VOLUME SIX HUNDRED AND SIXTY TWO

METHODS IN ENZYMOLOGY

Selenoprotein Structure and Function

METHODS IN ENZYMOLOGY

Editors-in-Chief

ANNA MARIE PYLE
Departments of Molecular, Cellular and Developmental Biology and Department of Chemistry
Investigator, Howard Hughes Medical Institute
Yale University

DAVID W. CHRISTIANSON
Roy and Diana Vagelos Laboratories
Department of Chemistry
University of Pennsylvania
Philadelphia, PA

Founding Editors

SIDNEY P. COLOWICK and NATHAN O. KAPLAN

VOLUME SIX HUNDRED AND SIXTY TWO

METHODS IN ENZYMOLOGY

Selenoprotein Structure and Function

Edited by

ERANTHIE WEERAPANA
Department of Chemistry,
Boston College,
Chestnut Hill, MA,
United States

ELSEVIER

ACADEMIC PRESS
An imprint of Elsevier

Academic Press is an imprint of Elsevier
50 Hampshire Street, 5th Floor, Cambridge, MA 02139, United States
525 B Street, Suite 1650, San Diego, CA 92101, United States
The Boulevard, Langford Lane, Kidlington, Oxford OX5 1GB, United Kingdom
125 London Wall, London, EC2Y 5AS, United Kingdom

First edition 2022

ISBN: 978-0-323-90735-4
ISSN: 0076-6879

For information on all Academic Press publications
visit our website at https://www.elsevier.com/books-and-journals

Publisher: Zoe Kruze
Developmental Editor: Federico Paulo S. Mendoza
Production Project Manager: James Selvam
Cover Designer: Alan Studholme

Typeset by STRAIVE, India

Contents

Contributors

Elias S.J. Arnér
Division of Biochemistry, Department of Medical Biochemistry and Biophysics, Karolinska Institutet, Stockholm, Sweden; Department of Selenoprotein Research, National Institute of Oncology, Budapest, Hungary

Clara A. Bailey
Department of Chemistry and Biochemistry, University of California, Santa Barbara, CA, United States

Daniel W. Bak
Department of Chemistry, Boston College, Chestnut Hill, MA, United States

Max J. Bedding
School of Chemistry; Australian Research Council Centre of Excellence for Innovations in Peptide and Protein Science, The University of Sydney, Camperdown, NSW, Australia

Mikael Björnstedt
Division of Pathology, Department of Laboratory Medicine, Karolinska Institutet, Karolinska University Hospital Huddinge, Stockholm, Sweden

Sofia Björnstedt
Division of Pathology, Department of Laboratory Medicine, Karolinska Institutet, Karolinska University Hospital Huddinge, Stockholm, Sweden

Juan Carlos Cáceres
Department of Chemistry and Biochemistry, University of California, Santa Barbara, CA, United States

Ye Cao
Interdisciplinary Research Center on Biology and Chemistry, Shanghai Institute of Organic Chemistry, Chinese Academy of Sciences, Shanghai, China

Anne E. Carlisle
Department of Molecular, Cell and Cancer Biology, University of Massachusetts Medical School, Worcester, MA, United States

Qing Cheng
Division of Biochemistry, Department of Medical Biochemistry and Biophysics, Karolinska Institutet, Stockholm, Sweden

Rujin Cheng
Department of Chemistry and Biochemistry, University of Delaware, Newark, DE, United States

Christina Z. Chung
Department of Molecular Biophysics & Biochemistry, Yale University, New Haven, CT, United States

Vidyadhar Daithankar
Department of Chemistry and Biochemistry, University of Delaware; Syva formulation, Siemens Healthineers, Newark, DE, United States

Jianguo Fang
State Key Laboratory of Applied Organic Chemistry and College of Chemistry and Chemical Engineering, Lanzhou University, Lanzhou, China

Tianyu Feng
Synthetic and Functional Biomolecules Center, Beijing National Laboratory for Molecular Sciences, Key Laboratory of Bioorganic Chemistry and Molecular Engineering of Ministry of Education, College of Chemistry and Molecular Engineering, Peking University, Beijing, China

Jinjun Gao
Peking-Tsinghua Center for Life Sciences, Academy for Advanced Interdisciplinary Sciences, Peking University, Beijing, China

Kei Goto
Department of Chemistry, School of Science, Tokyo Institute of Technology, Tokyo, Japan

Brandon L. Greene
Department of Chemistry and Biochemistry, University of California, Santa Barbara, CA, United States

Robert J. Hondal
Department of Biochemistry, University of Vermont, Burlington, VT, United States

Guogeng Jia
Synthetic and Functional Biomolecules Center, Beijing National Laboratory for Molecular Sciences, Key Laboratory of Bioorganic Chemistry and Molecular Engineering of Ministry of Education, College of Chemistry and Molecular Engineering, Peking University, Beijing, China

Dohoon Kim
Department of Molecular, Cell and Cancer Biology, University of Massachusetts Medical School, Worcester, MA, United States

Natalie Krahn
Department of Molecular Biophysics & Biochemistry, Yale University, New Haven, CT, United States

Sameer S. Kulkarni
School of Chemistry; Australian Research Council Centre of Excellence for Innovations in Peptide and Protein Science, The University of Sydney, Camperdown, NSW, Australia

Namgyu Lee
Department of Molecular, Cell and Cancer Biology, University of Massachusetts Medical School, Worcester, MA, United States

Jun Liu
Department of Chemistry and Biochemistry, University of Delaware, Newark, DE; Department of Pharmaceutical Chemistry, University of California, San Francisco, CA, United States

Ryosuke Masuda
Department of Chemistry, School of Science, Tokyo Institute of Technology, Tokyo, Japan

Meghomukta Mukherjee
Department of Physiology, University of Texas Southwestern Medical Center, Dallas, TX, United States

Richard J. Payne
School of Chemistry; Australian Research Council Centre of Excellence for Innovations in Peptide and Protein Science, The University of Sydney, Camperdown, NSW, Australia

Jennifer C. Peeler
Department of Chemistry, Colgate University, Hamilton, NY, United States

Ali Razaghi
Division of Pathology, Department of Laboratory Medicine, Karolinska Institutet, Karolinska University Hospital Huddinge, Stockholm, Sweden

Sharon Rozovsky
Department of Chemistry and Biochemistry, University of Delaware, Newark, DE, United States

Arun Kumar Selvam
Division of Pathology, Department of Laboratory Medicine, Karolinska Institutet, Karolinska University Hospital Huddinge, Stockholm, Sweden

Chenfang Si
Interdisciplinary Research Center on Biology and Chemistry, Shanghai Institute of Organic Chemistry, Chinese Academy of Sciences, Shanghai, China

Dieter Söll
Department of Molecular Biophysics & Biochemistry; Department of Chemistry, Yale University, New Haven, CT, United States

Anju Sreelatha
Department of Physiology; Center for Mineral Metabolism and Clinical Research, University of Texas Southwestern Medical Center, Dallas, TX, United States

Emma J. Ste. Marie
Department of Chemistry and Biochemistry, Norwich University, Northfield, VT, United States

Tímea Szekerczés
Division of Pathology, Department of Laboratory Medicine, Karolinska Institutet, Karolinska University Hospital Huddinge, Stockholm, Sweden

Chu Wang
Synthetic and Functional Biomolecules Center, Beijing National Laboratory for Molecular Sciences, Key Laboratory of Bioorganic Chemistry and Molecular Engineering of Ministry of Education, College of Chemistry and Molecular Engineering; Peking-Tsinghua Center for Life Sciences, Academy for Advanced Interdisciplinary Sciences, Peking University, Beijing, China

Eranthie Weerapana
Department of Chemistry, Boston College, Chestnut Hill, MA, United States

Fan Yang
Synthetic and Functional Biomolecules Center, Beijing National Laboratory for Molecular Sciences, Key Laboratory of Bioorganic Chemistry and Molecular Engineering of Ministry of Education, College of Chemistry and Molecular Engineering, Peking University, Beijing, China

Kenichi Yokoyama
Department of Biochemistry, Duke University, Durham, NC, United States

Baoxin Zhang
State Key Laboratory of Applied Organic Chemistry and College of Chemistry and Chemical Engineering, Lanzhou University, Lanzhou, China

Yaoyang Zhang
Interdisciplinary Research Center on Biology and Chemistry, Shanghai Institute of Organic Chemistry, Chinese Academy of Sciences, Shanghai, China

Preface

Selenoproteins contain the amino acid selenocysteine, a cysteine analog with selenium in the place of sulfur. Selenoproteins are observed in all domains of life, with 25 selenoproteins present in humans. Many selenoproteins are poorly characterized, especially with regard to their endogenous function and cellular regulation. The majority of functionally characterized selenoproteins display antioxidant activity, with others playing roles in thyroid hormone production, and selenoprotein synthesis. The characterization of selenoprotein function is complicated by the low abundance of selenoproteins, and the intricacies of the endogenous translational machinery required for selenoprotein production. Over the past decade, our understanding of selenoprotein structure and function has been accelerated by the development of innovative methods for the detection and production of selenoproteins. In Volume 662 of *Methods in Enzymology*, we compile a diverse collection of detailed protocols for the study of selenoproteins. The techniques presented utilize tools from chemistry, biochemistry, and molecular, cellular, and organismal biology to enable the investigation of selenoproteins both *in vitro* and *in vivo*.

Chapters 1 and 2 focus on the effect of exogenous selenium sources, primarily in the form of selenite, on selenoprotein expression and cellular function. In Chapter 1, Kim and colleagues describe methods to monitor selenium uptake into cells through the xCT transporter and to measure selenoprotein production upon perturbation of intracellular selenium levels. In Chapter 2, Björnstedt and colleagues discuss methods to monitor selenium cytotoxicity in cells, and assess how cytotoxicity is affected by diverse factors including cell culture conditions.

Chapters 3–7 describe recombinant protein expression methods for optimized selenoprotein production. In Chapter 3, Söll and colleagues describe methods to generate and screen tRNA libraries to identify engineered tRNAs with improved EF-Tu-mediated selenocysteine incorporation into a protein of interest in *Escherichia coli*. In Chapter 4, Arnér and colleagues describe methods to incorporate selenocysteine at a UGA codon in an *E. coli* strain that lacks other UGA codons as well as the release factor RF1. In Chapter 5, Greene and colleagues describe optimized methods for incorporating selenocysteine into thiyl radical enzymes to enable mechanistic studies into radical catalysis. In Chapter 6, Peeler and colleagues detail genetic code expansion methods to incorporate an unnatural photocaged selenocysteine into proteins for subsequent uncaging directly in mammalian cells. In Chapter 7, Rozovsky and colleagues describe the use of selenocysteine-mediated expressed protein ligation to produce selenoprotein S: an ER membrane-localized selenoprotein.

Chapters 8–11 outline methods to monitor the presence of selenoproteins in complex biological samples. In Chapter 8, Bak and colleagues adapt a cysteine-reactivity profiling platform, isoTOP-ABPP, for selenoprotein identification by performing the initial thiol/selenol-labeling step at low pH. Likewise, in Chapter 9, Zhang and colleagues apply a selenocysteine-specific mass spectrometry (SecMS) platform in cell lysates to identify novel selenoproteins. In Chapter 10, Wang and colleagues describe a computational algorithm, termed SESTAR, for identifying selenopeptides from mass-spectrometry data based on the unique isotopic signature of selenium. In Chapter 11, Fang and colleagues describe the use of a fluorescent probe, Sel-green, to detect selenol species in cells and discriminate from the more abundant thiol species.

Chapters 12–15 (the last group of submissions) offer an array of methods for studying and utilizing selenocysteine or selenoproteins for diverse biological applications. In Chapter 12, Sreelatha and colleagues describe methods to study selenoprotein O, a human selenoprotein that was recently discovered to catalyze AMPylation reactions. In Chapter 13, Hondal and colleagues describe the use of α-methyl selenocysteine as a more stable analog of selenocysteine that does not undergo oxygen-mediated β-elimination to dehydroalanine. In Chapter 14, Goto and colleagues describe nanosized molecular cavities that act as a protective cradle to enable mechanistic studies of reactive intermediates derived from selenocysteine. Last, in Chapter 15, Payne and colleagues adapt selenocysteine for chemical synthesis of proteins through a diselenide-selenoester ligation reaction.

I am very grateful to all the authors of this special issue on selenoproteins for their timely contribution and graciousness throughout the submission and review process. I acknowledge Dr. Daniel Bak for his help with identifying contributors for this issue. I also thank Dr. Daniel Bak and Dr. Jennifer Peeler for their help with the editing and reviewing process. The staff at Elsevier, especially Paulo Mendoza, deserve much recognition for all the editorial assistance with this special issue. Finally, I am very grateful for the Editors-in-Chief of *Methods in Enzymology*, Dr. David Christianson and Dr. Anna Pyle, for the invitation to compile this exciting collection of methods. I am excited for the broader community to adopt these methods for the study of selenoproteins, and look forward to the next generation of methods that will further propel our understanding of selenoprotein biology.

Eranthie Weerapana

CHAPTER ONE

Examining xCT-mediated selenium uptake and selenoprotein production capacity in cells

Namgyu Lee, Anne E. Carlisle, and Dohoon Kim*
Department of Molecular, Cell and Cancer Biology, University of Massachusetts Medical School, Worcester, MA, United States
*Corresponding author: e-mail address: dohoon.kim@umassmed.edu

Contents

Abstract

The production of selenoproteins in cancer cells is dependent on uptake of selenium and processing via the selenocysteine biosynthesis pathway. Both the uptake and processing of selenium has recently shown to be upregulated in subsets of cancer cells due to their increased expression of xCT transporter, and the resulting increased expression of selenoproteins such as GPX4 can play multiple roles in cancer cells such as providing protection against ferroptotic insults. Here, we describe a set of protocols designed to measure this process in cancer cell culture—the measurement of xCT transporter expression and activity, the intracellular uptake of selenium in cancer cells, and the expression of selenoproteins as the final functional readout of this process. The successful measurement of xCT requires non-denaturing western blotting of xCT subunits, while its activity is determined by the measurement of reduced thiol groups that accumulate over time, as determined by Ellman's reagent. Selenium uptake is determined by supplementing a selenium source and then measuring total intracellular selenium levels, which is determined from digested cellular material using a reactive fluorescent

Methods in Enzymology, Volume 662
ISSN 0076-6879
https://doi.org/10.1016/bs.mie.2021.10.002

probe or via inductively coupled plasma mass spectrometry. Finally, specific tips for efficiently determining the expression level of a set of "indicator" selenoproteins is provided. These parameters allow one to determine the "selenophilicity" of cells, i.e., the ability of cells to utilize selenite to upregulate their selenoprotein production and thus antioxidant defenses.

1. Introduction

Selenium is a trace nutrient that is used in the selenocysteine biosynthesis pathway which produces selenocysteinyl-tRNA, which is in turn required in the translation of 25 selenoproteins in humans (Gonzalez-Flores, Shetty, Dubey, & Copeland, 2013; Labunskyy, Hatfield, & Gladyshev, 2014). While functions of several selenoproteins are yet unknown, characterized selenoproteins include those involved in redox homeostasis, selenium transport and metabolism, thyroid hormone regulation, and endoplasmic reticulum function (Labunskyy et al., 2014). Thus, selenocysteine is referred to as the 21st amino acid.

The selenocysteine biosynthesis pathway involves (1) provision of a specialized tRNA precursor, as well as (2) multistep formation of selenophosphate before the two components are used to form selenocysteine (Gonzalez-Flores et al., 2013; Turanov et al., 2011) (Fig. 1). In the first part, a transfer RNA (tRNA) is aminoacylated by seryl-tRNA synthetase (SERS) with serine and then O-phosphoseryl-tRNA kinase (PSTK) phosphorylates the seryl-tRNA to form O-phosphoseryl-tRNA. For the second part, dietary selenate or selenite can be converted to selenide, which is in turn metabolized by the enzyme selenophosphate synthetase 2 (SEPHS2) to form selenophosphate. An alternative source of selenide is conversion of selenocysteine - either formed from selenoprotein degradation or metabolism of selenomethionine - to selenide by selenocysteine lyase (SCLY; not shown in the diagram for simplification). Selenide is converted to selenophosphate by selenophosphate synthetase 2 (SEPHS2). Finally, the O-phosphoseryl-tRNA and selenophosphate components are processed by selenocysteine synthase (SEPSECS) to yield selenocysteinyl-tRNA. Thus, both the initial availability of selenium and the metabolic processes for selenocysteinyl-tRNA formation can be determinants for the expression levels of selenoproteins.

Selenoproteins include multiple glutathione reductases (GPX1,2,3,4,6) and thioredoxin reductases (TXNRD1/2/3), and thus the importance of selenium in a cell's redox homeostasis is well appreciated (Labunskyy et al., 2014). On the other hand, there has been renewed interest in selenium

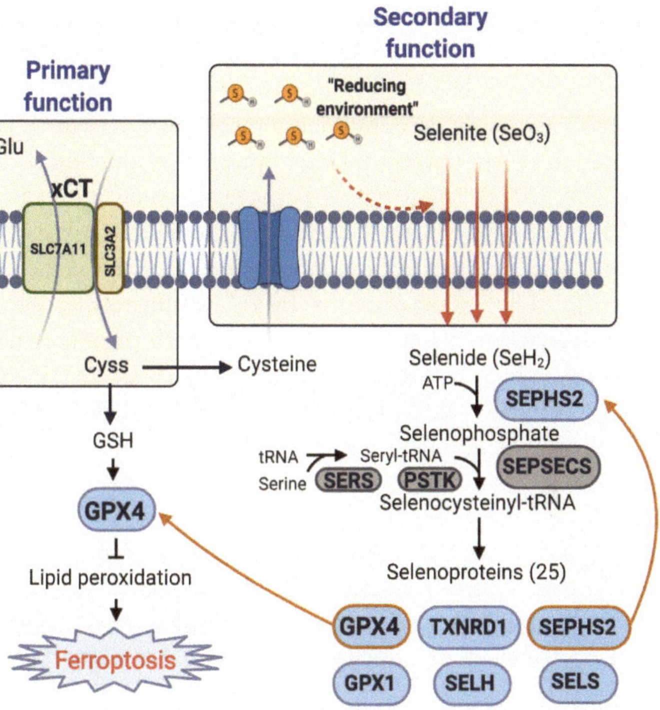

Fig. 1 Schematic diagram of functions of xCT on cystine and selenium uptake, and Sec biosynthesis pathway. xCT primarily has a role in cystine import and glutamine export. When xCT is active, cystine (the oxidized form of cysteine) is imported and reduced to cysteine. The subsequent export and accumulation in the media of cysteine provides thiol groups with reducing potential which results in the reduction of selenite to the volatile selenide. The diffusion of selenide into cells, and its further metabolism via SEPHS2 and SEPSECS results in the production of selenocysteinyl-tRNA which is incorporated into one of 25 selenoproteins. These include GPX4, a well-known negative regulator of ferroptosis which works by repairing peroxidized lipids. Thus, xCT has a secondary function in the import and processing of selenium beginning from the dietary form selenite. Created with Biorender (https://app.biorender.com/).

metabolism as some of these proteins, and in particular GPX4, have recently been identified as important regulators of ferroptosis, an iron-dependent form of regulated cell death (Dixon et al., 2012; Yang et al., 2014). Selenium-mediated protection against ferroptosis has been recognized as a protective mechanism in neurodegeneration (Dominiak, Wilkaniec, Wroczynski, & Adamczyk, 2016). Furthermore, protection against ferroptosis is important for cancer cell survival (Yang et al., 2014), and ferroptosis is engaged by chemotherapeutics (Su et al., 2020). For these reasons targeting of GPX4 has been of high interest as a potential cancer therapy.

Strikingly, selenium was long known to preferentially be taken up by tumors (Cavalieri & Scott, 1968; Cavalieri, Scott, & Sairenji, 1966), the mechanisms of which had been poorly understood. Providing an explanation, our group recently demonstrated a mechanism by which cancer cells have elevated selenium uptake and selenoprotein production (Carlisle et al., 2020; Lee et al., 2021). Many types of cancer cells such as breast cancer cells employ a mechanism for increased selenium (in the form of selenite) uptake that is dependent on their increased expression of the glutamate/cystine antiporter xCT. The import of cystine by xCT ultimately leads to formation and export of cysteine, the accumulation of which promotes reduction of selenite to selenide (Carlisle et al., 2020; Lee et al., 2021; Olm et al., 2009) (Fig. 1). Thus, xCT formation can drive selenium uptake and selenoprotein production in cancer cells, and thus impact their biology such as in increasing their antioxidant capacity (Carlisle et al., 2020; Lee et al., 2021).

2. Measurement of selenium uptake and selenoprotein production as mediated by xCT transporter

Here, we describe a set of methods to determine various aspects of the xCT-mediated selenium uptake process in the cell culture context. First, we describe the determination of xCT levels, which involves non-heat-denatured western blot detection of SLC7A11 and SLC3A2 subunit protein levels which comprise xCT. Second, as the mechanism of xCT promotion of selenium uptake involves the export of cysteine which provides extracellular thiol groups, we describe the determination of reactive thiols from cell conditioned media as a readout of xCT activity. Third, we describe two procedures which can be used to determine total intracellular selenium, which can be used to determine the selenium uptake of cancer cells following media supplementation of selenium sources. Finally, we describe details for optimizing western blotting-based detection of various selenoproteins as an indirect, ultimate readout of selenium uptake/selenocysteine biosynthesis pathway activity.

2.1 Evaluation of xCT expression of SLC7A11 and SLC3A2

In many cases, protein expression is correlated with the degree of protein activity. Indeed, we have shown that the protein expression of the xCT subunits SLC7A11 and SLC3A2 is associated with the activity of xCT in breast normal and cancer cell lines (Carlisle et al., 2020; Lee et al., 2021). Thus, confirming the expression of xCT by western blotting or immunohistochemistry is a relatively straightforward method for predicting xCT activity.

2.1.1 Technical tips for the non-heat-denaturing Western blot of xCT

Due to the tendency of transmembrane proteins to self-aggregate following heat denaturation, standard western blotting which usually involves boiling of samples before gel loading can result in the failure of detection of the transmembrane proteins (Tsuji, 2020). Since both subunits of xCT are transmembrane proteins, we compared the effect of boiling on the blotting of SLC3A2 and SLC7A1. The SLC7A11 protein was not detected at the predicted size (35 kDa) within the heated samples, instead showing up as a band of much larger size than predicted (Fig. 2C, blue arrow). In the non-heated samples, the proteins shows up at the correct size, and only in the MDAMB231, CAL120, and Hs578T breast cancer cells which were previously defined as high xCT cancer cells (Carlisle et al., 2020) (Fig. 2C, red arrow). Since other bands (~135 kDa) were also present in the blot using the SLC7A11 antibody, we validated the specificity of SLC7A11 antibody using lysates of CTRL and SLC7A11 KO cells, which were generated by lentiviral transduction of CRISPR/Cas9, using the pLENTICRISPR V2 system (Carlisle et al., 2020; Sanjana, Shalem, & Zhang, 2014). The band at the predicted size disappeared in the SLC7A11 KO cell lysates, confirming that the band at 35 kDa in Fig. 2C is the correct one. In the case of SLC3A2, the band is correctly detected in either heated or unheated samples (Fig. 2D). The antibody specificity was also confirmed as the band at the predicted size (85–95 kDa) was not detected in SLC3A2 KO cells (Carlisle et al., 2020; Sanjana et al., 2014) (Fig. 2D). Overall, the heating seems to make SLC7A11 proteins aggregated, which impairs their electrophoresis in the SDS-PAGE and cause them to remain at top of the gel (Fig. 2C, blue arrow). Thus protein samples should not be boiled, to preserve SLC7A11 detection using western blotting. Importantly, as we have shown (Lee et al., 2021), robust detection of both SLC7A11 and SLC3A2 is correlated with active xCT in breast cancer patient tumor samples, as inferred by elevated expression of GPX4. Expression of one subunit but not the other is correlated with weak/negligible GPX4 expression (Lee et al., 2021).

- Transfer condition: proteins were transferred to 0.45 μm PVDF membrane at 390 mA for 2 h in the 1 × transfer buffer with 20% EtOH.
- Washing condition: three times vigorous washing for 10 min on a rocking apparatus with 15–20 mL of 1xTBS-T buffer.

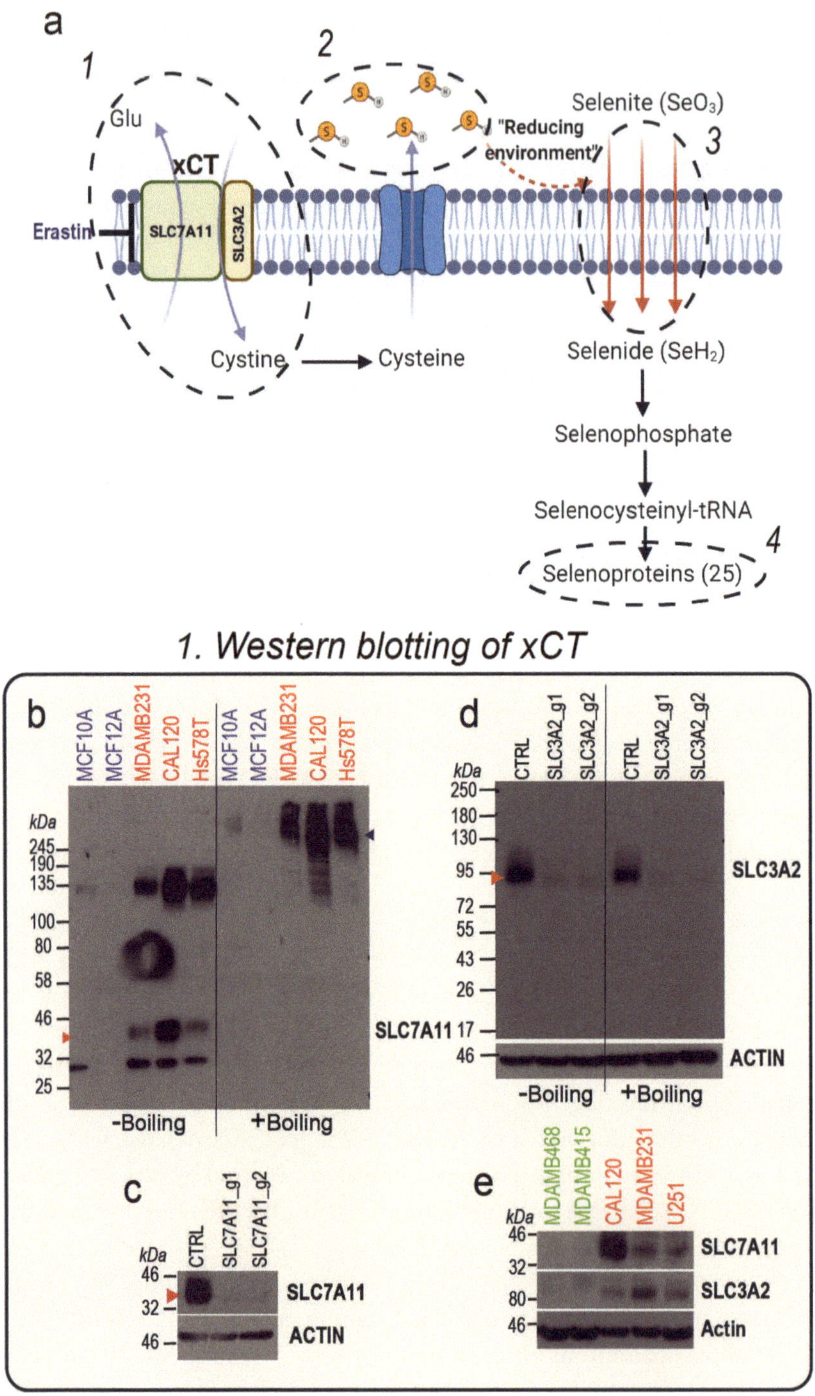

Fig. 2 See figure legend on opposite page.

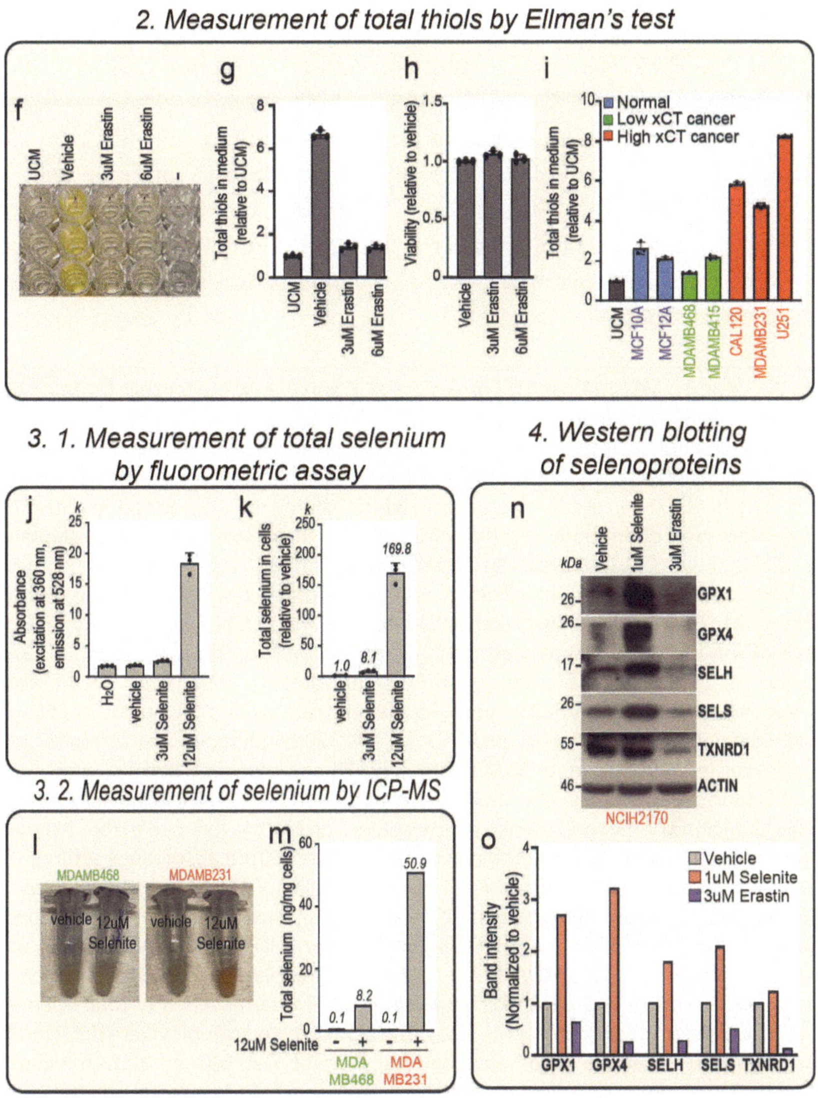

Fig. 2 Various approaches to assess the xCT function on selenium uptake. Continue to next page. (A) Various approaches to assess the xCT function on selenium uptake. Four different steps in the selenocysteine biosynthesis pathway, which could be measured by separate experiments, for evaluating the xCT activity and subsequent processing to selenoproteins. The xCT activity could be measured at the indicated steps of the pathway by (1) Western blotting of xCT, (2) Ellman's test, (3) measurement of total selenium by fluorometric assay, or ICP-MS, and (4) Western blotting of selenoproteins. Created with Biorender (https://app.biorender.com/). (B) The effect of boiling samples on the Western blotting of SLC7A11. Lysates of normal immortalized breast cells (MCF10A,

(Continued)

- Protein quantities and detection sensitivity: around 30 μg of protein is required to get a good signal for SLC7A11 with diluted Femto substrate and SLC3A2 with ECL substrate (refer to Section 2.4 regarding substrates).
- Dilution factors and incubation time for antibodies: anti-SLC3A2 and anti-SLC7A11 antibodies were diluted in 5% BSA or skim milk in TBS-T buffer at 1:1000 dilution. Secondary antibody was diluted in 5% skim milk in TBS-T buffer at 1:2000 dilution. Primary antibodies were incubated overnight (12–16 h) and secondary antibodies were incubated for 2 h.

Fig. 2—Cont'd MCF12A) which express low xCT, and breast cancer cells (MDAMB231, CAL120, Hs578T) which express high xCT (Carlisle et al., 2020) were used as a negative and positive control, respectively. The SLC7A11 was correctly detected in non-boiled high xCT cancer cells as indicated by a red arrowhead. The blue arrowhead indicates a potential SLC7A11 protein aggregate which seems to be formed by heating. (C) Confirmation of antibody specificity for SLC7A11 in control and SLC7A11 depleted by CRISPR/Cas9 CAL120 cells. Predicted and confirmed size of SLC7A11, 35 kDa. (D) Verification of antibody specificity for SLC3A2 and effect of boiling on the western blotting of SLC3A2. Non-boiled and boiled control and SLC3A2 depleted by CRISPR/Cas9 MDAMB231 cell lysate was used. Predicted and confirmed size of SLC3A2, 85–95 kDa. (E) Western blotting of xCT in the breast cancer cells (MDAMB468, MDAMB415, CAL120, MDAMB231) and glioblastoma cells (U251). (F) A picture of 96 well plate containing unconditioned media (UCM), CAL120 conditioned media treated with either vehicle, 3 μM Erastin, or 6 μM Erastin, plus Ellman's reagent; and (−) which indicates Ellman's reagent only. The media was conditioned for 24 h. (G) Relative total thiols in the UCM and conditioned media. Yellow-colored product in the media from the plate "f" after mixing with Ellman's solution was measured at 450 nm absorbance setting. The values are relative to UCM which was set as 1. (H) Viability of CAL120 which was used for media conditioning for Ellman's test. (I) Relative total thiols in UCM and conditioned media of normal (MCF12A, MCF10A), low xCT cancer (MDAMB468, MDAMB415), and high xCT cancer (CAL120, MDAMB231, U251). The media was conditioned for 22 h. The values are relative to UCM which was set as 1. (J) Quantification of total selenium in U251 cell pellets by a fluorometric assay. The acid digested samples were detected by microplate reader at the absorbance setting, excitation at 360 nm, and emission at 528 nm. (K) Relative total selenium in vehicle, 3 μM, and 12 μM selenite treated U251. All background (H_2O condition) subtracted values were normalized to the vehicle-treated condition, set as 1. (L) Picture of MCAMB486 and MDAMB231 pellets treated with vehicle and 12 μM selenite for 2 h. (M) Quantification of total selenium in cell pellets measured by ICP-MS. The total selenium was normalized to weight of pellets. The cell pellets presented in "l" were subjected to acid digestion and ICP-MS analysis. The detected selenium was normalized to the weight of a cell pellet. (N) Western blotting of selenoproteins in NCIH2170 cells treated with vehicle, 1 μM selenite, or 3 μM Erastin. (O) Quantification of band intensity of blots of selenoproteins. Values are normalized to the band intensity value from vehicle-treated sample which was set as 1.

Reagent or resource	Source	Identifier
Rabbit polyclonal anti-SLC3A2	Sigma-Aldrich	Cat#HPA017980
Rabbit monoclonal anti-xCT/ SLC7A11	Cell Signaling Technology	Cat#12691
4–20% Precast Protein Gels	BioRad	Cat#4561096
Transfer or Electro Blotting Buffer (10 ×)	Boston Bioproducts	Cat#3BP190
TBST (10 ×, with 0.5% Tween-20, pH 7.4)	Boston Bioproducts	Cat#IBB181
mouse anti-rabbit IgG-HRP	Santa Cruz	Cat#sc2357
ECL Western Blotting Substrate	Thermo Fisher	Cat#32106
West Pico PLUS Chemiluminescent Substrate	Thermo Fisher	Cat#34580
West Femto Maximum Sensitivity Substrate	Thermo Fisher	Cat#34095

2.2 Measurement of extracellular thiols as readout of xCT function leading to uptake

When cystine is imported into cells by xCT, cysteine which has a thiol group is exported out of cells, which results in an extracellular reducing environment (Fig. 1). Thus, measuring total thiols in the conditioned media could reflect the activity of xCT of the cells. The total thiols can be measured by Ellman's reagent, 5,5′-dithio-bis-(2-nitrobenzoic acid), also known as DTNB (Tietze, 1969). Ellman's reagent reacts with a free sulfhydryl (thiol) group to yield a mixed disulfide and 5-nitro-2-thiobenzoic acid (TNB; λ max at 412 nm), a measurable, yellow-colored product in a short reaction time. Using this method, we previously revealed that high xCT cancer cells have high extracellular reducing capacity (Carlisle et al., 2020). Consistent with our previous reports (Carlisle et al., 2020), a yellow-colored product was observed in the media that CAL120 cells, breast cancer cells expressing high xCT, were cultured in for 12–24 h (= "conditioned media") (Fig. 2F and G). This signal is not present in unconditioned media (media incubated in the absence of any cells). Furthermore, signal is lost in conditioned media from CAL120 cells treated with Erastin, an xCT inhibitor.

Consistently, the levels of total thiols were high in high xCT cancer cells but not in low xCT cancer or normal cells (Fig. 2I). Thus, Ellman's test provides a readout of xCT activity which is tightly linked to selenium uptake. The detailed procedure is described in the following protocol section.

2.2.1 Protocol for Ellman's test

2.2.1.1 Ellman's test for measuring xCT activity on extracellular reduction

Before you begin

Timing: 10 min

1. Prepare 10 mM DTNB stock solution
   - **(a)** Dissolve 40 mg DTNB in 10 mL DMSO
   - **(b)** The stock solution can be stored at 4 °C for 3 months

2.2.1.2 Materials and equipment

Reagent or resource	**Source**	**Identifier**
Reagent		
5,5′-dithio-bis-(2-nitrobenzoic acid)	Sigma-Aldrich	Cat#HPA017980
Erastin	Cayman Chemical	Cat#17754
Phenol red free media Ex: DMEM, no phenol red DMEM/F-12, no phenol red RPMI 1640 Medium, no phenol red	 Thermo Fisher Thermo Fisher Thermo Fisher	 Cat#21063029 Cat#21041025 Cat#11835030
CellTiterGlo	Promega	Cat#G7570
Instrument		
GloMax Discover Microplate Reader (Any brand/model having a filter around 412 nm)	Promega	Cat#GM3000

2.2.1.3 Step-by-step method details

Day 1

Timing: 1 h

1. Seed around 10,000 cells per well in a 96 well plate for biological triplicates in 100 μL of phenol red free media per well
   - **(a)** Phenol red can interfere with the colorimetric signal. Thus, phenol red free media is recommended

(b) Set up 3 wells of media only without cells which will be used as an unconditioned media (UCM), the signal of which will used for normalization

Day 2

Timing: vary depending on how long the media will be conditioned (recommended conditioning time-12–16 h)

2. Change media with fresh phenol red free media using multichannel pipette
   (a) Premix drug treatments (e.g., Erastin) with media where there are drug treatment conditions
   (b) Non-toxic doses of Erastin (e.g., from 3 to 6 μM in MDAMB231 and CAL120; PMID: 33672555) are recommend for Ellman's test as decreased cell numbers due to drug toxicity can affect the result
3. After a determined incubation time, collect the conditioned media with multichannel pipette and transfer into a new 96 well plate containing 50 μL of 10 mM DTNB solution
   (a) Prepare a new 96 well plate having 50 μL of 10 mM DTNB solution in each well and turn on the plate reader and the program for ready to use in advance to read plate as soon as the conditioned media is mixed with Ellman's reagent
   (b) As you transfer the conditioned media to the DTNB containing plate, mix the contents in the same action by multichannel pipetting up and down a few times, while avoiding making bubbles which can impact the reading
4. Read the plate at 412 nm absorbance setting
   (a) The yellow TNB formed can be measured around 412 nm (ideal). But other filters that are close to 412 nm can be used
5. Data analysis

Critical tips: (1) Do not leave out a plate which has conditioned media out of the incubator for too long, which can lead to oxidation of media and formation of yellow color in the Ellman containing media; measure samples as soon as possible from endpoint. (3) Different cell confluencies also can affect the values. Thus, make sure cell numbers in each well at endpoint (which can be estimated from CellTiterGlo total cell viability, described below) is not significantly different between the groups at experimental endpoint, which can skew values.

Optional: When the xCT activity is compared across different cell lines, fresh media can be added onto the wells after step 3 and then the plate can be subjected to CellTiterGlo. (CTG) (Promega) luminescent cell viability assays, which will provide total cell viability in each well, a proxy for total cell numbers. In this manner one can normalize the Ellman's test values to total viable cell amount.

2.2.1.4 Expected outcomes

The final values will be higher from conditioned media of high xCT-expressing cells compared to low xCT-expressing cells. The values will be greatly diminished by KO of SLC7A11/SLC3A2 or pharmacological inhibition of xCT (Erastin, Sulfasalazine, Sorafenib), serving as an important control.

2.2.1.5 Quantification and statistical analysis

First, subtract all values to the values from Ellman's reagent only. Then normalize all values to that of unconditioned media (UCM, set as 1.0). Optional: when the values are compared across different cell lines, the subtracted values could be normalized to cell number or CellTiterGlo value as described to minimize the noise from cell seeding number errors.

2.3 Measurement of selenium uptake

The xCT mediated selenium uptake could be measured by directly quantifying the total selenium levels in cells. In order to determine the selenium uptake capacity of cells, they should be treated with a relatively high dose of exogenous selenium source such as sodium selenite at a given time prior to endpoint (e.g., 2 h). Following incubation, cells are washed, and then total selenium levels following this provides a readout of selenium uptake. Uptake was confirmed to be increased in selenite-treated samples by both fluorometric and ICP-MS based methods (Fig. 2J–M). Consistent with our previous report on the correlation of xCT expression and cellular capacity of selenium uptake (Carlisle et al., 2020), MDAMB231, a high xCT breast cancer cell line, was confirmed to have higher selenophilicity compared with MDAMB468, a low xCT breast cancer cell line (Fig. 2M). Interestingly, a reddish coloring of the cell pellet, which appears to arise from high selenium content in the cells, was observed in the high xCT MDAMB231 cell line in the selenite treated condition (Fig. 2I).

As selenium is a metalloid element that cannot be readily detected by conventional liquid chromatography–mass spectrometry (LC-MS), alternative methods are required to measure selenium in a cell. Two different methods, fluorometric and ICP-MS (inductively coupled plasma-mass spectrometry) based measurements, are available for quantifying total selenium

(Delafiori, Ring, & Furey, 2016; Sheehan & Gao, 1990). Both these methods first require treatment with strong acids at high temperatures to fully digest compounds and break selenium species down to their base atomic form. The fluorometric method utilizes a compound that reacts with selenium to provide a fluorescent signal (Sheehan & Gao, 1990). Briefly, the reagent 2,3-diaminonaphthalene (DAN) reacts with the free selenium formed from the acid digestion, to form a fluorophore (piazselenol). This fluorophore is extracted with cyclohexane and then can be quantified at optimal excitation wavelength 366 nm, and emission wavelength 544 nm. The second method utilizes ICP-MS which can detect various metal and metalloid atoms not detectable by LC-MS (Delafiori et al., 2016). Both methods provide a readout of the overall selenium levels in a cell, but by itself cannot provide information on the species of selenium. However, in conjunction with treatment of a selenium source in the media for a determined period, these measurements provide a readout of selenium uptake rate for a given cell line.

2.3.1 *Protocol for fluorometric assay*

2.3.1.1 Measurement of cellular selenium uptake capacity using fluorometric assay

Before you begin

Timing: 30 min

1. Low xCT cell lines (e.g., MCF10, MDAMB468) and high xCT cell lines (CAL120, U251) can be used as negative and positive controls for selenite uptake. Erastin treated cells or cells with KO of xCT subunits can also use as a negative control for selenite uptake
2. Make 1 mL of 12 mM sodium selenite stock solution
   (a) Dissolve 2.075 mg sodium selenite in 1 mL Milli-Q water
3. Make 10 mL of a 4:1 mixture of Nitric acid (HNO_3) and Perchloric acid ($HClO_4$) in safety cabinet, behind transparent plastic shielding, as these acids are both caustic and can explosively react
   (a) Mix 8 mL HNO_3 and 2 mL $HClO_4$
   (b) Make solution fresh for every assay
4. Make 20 mL of a 2.5 mM solution of Ethylenediaminetetraacetic acid (EDTA).
   (a) Dissolve 14.612 mg EDTA in 20 mL Milli-Q water
5. Make 10 mL of a 6.3 mM 2,3-diaminonapthalene HCl (DAN) solution
   (a) Dissolve 9.967 mg DAN powder in 10 mL Milli-Q water

(b) Sonicate on medium power for 10 rounds, 30 s on and 30 s off, until DAN reagent dissolves. Appearance of some particulates is normal
(c) Make solution fresh for every assay.

2.3.1.2 Materials and equipment

Reagent or resource	**Source**	**Identifier**
Reagent		
Sodium selenite	Sigma-Aldrich	Cat#S5261
Milli-Q water	N/A	
Nitric acid (HNO_3)	Sigma-Aldrich	Cat# 438073
Perchloric acid ($HClO_4$)	Sigma-Aldrich	Cat# 311413
Hydrochloric acid (HCl)	Sigma-Aldrich	Cat# 320331
Ethylenediaminetetraacetic acid (EDTA)	EMD Millipore	Cat# 324503
2,3-diaminonapthalene-HCl (DAN)	ChemSavers	Cat# 23DANH1G
Cyclohexane solution	Sigma-Aldrich	Cat# 650455
Instrument		
GloMax Discover Microplate Reader	Promega	Cat#GM3000
2 Heat blocks	Any brand/model which can reach up to 130 °C	
Plasticware or glassware		
Borosilicate glass, screw top with caps, flat bottom, 16x100mm, 10 mL vol	Karter Scientific	Cat# 226N8
96 well plate, transparent	Any brand	
10 cm^2 plate, cell culture treated	Any brand	

2.3.1.3 Step-by-step method details

Day 1

Timing: 1 h

1. Seed 7–15 M cells into 10 cm plates which will result in approximately 75% confluency the next day

(a) Plates cells for biological triplicate for each condition. For example, seed cells in six 10 cm plates, three will be treated with 12 μM sodium selenite and 3 with vehicle to create biological triplicate for each condition

Note: Depending on cell growth rate and cell size, the cell seeding number can be optimized.

Day 2

Timing: 10 min

2. When cells reach approximately 75% confluency, change them into fresh media. This is key, as selenite uptake capacity is directly tied to extracellular accumulation of thiols resulting from xCT activity. Changing to fresh media here will prevent over-conditioning—the saturation of thiol groups - which could negate the ability to discern differences in selenium uptake capacity between conditions.

Day 3

Timing: 3 h

3. After the cells have conditioned the media for 24 h, treat cells with 12 μM selenite or vehicle and incubate at 37 °C for 2 h

(a) When selenite/vehicle is added, just add 10 μL of 1000 × stock to the media without media change. The thiol accumulation in the media in the preceding 24 h is a key factor in selenite uptake rate. Conditioning time at this step may need to be optimized to prevent over-conditioning, or even under-conditioning which will may not allow enough selenium import for robust detection.

(b) Before proceeding to the next step, turn on the heat block and set to 130 °C.

4. Harvest cells by trypsinization or scraping and centrifugation at 1500 rpm, 4 °C for 3 min.

5. Aspirate the media and wash cell pellets with 1 mL cold PBS for three times, then count to 5 million and resuspend in 100 μL Milli-Q H_2O and add to the borosilicate glass tubes.

6. Add 500 μL of the 4:1 mixture of HNO_3 and $HClO_4$ to each tube and incubate on the heat block overnight.

(a) Set the heat block behind a plastic shield; mix and add acids behind the plastic shield.

(b) Add a sample for background subtraction, which is without any cells but includes all reagents.

Critical: Do not expose the conditioned media out of incubator too long when selenite is supplemented. It can affect the level of media conditioning.

Take the plate out one by one from the incubator right before adding the drug.

Day 4

Timing: 3 h

7. Add 500 μL HCl and incubate at 130 °C for 1 h.
   a. Perform in the hood.
8. Remove the tubes to a rack and allow them to cool to room temperature (approximately 1 h).
   a. This time can be used to set the second heat block to 60 °C.
9. Add 2 mL of EDTA and 500 μL of DAN reagent and incubate for 30 min at 60 °C.
   a. Do not cap tightly.
10. Add 1 mL of cyclohexane solution to each tube, recap, and place samples on a shaker for 5 min.
11. Aliquot 150 μL of each sample, in triplicate, to a 96 well plate.
   a. Also add 3 wells of 150 μL cyclohexane as a blank.
12. Read the plate in a fluorimeter at 360 nm excitation and 528 nm emission.
   a. The optimal excitation wavelength is 366 nm and emission wavelength is 544 nm, but plate readers able to read similar wavelength peaks will work.
13. Data analysis.

2.3.1.4 Expected outcomes

First, the cells that were supplemented with extracellular selenium will have higher measurements than those with basal selenium levels from the culture media. The normalized fluorescence values will be higher in cell lines having high expression and activity of xCT compared to those which are low xCT. The values will be diminished by KO of SLC7A11 or SLC3A2 or xCT inhibitor treated.

2.3.1.5 Quantification and statistical analysis

First, subtract from each of the values the average of the reagent blank (from step 6(b)) values to normalize out any background. Then, normalize the values for the selenite supplemented cells to the basal intracellular selenium values. This will give a relative selenium uptake measurement.

2.3.1.6 Advantages

Selenium uptake can be measured relatively easily and quickly, and across multiple cell lines or genetic/treatment conditions.

2.3.1.7 Safety considerations and standards

All steps involving acids must be carried out in the hood until taking out and aliquoting into plate. Caustic, explosively reactive acids are used and must be handled with care. Acids should be mixed behind a plastic shield within the safety hood.

2.3.2 Protocol for sample prep for ICP-MS analysis

2.3.2.1 Measurement of cellular selenium uptake capacity using ICP-MS

Before you begin

Timing: 30 min

1. Low xCT cell lines (e.g., MCF10, MDAMB468) and high xCT cell lines (CAL120, U251) can be used as negative and positive controls for selenite uptake. Erastin treated cells or cells with KO of xCT subunits can also used as a negative control for selenite uptake
2. Make 1 mL of 12 mM sodium selenite stock solution
   1) Dissolve 2.075 mg sodium selenite in 1 mL Milli-Q water
3. Make 1:1 H_2O_2/HNO_3 mixture

2.3.2.2 Materials and equipment

Reagent or resource	Source	Identifier
Reagent		
Sodium selenite	Sigma-Aldrich	Cat#S5261
Milli-Q water	N/A	
Nitric acid (HNO_3)	Sigma-Aldrich	Cat# 438073
Instrument		
ICP-Mass spectrometry	Agilent 7800 ICP-MS system	
Heat block	Any brand/model which can reach up to 130 °C	
Sonicator	Any brand/model	
Plasticware or glassware		
Glass tube with Screw Cap, 16 × 100 mm	HACH	Cat# 2517600

2.3.2.3 Step-by-step method details

Day 1

Timing: 30 min

4. Seed 15–30 M cells with 20 mL growth media in 150 dish to form around 80% confluency the next day.

Day 2

Timing: 4 h

5. Add 20 μL of 12 mM selenite stock solution directly into the conditioned media, without changing media for 24 h

(a) As with the DAN method, an optimized amount of media conditioning where over-conditioning is prevented, is important. Here, a larger amount of cells are grown in 15 dishes to provide sufficient material for adequate ICP-MS detection. The timing is thus different, where the 24 h conditioning starts right from plating instead of after cells have reached 75% confluency. Conditioning time may need to be further optimized to prevent over-conditioning or under-conditioning as with the DAN method

(b) During the incubation time, set the temperature of centrifuge at 4 °C

(c) Measure the weight of Eppendorf tubes which will be used for collecting cell pellets in order to subtract the weight of tubes for quantifying weight of cell pellets later

6. After 2 h incubation, remove the media and detach cells from the plate by trypsinization or scraper and collect the cells in 15 mL tubes

(a) The incubation time can be optimized depending on the dosage of selenite and cellular capacity on extracellular reduction. This optimization step aims to obtain sufficient selenium import and sufficient cells to provide a satisfactory ICP-MS peak signal well above a detection limit. Insufficient cell quantity and/or insufficient selenium import will result in noisy values or lack of ICP-MS detection of selenium

7. Wash cell pellet with 1 mL cold PBS three times. For the third wash, move the cell mixture to the Eppendorf tubes whose weight were premeasured and recorded, and centrifuge at 2500 rpm and 4 °C for 5 min for pelleting

8. Aspirate the PBS and measure the weight of the tubes

(a) Do not disturb cell pellet when PBS is aspirated

(b) Subtract the weight of the tubes to the recorded weight of each tube

Pause Point: The cell pellets can be stored at −80 °C prior to the next step.

Day 3–4

Timing: 2 days

9. Treat all cell pellets with 1000 μL of 1:1 H_2O_2/HNO_3 mixture and allow to stand for 12 h at room temperature with periodic venting

10. Sonicate samples at 35 kHz and 40 °C for 1 h with periodic venting for 1 h

11. Transferred the contents of the Eppendorf tube to a glass digestion tube and rinse the same Eppendorf tube with 500 μL of Milli-Q water for three times and also transfer the Milli-Q water used for rinse to the glass tube

12. Seal the tubes and heat at 100°C for 8h

13. Cool and dilute to 2mL with ultrapure water and heat at 100°C for 4h

2.3.2.4 Example condition for ICP-MS run

The machine used is an Agilent 7800 ICP-MS system fitted with a standard concentric nebulizer, a Peltier-cooled double-pass Scott-type spray chamber, a torch shield and standard Ni interface cones and a reaction gas chamber operating at 4.3mL/min He.

The peri-pump should be set to deliver 200μg/L yttrium as an internal standard and masses at 77, 78, 82, and 89 (Y-89) are scanned using a spectrum analysis with 3 points per mass, with 100 repetitions.

Following each analytical sample, the probe sample introduction system is rinsed with 10% nitric acid for 120s to minimize sample carryover. Calibration curves were prepared utilizing standard solutions of Se (Ultra Scientific) and QC samples utilizing a multi-element standard (Environmental Express). Additionally, mock samples were prepared for each calibration level and subjected to the sample preparation to allow adequate recovery. In all of the cases the recoveries were 100±10% (data not shown). The calibration solutions were analyzed without background subtraction using the 89 internal standard peaks. In all cases the linear regression was $r \geq 0.999$. The relative concentrations based on the use of the 77 and 82 isotopes were examined agreed within ±10%, so the average concentration was reported.

2.3.2.5 Expected outcomes

Higher peak measurements are expected to be observed from selenite treated samples and samples from high xCT cells compared with vehicle treated samples and samples from low xCT cell condition. Inhibition or KO of xCT is expected to prevent intracellular selenium.

2.3.2.6 Quantification and statistical analysis

Analyze samples using the 82 isotope of selenium. Construct a curve from 0 to 100ppb and any concentrations outside the dynamic range can be diluted to be within range. Next, calculate the results for amount selenium extracted from sample (ng/sample).

* Calculation: Quantity of selenium extracted from sample (ng/sample) = (ICP-MS concentration result/200) × 1000.

2.3.2.7 Advantages

Absolute quantification of total selenium is possible by comparing with peaks from selenium standards.

2.3.2.8 Limitations

Due to the detection limit of selenium, the basal level of selenium in low xCT cells sometimes leads to failure of detection and relatively large quantity of pellets are required for the quantification.

2.3.2.9 Safety considerations and standards

Exercise caution and utilize protective equipment such as gloves and eye protection against Nitric acid. Process samples in a safety hood.

2.4 Western blotting of selenoproteins as final readout

It is known that the level of amino acid pools can affect the expression of proteins by limiting or activating the translation of mRNA (Kimball, 2002). Active selenocysteine synthesis pathway can lead to an increased selenocysteine pool and thus increased selenoprotein production (Chen et al., 2017; Lee et al., 2021). In addition, the translational control mechanism for the expression of selenoproteins becomes more critical upon the oxidative stress by up-regulation of UGA-selenocysteine recoding efficiency (Touat-Hamici, Legrain, Bulteau, & Chavatte, 2014). Thus, regulation of selenium availability can affect the expression of selenoproteins. Supporting this notion, we previously confirmed that xCT inhibition decreases expressions of selenoproteins, and selenite supplement increases expression of GPX4 (Carlisle et al., 2020; Lee et al., 2021). Consistent with this, we show xCT inhibition by the treatment of Erastin decreases, and selenite supplement increases, the expression levels of various selenoproteins (Fig. 2N and O) in CAL120 breast cancer cells. Therefore, evaluation of the expression of selenoproteins after selenite supplements is an indirect but practical downstream readout for the selenophilicity of cells. Here we describe technical tips for western blotting of selenoproteins.

2.4.1 Technical tips for the Western blotting of selenoproteins

We confirmed that boiling did not have a major impact on the SDS-PAGE for the selenoproteins (data not shown). However, due to the low basal levels of some selenoproteins or sensitivity issue of antibodies, some targets are not easy to detect by western blotting with an relatively weak chemiluminescent substrate such as ECL. Here, we suggest taking advantage of chemiluminescent substrates which have different sensitivities—ECL provides the weakest chemiluminescent signal, Pico substrate provides moderate luminescence, while Femto substrate is so potent as to easily "burn" the blot, and thus should always be diluted 1:10 in Pico substrate. In the case of probing GPX4, which is easily detected, ECL or Pico substrate can be used depending on the cell line. For probing the more difficult to detect GPX1 and SELH, it is recommended to use the Femto:Pico mixture. Furthermore, when using Pico or Femto:Pico substrate, all the prior washes should be thorough, with extra volume of TBS-T buffer, as any small amount of background secondary signal will be amplified.

An additional tip for probing selenoproteins is to reuse the membranes and horizontally cutting membranes to save protein samples. The predicted and detected sizes of TXNRD1 (55 kDa) and SEPHS2 (48 kDa) are similar, and GPX1 (23 kDa), GPX4 (21 kDa), SELH (13.5 kDa), and SELS (21 kDa) are similar. Thus, the membrane could be cut around 30 kDa, so the upper region can be used for probing TXNRD1 or SEPHS2, and a lower part of the membrane could be used for small selenoproteins such as GPX1/4, SELH, and SELS. Another strategy is reprobing the targets. As detection of GPX1 or SELH requires Femto: Pico substrate which is the most sensitive substrate for the detection, the membrane used for detection of GPX1 and SELH could be re-used for re-probing other more robustly detected targets such as GPX4 and SELS which could be detected by the weaker ECL (or Pico), even without membrane stripping steps. Because GPX1 and SELH cannot be detected by ECL or Pico substrate, the signal from the secondary antibody bound to GPX1 or SELH antibodies will not affect the signals from the secondary antibody bound to GPX4 or SELS antibodies with ECL (or Pico) substrates. The membrane used for probing GPX1 or SELH can be just washed with Milli-Q water three times for 5 min, blocked with blocking buffer for 30 min, and then incubated with a primary antibody for GPX4 or SELH.

Lastly, to obtain a downstream readout for selenium uptake and selenocysteine biosynthesis, selenoproteins whose levels are tightly regulated by the availability of selenium should be examined via Western blot. It has been observed that limiting a selenium availability in animals decreased the expression of some selenoproteins more than others (Burk & Hill, 2015). From this finding, the concept of a selenoprotein hierarchy implying that when selenium supply is limited to support full expression of all the selenoproteins, the synthesis of some of them is decreased so that selenium can be used for others expressed more fully was introduced (Burk & Hill, 2015). For example, the MDAMB231 and NCIH2170 lung cancer cell lines that we tested, selenium restriction tightly curbs the expression of GPX1 and GPX4 (Fig. 2N and O; Carlisle et al., 2020). Also, the expressions of GPX1 and GPX4 are dramatically increased by selenium supplements. However, the expression of TXNRD1 was less affected by selenium availability in NCIH2170 cancer cells. Thus, a proper selection of selenoproteins is required for use as a final readout, and we suggest GPX4 and GPX1 as good downstream indicators of selenium uptake and selenocysteine biosynthesis leading to selenoprotein production.

- Transfer condition: proteins were transferred to 0.45 μm PVDF membrane at 390 mA for 2 h in the 1 × transfer buffer with 20% EtOH
- Washing condition: three vigorous washes for 10 min each on a rocking apparatus with 15–20 mL of 1xTBS-T buffer
- Amount of protein and detection condition: around 30μg of protein is required to get a good signal for GPX4, GPX1 and SELH with diluted Femto substrate and TXNRD1 and SELS with ECL substrate

  Dilution factors and incubation time for antibodies: all primary antibodies were diluted in 5% BSA or skim milk in TBS-T buffer with 1:1000 dilution. Secondary antibody was diluted into 5% skim milk in TBS-T buffer with 1:2000 dilution. Primary antibodies were incubated overnight (12–16 h) and secondary antibodies were incubated for 2 h.

Reagent or resource	**Source**	**Identifier**
Antibodies		
Rabbit polyclonal anti-GPX1	Cell Signaling Technology	Cat#12691
Rabbit monoclonal anti-GPX4	Abcam	Cat#ab125066
Rabbit polyclonal anti-SELH (VIMP)	Sigma-Aldrich	Cat#HPA010025
Rabbit polyclonal anti-SELS (C11ORF31)	Sigma-Aldrich	Cat#HPA048362
Mouse monoclonal anti-TXNRD1	Santa Cruz Biotechnology	Cat#sc28321

3. Summary

We have described a series of protocols which allow an investigator to determine the capacity of xCT-mediated selenium uptake and selenoprotein production of cells, which can be used in a cell culture context, either in cancer or nontransformed cells. We believe measuring these parameters are highly useful for determining which cells are "selenophilic," i.e., able to readily utilize selenite to upregulate their selenoprotein production and thus antioxidant defenses. We note that there are additional mechanisms for a cell to obtaining selenium that is likely independent of xCT, such as import of the selenocysteine carrier protein SEPP1, or via conversion of selenomethionine. Even in these cases, many of our protocols such as measurement of total selenium levels or selenoprotein expression can be easily modified to determine selenium levels or selenoprotein production following supplementation of these other selenium sources.

Acknowledgment

This work was supported by the Suh Kyungbae Foundation (SUHF) Young Investigator Award to D.K.

References

Burk, R. F., & Hill, K. E. (2015). Regulation of selenium metabolism and transport. *Annual Review of Nutrition, 35*, 109–134.

Carlisle, A. E., Lee, N., Matthew-Onabanjo, A. N., Spears, M. E., Park, S. J., Youkana, D., et al. (2020). Selenium detoxification is required for cancer-cell survival. *Nature Metabolism, 2*(7), 603–611.

Cavalieri, R. R., & Scott, K. G. (1968). Sodium selenite Se 75. A more specific agent for scanning tumors. *Journal of the American Medical Association (JAMA), 206*(3), 591–595.

Cavalieri, R. R., Scott, K. G., & Sairenji, E. (1966). Selenite (75Se) as a tumor-localizing agent in man. *Journal of Nuclear Medicine*, 7(3), 197–208.

Chen, Y. F., Lin, H. C., Chuang, K. N., Lin, C. H., Yen, H. S., & Yeang, C. H. (2017). A quantitative model for the rate-limiting process of UGA alternative assignments to stop and selenocysteine codons. *PLoS Computational Biology, 13*(2), e1005367.

Delafiori, J., Ring, G., & Furey, A. (2016). Clinical applications of HPLC-ICP-MS element speciation: A review. *Talanta, 153*, 306–331.

Dixon, S. J., Lemberg, K. M., Lamprecht, M. R., Skouta, R., Zaitsev, E. M., Gleason, C. E., et al. (2012). Ferroptosis: An iron-dependent form of nonapoptotic cell death. *Cell, 149*(5), 1060–1072.

Dominiak, A., Wilkaniec, A., Wroczynski, P., & Adamczyk, A. (2016). Selenium in the therapy of neurological diseases. Where is it going? *Current Neuropharmacology, 14*(3), 282–299.

Gonzalez-Flores, J. N., Shetty, S. P., Dubey, A., & Copeland, P. R. (2013). The molecular biology of selenocysteine. *Biomolecular Concepts, 4*(4), 349–365.

Kimball, S. R. (2002). Regulation of global and specific mRNA translation by amino acids. *The Journal of Nutrition, 132*(5), 883–886.

Labunskyy, V. M., Hatfield, D. L., & Gladyshev, V. N. (2014). Selenoproteins: Molecular pathways and physiological roles. *Physiological Reviews, 94*(3), 739–777.

Lee, N., Carlisle, A. E., Peppers, A., Park, S. J., Doshi, M. B., Spears, M. E., et al. (2021). xCT-driven expression of GPX4 determines sensitivity of breast cancer cells to ferroptosis inducers. *Antioxidants (Basel), 10*(2).

Olm, E., Fernandes, A. P., Hebert, C., Rundlof, A. K., Larsen, E. H., Danielsson, O., et al. (2009). Extracellular thiol-assisted selenium uptake dependent on the x(c)-cystine transporter explains the cancer-specific cytotoxicity of selenite. *Proceedings of the National Academy of Sciences of the United States of America, 106*(27), 11400–11405.

Sanjana, N. E., Shalem, O., & Zhang, F. (2014). Improved vectors and genome-wide libraries for CRISPR screening. *Nature Methods, 11*(8), 783–784.

Sheehan, T. M., & Gao, M. (1990). Simplified fluorometric assay of total selenium in plasma and urine. *Clinical Chemistry, 36*(12), 2124–2126.

Su, Y., Zhao, B., Zhou, L., Zhang, Z., Shen, Y., Lv, H., et al. (2020). Ferroptosis, a novel pharmacological mechanism of anti-cancer drugs. *Cancer Letters, 483*, 127–136.

Tietze, F. (1969). Enzymic method for quantitative determination of nanogram amounts of total and oxidized glutathione: Applications to mammalian blood and other tissues. *Analytical Biochemistry, 27*(3), 502–522.

Touat-Hamici, Z., Legrain, Y., Bulteau, A. L., & Chavatte, L. (2014). Selective up-regulation of human selenoproteins in response to oxidative stress. *The Journal of Biological Chemistry, 289*(21), 14750–14761.

Tsuji, Y. (2020). Transmembrane protein western blotting: Impact of sample preparation on detection of SLC11A2 (DMT1) and SLC40A1 (ferroportin). *PLoS One, 15*(7), e0235563.

Turanov, A. A., Xu, X. M., Carlson, B. A., Yoo, M. H., Gladyshev, V. N., & Hatfield, D. L. (2011). Biosynthesis of selenocysteine, the 21st amino acid in the genetic code, and a novel pathway for cysteine biosynthesis. *Advances in Nutrition, 2*(2), 122–128.

Yang, W. S., SriRamaratnam, R., Welsch, M. E., Shimada, K., Skouta, R., Viswanathan, V. S., et al. (2014). Regulation of ferroptotic cancer cell death by GPX4. *Cell, 156*(1–2), 317–331.

CHAPTER TWO

Methods for accurate and reproducible studies of pharmacological effects of selenium in cancer

Arun Kumar Selvam, Tímea Szekerczés, Sofia Björnstedt, Ali Razaghi, and Mikael Björnstedt*
Division of Pathology, Department of Laboratory Medicine, Karolinska Institutet, Karolinska University Hospital Huddinge, Stockholm, Sweden
*Corresponding author: e-mail address: mikael.bjornstedt@ki.se

Contents

Methods in Enzymology, Volume 662
ISSN 0076-6879
https://doi.org/10.1016/bs.mie.2021.10.019

Abstract

Selenium compounds have pronounced effects on cell growth and proliferation. Nutritional levels induce selenoproteins. However, the antineoplastic effects of supra-nutritional selenium levels are not mediated by selenoproteins. The most studied compound, selenite, was shown in a clinical trial to possess extraordinary pharmacological properties. The uptake of selenite as for GS-Se-SG and selenocystine is dependent on the extracellular reducing environment maintained by the X_c^- cystine transporter (x_c^- antiporter) ensuring a high level of extracellular cysteine. The expression of the x_c^- antiporter is vital for selenium cytotoxicity and any xenobiotic or media constituents modulating the expression of this antiporter will greatly affect the cellular response. Cytotoxicity determinations are often difficult to interpret and repeat due to differences in culture conditions. In the current chapter, factors influencing the cellular response, e.g., media composition, cell culturing conditions, assays for key enzymes of importance for selenium metabolism and effects, along with selenium mediated modulation of microRNA expression and immune responses are treated.

1. Introduction

Selenium compounds have dual physiological effects depending on concentration, chemical species, and the nature of the exposed cells (Selenius, Rundlöf, Olm, Fernandes, & Björnstedt, 2010). Several epidemiological studies show that low basal selenium levels are associated with an increased risk for cancer. Furthermore, supplementation to populations with low basal selenium levels have shown beneficial effects by preventing tumor development, cardiac death, and overall mortality (Alehagen et al., 2016; Alehagen, Johansson, Bjornstedt, Rosen, & Dahlstrom, 2013; Bjornstedt & Fernandes, 2010). Supra-nutritional, pharmacological levels are specifically cytotoxic to tumor cells, especially tumors highly resistant to standard chemotherapy (Misra, Boylan, Selvam, Spallholz, & Bjornstedt, 2015). Cytostatic/cytotoxic effects of selenium on tumor cells have been shown in numerous cell lines, animal models and primary cells (Misra et al., 2015). The concentration of selenium is critical and decides if selenium will act as an antioxidant or a prooxidant. Redox-active selenium compounds including selenite, selenodiglutathione and selenocysteine are of particular interest since they, at levels exceeding the dose required for the saturation of selenoproteins, will redox cycle with oxygen and thiols giving rise to a non-stoichiometric ROS production which will continue until the exhaustion of oxygen or thiols (Bjornstedt, Kumar, Bjorkhem, Spyrou, & Holmgren, 1997; Kumar, Björnstedt, & Holmgren, 1992). The redox cycles will thus cause not only oxidative stress due to ROS formation but also depletion of NADPH and thiols, leading to collapse of the defense system and the synthetic machinery of vital functions for the survival and growth of the exposed cell. The prooxidant/cytotoxic effects are surprisingly specific to cancer cells, in particular cells that are highly resistant to cytostatic drugs (Bjorkhem-Bergman et al., 2002; Madeja et al., 2005). A great body of evidence clearly shows that there thus is a therapeutic window in which levels of selenium that do not affect the viability and growth of benign/normal cells are highly cytotoxic and inhibitory to malignant, resistant cells (Selenius et al., 2010). The opposite effects of standard cytostatic drugs as compared to selenium compounds are explained by the properties of the resistant phenotype expressed by these cells. One striking feature of the resistant phenotype is the induction of intra- and extracellular thiols, the expression of membrane pumps including MRP:s and the x_c^- antiporter (Misra et al., 2015; Olm et al., 2009; Wallenberg, Misra, & Bjornstedt, 2014). The latter ensures the supply of cysteine for the synthesis of glutathione and proteins. We have previously shown that the cytotoxicity of the above-mentioned selenium compounds is completely dependent on the extracellular redox environment. Resistant cells benefit from a reducing extracellular environment since thiols will inactivate and disarm xenobiotics. The leading extracellular thiol is cysteine and the extracellular level is maintained by the x_c^- antiporter, which will mediate uptake of the oxidized dimer cystine at the expense of a glutamate molecule. Intracellularly, cystine is reduced by redox enzymes, in particular the thioredoxin system and the reduced cysteine is excreted by, e.g., MRP:s (Olm et al., 2009). The selenium metabolite selenide, a very reactive anion, is the main mediator of redox cycles causing ROS formation and cytotoxicity, but it is also the key species for selenoprotein synthesis (Bjornstedt et al., 1997). Selenide is readily taken up by the cells as compared to selenite and selenodiglutathione. Extracellular reduction of these species will thus facilitate uptake and a reducing extracellular environment will cause specific and high accumulation of selenium.

A current trend within Oncology is the concept of targeted therapy based on single mutations in the tumor cells. However, although an interesting approach, single targets often will result in the development of resistance. This is also true for several traditional, more blunt cytostatic formulations, where only one or a few mechanisms mediate the cytotoxic effects. In this context, selenium compounds are highly relevant and interesting due to multimodal effects including induction of apoptosis, cell cycle arrest, transcription factor regulation, inhibition of DNA synthesis and DNA damage (Wallenberg et al., 2014). The great number of diverse mechanisms explain why resistance to selenium is not reported and thus not considered a major challenge.

Despite a great body of evidence from in vitro and animal experiments clearly showing remarkable antineoplastic efficacy of selenium, formulations including selenium are not yet in clinical practice. The reason for this is the shortage of human clinical trials. We published in 2015 the first-in-man clinical trial on iv selenite in end-stage cancer patients (Brodin et al., 2015). The results showed an unexpectedly high tolerance with an MTD of 10.2 mg/square meter body surface, a short half-life (18 h) and no signs of accumulation or selenosis. The results are very encouraging since the pharmacokinetic properties thus will allow a rather aggressive regimen with the possibility of repeated infusions without the risk of bioaccumulation and chronic selenosis.

Selenite is, as demonstrated above, very reactive and the only route of administration, possible to achieve high systemic levels, is by intravenous infusion. There are several peroral formulations available including yeast extracts (usually with selenomethionine as the main constituent) and selenomethionine. The latter is inserted into proteins randomly in place of methionine and as a result, there is a substantial risk of selenosis after prolonged and high-level consumption. Selenomethylselenocysteine (MSC) is a very promising peroral drug due to low inherent toxicity and high bioavailability. MSC is to be considered a prodrug, since it is not toxic or accessible for selenoprotein synthesis unless metabolized by kynurenine aminotransferases (KYATs) (Selvam & Bjornstedt, 2020). These enzymes do not exist in plants but are expressed by mammalian cells and tissues, where they are important for sulfur amino acid metabolism (Cooper et al., 2011; Eliot & Kirsch, 2004; McGoldrick, Lock, Rodilla, & Hawksworth, 2003). KYAT:s catalyze MSC either to β-methylselenopyruvate (MSP) by transamination or to monomethylselenol (MMS) by β-elimination (Han, Cai, Tagle, & Li, 2010). The route is determined by the nature of the substrate, where selenium substrates favor β-elimination as compared to sulfur substrates and also by the presence of regulatory α-ketoacids. The β-elimination product, MMS, is highly reactive and considered one of the most cytotoxic selenium metabolites (Gabel-Jensen, Lunoe, & Gammelgaard, 2010; Vadhanavikit, Ip, & Ganther, 1993). For studies concerning the effects of MSC on cancer cells, knowledge of the levels and activities of KYAT:s are absolutely essential.

Herein, important factors for the accurate assessment of selenium cytotoxicity in cell lines are treated, with special emphasis on culture conditions, critical assays for thiol measurements, microRNA changes, enzymes important for selenium metabolism and cytotoxic mechanisms. Due to the chemical and pharmacological properties, the studies performed in cancer cell lines and the potential for therapeutic use, we focus the present chapter on selenite and selenomethylselenocysteine (MSC).

2. Selenium cytotoxicity in cancer cell lines

2.1 Effects of cell culture media composition on IC_{50} determinations of selenium compounds in cancer cell lines

Numerous studies report cytotoxicity of selenium compounds in cell cultures. However, the reported IC_{50} values vary greatly and experimental results in cancer cell lines are often difficult to reproduce (Arodin Selenius, Wallenberg Lundgren, Jawad, Danielsson, & Björnstedt, 2019). Table 1 summarizes the deviations in IC_{50} with different media used. Depending on media and growth conditions, the IC_{50} of the same cell line may vary by a factor of three, making comparisons impossible.

The basal constituents of commercial cell culture media are sugar, amino acids, minerals (including iron, potassium and calcium) and vitamins, all of which may affect cell proliferation. Furthermore, several media also contain trace amounts of selenium, which may disturb and alter baseline conditions. Addition of serum, usually fetal bovine serum, is a common source for growth factors and hormones, including insulin. The contents of glucose vary greatly among different media and a span of 5.5–25 mM is not unusual

Table 1 Variations in IC_{50} by using different cell culture media.

Selenite cell line	Medium	FBS	Antibiotics	IC_{50}	Evaluated by
A549	RPMI 1640	10%	Yes	~6 μM	MIT
A549	DMEM DMEM + non ess.	10%	No	8.2 μM	Sulforhodamine
A549	AA	2%	Yes	5 μM = 85% viability	MIT
HepG2	1:1DMEM:F12	10%	Yes	25.7 μM	MIT
HepG2	DMEM	10%	Yes	~10 μM	MIT (48 h)
HepG2	MEM	10%	Yes	~7.5 μM	Tunnel-assay
Huh7	DMEM	10%	Yes	~20 μM	XTT
HL-60	RPMI 1640	10%	Yes	20 μM	DNA fragmentation
NB4	RPMI 1640	10%	Yes	20 μM	Trypan blue
MSC cell line	**Medium**	**FBS**	**Antibiotics**	**IC_{50}**	**Evaluated by**
HT 29/SW480/ SW620	DMEM	10%	Yes	~64/~32/~90 μM	MTT (48 h)
HL-60	RPMI 1640	10%	Yes	50 μM	DNA fragmentation

AA, amino acids; DMEM, Dulbecco's Modified Eagle's Medium; non ess, non-essential; F12, Ham's F-12 Nutrient Mixture; FBS, Fetal bovine serum; IC_{50}, half-maximal inhibitor concentration; MEM, minimum essential medium; MIT, 3-(4,5-dimethylthiazol-2-yl)-2,5-diphenyltetrazolium bromide assay; RPMI 1640, Roswell Paris Memorial Institute 1640 Medium; XTT, (2,3-bis-(2-methoxy-4-nitro-5-sulfophenyl)-2H-tetrazolium-5-carboxanilide) assay.

Adapted from Arodin Selenius, L., Wallenberg Lundgren, M., Jawad, R., Danielsson, O., & Björnstedt, M. (2019). The cell culture medium affects growth, phenotype expression and the response to selenium cytotoxicity in A549 and HepG2 cells. *Antioxidants (Basel), 8*(5). doi:10.3390/antiox8050130.

(Arodin Selenius et al., 2019). Since glucose generally stimulates growth differences in glucose levels may influence the effect of selenium compounds to a large extent and any comparisons of IC_{50} for a particular selenium compound in a specific cell line must be done with consideration to the glucose levels in the particular experiments. Several selenium compounds are highly reactive, and constituents of media may interact with selenium as exemplified by thiols and vitamins, in particular vitamin C. In Fig. 1, a comparison between the obtained IC_{50} values for selenite and MSC in lung cancer and hepatocellular carcinoma cell lines cultured in different media shows marked variability dependent on the media used.

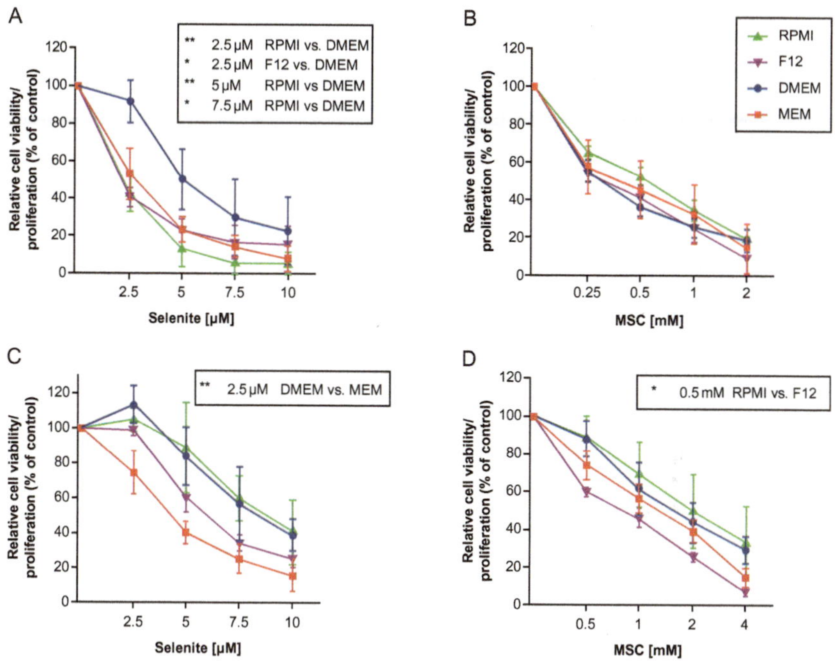

Fig. 1 Selenite and MSC cytotoxicity in human lung cancer-(A549), panels (A) and (B), and hepatocellular carcinoma cell lines (HepG2), panels (C) and (D). *Adapted from Arodin Selenius, L., Wallenberg Lundgren, M., Jawad, R., Danielsson, O., & Björnstedt, M. (2019). The cell culture medium affects growth, phenotype expression and the response to selenium cytotoxicity in A549 and HepG2 cells.* Antioxidants (Basel), 8*(5). doi:10.3390/antiox8050130.*

2.2 Phenotype expression depending on cell culture media

An important factor to consider during investigations of selenium cytotoxicity is the cellular phenotype, i.e., epithelial or mesenchymal (Nilsonne et al., 2009). This is of particular importance since the sensitivity of cells to redox-active selenium compounds has been shown to increase after the transition to mesenchymal phenotype (Nilsonne et al., 2006, 2009). Different media may affect the phenotype of cells, as demonstrated for the lung cancer cell line A549 (Fig. 2).

One simple and accurate method to monitor any shift in phenotype is by light microscopy, where the morphology will reveal a phenotypic shift from cuboidal epithelial to cigar-shaped mesenchymal. The result is further supported by immunohistochemical staining for CK18 and Vimentin (Fig. 2).

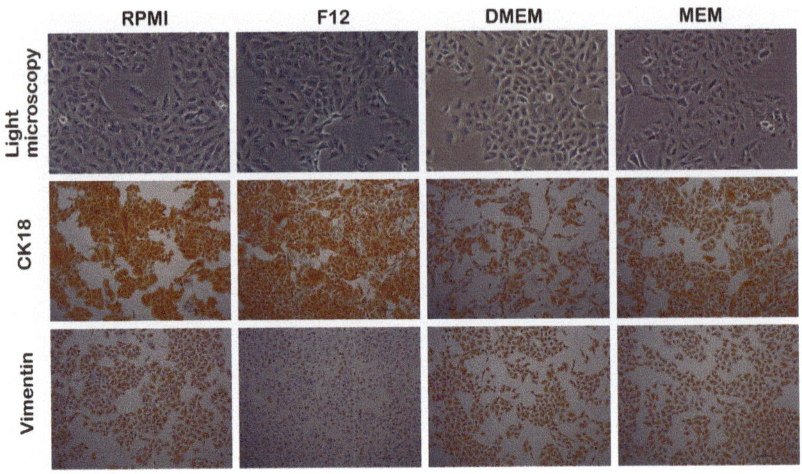

Fig. 2 Phenotypic characterization of human lung cancer cells (A549), by light microscopy (40×) and immunohistochemistry using antibodies to CK18 and vimentin. *Adapted from Arodin Selenius, L., Wallenberg Lundgren, M., Jawad, R., Danielsson, O., & Björnstedt, M. (2019). The cell culture medium affects growth, phenotype expression and the response to selenium cytotoxicity in A549 and HepG2 cells.* Antioxidants (Basel), *8(5). doi:10.3390/antiox8050130.*

2.3 Protocol for adherent cells

2.3.1 Materials

Superfrost plus glasses

Rectangular multi-well dishes (Nunclon, Fisher Scientific, Rockford, IL, USA)

PBS

Polyethylene-glycol

Anti-Cytokeratin 18 and anti-Vimentin (VIM-V9 (Novocastra, Leica Biosystems, Nubloch, Germany))

2.3.2 Procedure

(1) Cells are seeded on superfrost plus glasses, placed in a rectangular multi-well dish and incubated for 48h in an incubator under atmospheric oxygen and 5% CO_2.

(2) The attached cells are washed by PBS and fixed by polyethylene-glycol (PEG).

(3) For immunohistochemical staining to detect CK18 and Vimentin the cells are incubated with anti-Cytokeratin 18 (1:200 dilution) and anti-Vimentin (VIM-V9, 205 mg/L, 1:2000 dilution) followed by standard IHC detection by secondary antibody.

2.4 Protocol for cells in suspension

2.4.1 Preparation of cell blocks for suspension cells

For cells in suspension, a fast and reliable method is to prepare cell blocks. Cell suspension cultures can be fixed, embedded in agarose as an inert material, and processed as tissue fragments. The agarose-based compact cell block technique enables the preparation of high-quality cell blocks without loss of cells during block preparation.

2.4.2 Materials and equipment

- Low-melting point agarose
- PBS
- Formalin
- Cold plate or ice
- 1.5 mL Eppendorf tubes
- Tissue sponges
- Tissue cassettes

2.4.3 Procedure

1. Prepare 1% agarose solution using a low-melting point agar in PBS. Heat the agarose until it is completely dissolved.
2. Approximately 20×10^6 cells are centrifugated in a 1.5 mL Eppendorf tube and remove the supernatant.
3. Resuspend the cell pellets in 0.5–1 mL of 1% agarose solution.
4. The closed Eppendorf tubes are placed with the cap down and place on a cold plate or ice to solidify the agar.
5. Open the caps of the Eppendorf tubes carefully and gently pull out the agarose cylinders.
6. Place two tissue sponges in the tissue cassettes and insert the cell-agarose cylinder between the two sponges and close the cassettes.
7. Place the cassettes in formalin for more than 2 h and after formalin fixation, the samples can be embedded in a paraffin block.

Note: After the cutting of the block and deparaffination of the sections, the standard immunohistochemical procedures on these samples can be carried out without modification.

2.5 The activity of the selenoprotein thioredoxin reductase (TXNRD) in cells cultured in different media

The expression and activity of Thioredoxin reductase, an important cellular selenoprotein is affected by the media composition as demonstrated in both

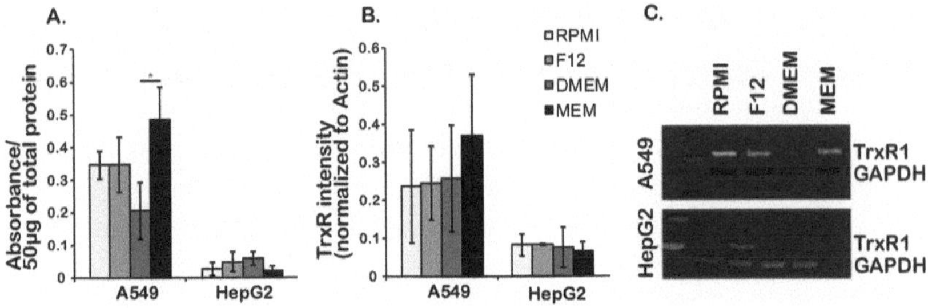

Fig. 3 The expression and activity of the selenoprotein Thioredoxin Reductase (TrxR, TXNRD) in A549 and HepG2 cells after incubation in different media. *Reproduced from Arodin Selenius, L., Wallenberg Lundgren, M., Jawad, R., Danielsson, O., & Björnstedt, M. (2019). The cell culture medium affects growth, phenotype expression and the response to selenium cytotoxicity in A549 and HepG2 cells.* Antioxidants (Basel), 8*(5). doi:10.3390/ antiox8050130.*

lung cancer and hepatocellular carcinoma cell lines Fig. 3. For methodological description see Section 4.

2.6 The extracellular redox milieu

The uptake of selenite, GS-Se-SG and selenocystine are all dependent on the x_c^- antiporter. Selenite and GS-Se-SG must be reduced to selenide, which is easily taken up by cells. The x_c^- antiporter is widely expressed by cancer cells and ensures the continuous supply of cysteine by the specific uptake of cystine at the expense of glutamate. Cystine is reduced intracellularly by the Thioredoxin system and other NADPH-dependent oxidoreductases. Due to the chemical similarities between cysteine and selenocystine, the antiporter cannot discriminate between the sulfur amino acid and its selenium analog, why this antiporter also is responsible for uptake and accumulation of selenocystine upon exposure. After reduction, excess cysteine is secreted to the extracellular space by several proteins known to confer resistance to cytostatic drugs, e.g., the MRP superfamily of proteins. Thus, the action of the x_c^- antiporter is of the utmost importance for selenium cytotoxicity (Fig. 4). Non-specific antibodies are often used to detect the x_c^- antiporter (xCT protein) in western blot and immunohistochemistry. The main reason is confusion concerning the predicted molecular weight of xCT, where the correct band in western blots is present at 35 kDa (Van Liefferinge et al., 2016). Our group has testes a wide variety of xCT antibodies from various sources and identified a reliable monoclonal antibody (clone, D2M7A, lot # 3, Ref: 40/2019). We have also verified the results by knock-down and over-expression experiments.

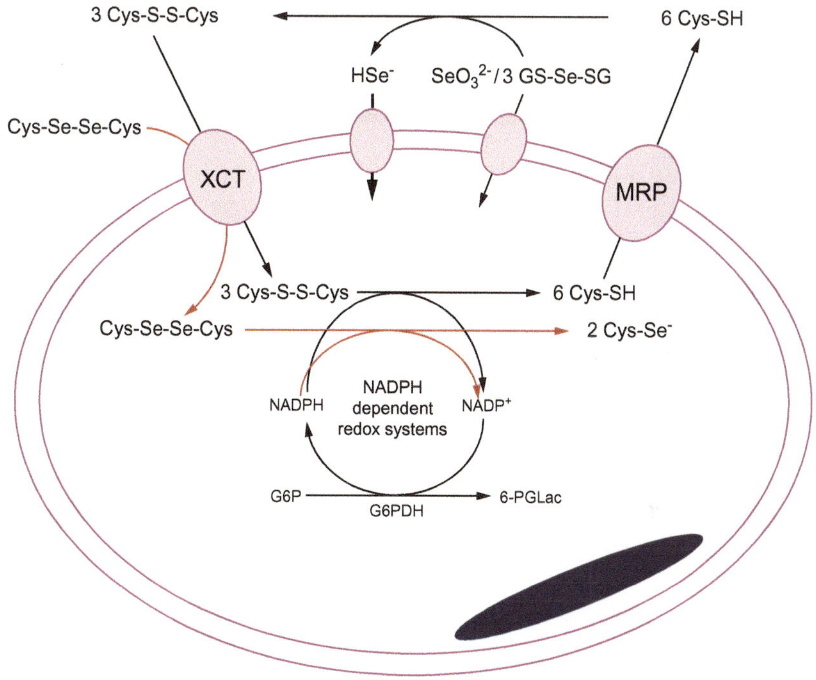

Fig. 4 Mechanism of cellular uptake of selenite, GS-Se-SG and selenocystine. The x_c^- antiporter mediates uptake of cysteine and selenocystine. The reduced cysteine is secreted by MRP:s and provides a reducing extracellular environment which reduces selenite and GS-Se-SG to selenide (HSe^-) which is readily taken up by cells. *Modified from Olm, E., Fernandes, A. P., Hebert, C., Rundlof, A. K., Larsen, E. H., Danielsson, O., et al. (2009). Extracellular thiol-assisted selenium uptake dependent on the x(c)-cystine transporter explains the cancer-specific cytotoxicity of selenite.* Proceedings of the National Academy of Sciences of the United States of America, 106*(27), 11400–11405. doi:10.1073/pnas.0902204106.*

The concentration of extracellular thiols is an important factor in any determinations of selenium cytotoxicity and the expression of the x_c^- antiporter together with the thiol concentration of the media must be measured. The concentration of extracellular thiols varies depending on the media used (Fig. 5).

Apart from the media composition, the cell density is critical for the extracellular thiol concentration and thereby for the cytotoxicity of selenite and GS-Se-SG. The thiol concentration varies depending on the cell density. A six-fold change in cell density will result in a nearly five-fold decrease in IC_{50} solely depending on the thiol secretion to the media (Fig. 6) (Misra et al., 2016). Therefore, information concerning the cell density must be provided in the description on selenium cytotoxicity in cell lines.

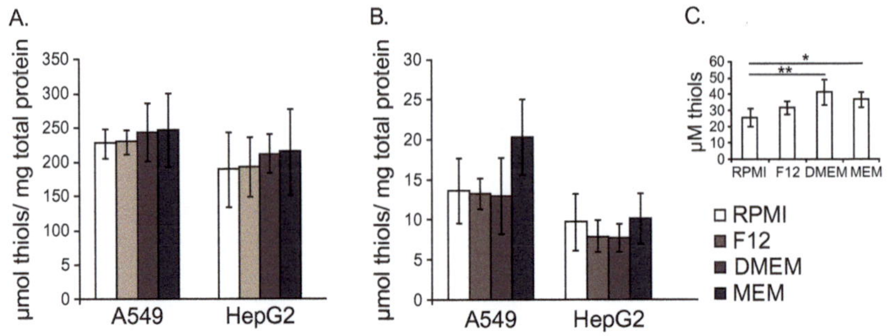

Fig. 5 Intracellular thiols (panel A) and extracellular thiols (panel B) in A549 and HepG2 cells cultured in different cell culture media and basal levels of thiols in fresh media (panel C). *Reproduced from Arodin Selenius, L., Wallenberg Lundgren, M., Jawad, R., Danielsson, O., & Björnstedt, M. (2019). The cell culture medium affects growth, phenotype expression and the response to selenium cytotoxicity in A549 and HepG2 cells.* Antioxidants (Basel), 8*(5). doi:10.3390/antiox8050130.*

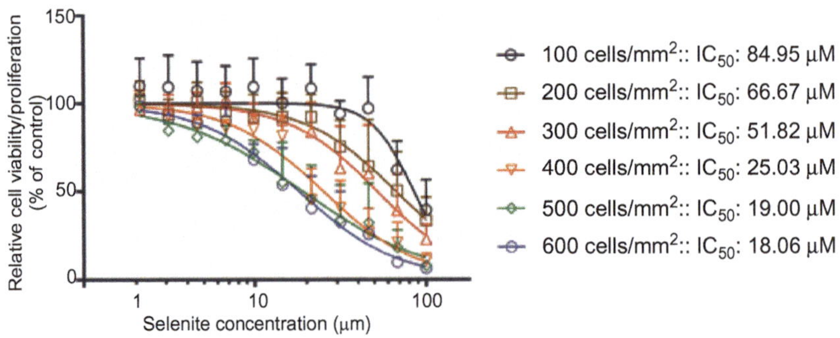

Fig. 6 The effect of cell density on IC_{50} of NB4 cells due to variable levels of extracellular thiols. *Modified from Misra, S., Selvam, A. K., Wallenberg, M., Ambati, A., Matolcsy, A., Magalhaes, I., … Björnstedt, M. (2016). Selenite promotes all-trans retinoic acid-induced maturation of acute promyelocytic leukemia cells.* Oncotarget, 7*(46), 74686–74700. doi: 10.18632/oncotarget.12531.*

2.7 Thiol measurements

2.7.1 Materials

- PBS
- Tris-EDTA buffer (50 mM/1 mM)
- Guanidine-HCL
- DTNB
- 96-Well UV flat-bottom plates
- 96-Well plate spectrophotometer

2.7.2 Procedure

(1) Culture cells for 24 or 48 h in complete cell culture media.
(2) Aspirate 1 mL of medium from cell culture and store on ice.
(3) The thiol measurements are preferably performed in a quartz 96-well plate.
(4) Samples of the medium are added to a reaction mixture in Tris-EDTA with Guanidine-HCL and DTNB in final concentrations of 2.75 M and 37.2 μg/mL respectively.
(5) The absorbance at 412 nm is recorded.
(6) Assay mixture with fresh media serves as a blank.

2.7.3 Quantification

The extinction coefficient for thionitrobenzoic acid is 13,600 M^{-1} cm^{-1}.

The values are calculated and denoted as total extracellular thiols in μM.

2.8 Alternative method—Thiol measurement by DTDP

2.8.1 Buffers and chemical preparation

1 M Monosodium phosphate (M.Wt:137.99 g mol^{-1}): Dissolve 2.07 g of NaH_2PO_4 into 12 mL of distilled water. Add distilled water to a final volume of 15 mL.

0.5 M EDTA: Dissolve 186.12 mg of EDTA into 0.8 mL of distilled water until everything gets dissolved. Add distilled water to a final volume of 1 mL.

To prepare 100 mM NaH_2PO_4 and 0.2 mM EDTA buffer (pH 4.5): Add 10 mL of 1 M NaH_2PO_4 from the stock solution and 40 μL of 0.5 M EDTA to 80 mL of distilled water. Adjust the pH to 4.5 using 10 N NaOH or concentrated HCl. Adjust the final volume to 100 mL with distilled water.

Aldrithiol/DTDP Stock preparation (Mol. Mass: 220.31):

18 mM of DTDP + 12 mM HCl: To 39.66 mg of DTDP add 3 mL of distilled water and 10 μL of concentrated HCl (37%, 12 M), stir until everything dissolved. Add distilled water to a final volume of 10 mL.

Working solution (1.8 mM): Dilute 1 mL of stock solution with 9.0 mL of dd H_2O. Store the aliquots in −20 °C.

2.8.2 Materials and equipment

100 mM Sodium phosphate EDTA buffer
Cell culture media
1.8 mM of DTDP

0.5 mL Eppendorf tubes
96-Well UV flat-bottom plates
96-Well plate spectrophotometer

2.8.3 Procedure

1 Culture cells for 24 or 48 h in complete cell culture media.
2 Aspirate 300 μL of cell culture media in the 0.5 mL Eppendorf tubes and keep it in the ice.
3 The reaction mixture contains 125 μL of buffer (pH 4.5), 75 μL of cell culture media and 25 μL of DTDP solution (1.8 mM).
4 Incubate at room temperature for 5 min.
5 Measure the absorbance at 324 nm.
6 Assay mixture with fresh media serves as a blank.

2.8.4 Expected outcomes

DTDP reacts with a thiol to form 4-thiopyridone (4-TP). The extinction coefficient is 21,400 M^{-1} cm^{-1}.

2.8.5 Quantification

The values are calculated and denoted as total extracellular thiols in μM.

2.8.6 Advantage

This assay method is rapid, simple, and 1.5-fold more sensitive than the commonly used Ellman method (Riener, Kada, & Gruber, 2002).

2.9 Interference of selenium compounds with cell viability tests and mycoplasma infection

The most commonly used viability assays are based on mitochondrial activity and may thus be influenced by redox-active selenium metabolites. When determining cytotoxic effects of a particular selenium compound, the assay results must be compared to visual determination and Trypan Blue exclusion. MSC, e.g., is interfering with the WST-1 assay and IC_{50} determinations using this assay cannot be performed. Instead, our experience is that the ATP-based assay provides a reliable method.

In cell cultures, Mycoplasma infection is a serious problem and the cells must be regularly monitored. The cytotoxic effects of selenium compounds are greatly affected by the presence of Mycoplasma and IC_{50} values are generally inconsistent in infected cells. Mycoplasma contamination in cell cultures may cause inhibition of proliferation, increase in cell death, and fragmentation of DNA. It is possible to detect such contamination by DNA staining with fluorochromes. To interpret the results, the use of indicator cell lines as positive control may be useful. The other most commonly used method is the PCR and ELISA technique.

Once Mycoplasma contamination is detected in the cell culture, the cell culture can be treated with antibiotics specific for mycoplasma. However, antibiotic resistance occurs frequently in Mycoplasma, why the effectiveness of this method is only 66%–85%. The best approach after detecting mycoplasma is to terminate the cell culture and start a fresh batch of Mycoplasma-free cells. The best and most effective way to prevent Mycoplasma contamination is to follow the strict rule of good laboratory practice (GLP), maintain aseptic techniques in cell cultures and carry out periodic and regular mycoplasma testing (Nikfarjam & Farzaneh, 2012).

3. Metabolites and selenoproteins in plasma and serum after administration of pharmacological high doses of selenite IV

Selenium metabolites and selenoproteins in plasma vary greatly depending on the dose, route of administration and selenium compound used. After high dose administration of iv selenite to cancer patients (Brodin et al., 2015) the detected low molecular weight (LMW) fraction was minor, between 1% and 28% with an average of 8%, indicating that the major part was either incorporated into proteins, loosely attached to proteins or present as elemental selenium. The leading LMW species were selenite and the selenosugar, *Se*-methylseleno-*N*-acetylgalactosamine (SeGal) (Flouda et al., 2016).

The major selenoprotein in plasma is Selenoprotein P (SELENOP). SELENOP has been considered an important biomarker for nutritional selenium status with a believed maximum concentration after intakes corresponding to 100–150 μg selenium/day. However, recently SELENOP was shown to be a reliable biomarker also for therapeutic/pharmacological high doses of selenium, greatly exceeding the recommended upper intake level, with a steep and parallel increase to total plasma selenium levels (Brodin et al., 2020). Thus, the previous hypothesis that nutritional levels will result in a maximum concentration may reflect an intermediate plateau instead of a peak value. There are several ELISA kits available for the detection of SELENOP with highly variable quality and sensitivity. Determinations of SELENOP must be accompanied by total selenium measurements and the probability of the results must be assessed in light of the total selenium level and the knowledge that the SELENOP response have an intermediate plateau in the region between the upper intake nutritional level and pharmacological high doses.

4. Mammalian thioredoxin reductase (TrxR, TR, TXNRD) assay

4.1 Buffers and chemical preparation

1 M HEPES-buffer pH 7.5: Dissolve 23.83 g of HEPES in 80 mL of distilled water. Adjust the pH to 7.5 using 10 N NaOH and add distilled water to a final volume of 100 mL.

6 M guanidine hydrochloride in 0.2 M Tris-Cl (pH 8.0) (6 M Gua-HCl): Dissolve 57.32 g of guanidine hydrochloride in 100 mL of 0.2 M Tris-HCl buffer and adjust the pH to 8.0 using 10 N NaOH or concentrated HCl.

NADPH: 40 mg/mL in distilled water. NADPH should be prepared fresh every time.

DTNB: Dissolve 6 mg of DTNB in 1.5 mL absolute ethanol (store at −20 °C).

TXNRD standard: Dilute TXNRD to 20 nM in TE-buffer with 100 μg/mL of BSA.

Stop solution: Add 1.5 mL of DTNB (4 mg/mL) to 13.5 mL of 6 M guanidine hydrochloride in 0.2 M Tris-Cl (pH 8.0).

4.2 Materials and equipment

1 M HEPES buffer pH 7.5.
20 and 50 μg of protein from crude-cell extract.
200 mM EDTA pH 8.0.
NADPH 40 mg/mL.
Insulin 10 mg/mL.
DTNB.
6 M guanidine hydrochloride in 0.2 M Tris-Cl (pH 8.0).
E.coli Trx (120 μM).
Mammalian Thioredoxin Reductase (rTXNRD) (20 nM).
2, 1.5 mL Eppendorf tubes.
37 °C Heating block.
96-Well UV flat-bottom plates.
96-Well plate spectrophotometer.

4.3 Procedure

1. Mix the following components in a 2 mL Eppendorf tube.
   400 µL of 1 M HEPES buffer pH 8.0
   160 µL 200 mM EDTA pH 8.0
   120 µL NADPH
   1000 µL insulin
2. Take 20, 1.5 mL Eppendorf tubes and separate 9 tube each, i.e., A1, A2 … A9 and B1, B2, … B9 as shown in Table 2, name tubes S1 and S2 as sample 1 and sample 2.
3. Add 40 µL of the reaction mixture to each tube in series A and B.
4. In tube series A, S1 and S2 add 10 µL of Trx (120 µM) (40 + 10 µL of Trx).
5. Prepare the standard according to the following table.

Series	Standard (TXNRD) (µL)	Tube A (with Trx) + DH_2O (µL)	Tube B (without Trx) + DH_2O (µL)
1	0	50 + 70	40 + 80
2	5	50 + 65	40 + 75
3	10	50 + 60	40 + 70
4	20	50 + 50	40 + 60
5	30	50 + 40	40 + 50
6	40	50 + 30	40 + 40
7	50	50 + 20	40 + 30
8	60	50 + 10	40 + 20
9	70	50 + 0	40 + 10

6. In the tube, S1 and S2 add 20 and 50 µg of protein from the crude-cell extract and add dH_2O to a final volume of 120 µL.
7. Seal the tubes with parafilm and incubate them at 37 °C for 20 min.
8. Place all the tubes on ice and add 500 µL of stop solution (6 M Gua-HCl + 4 mg/mL DTNB).
9. Mix by inverting the tubes a few times.
10. Load 200 µL of the samples in triplicate in a 96-well flat-bottom glass plate.
11. Read at 412 nm within 20 min.

4.4 Quantification and expected outcomes

Abs A (with Trx)—Abs B (without Trx).

A standard curve is prepared for TXNRD1 and the sample absorbance is compared to the standards.

5. KYAT determination

5.1 *Se*-methylselenocysteine

Se-methylselenocysteine (MSC) is a naturally occurring selenium compound that has very favorable pharmacokinetic properties with high peroral bioavailability in humans. MSC has been proven to have a chemopreventive property, which is shown by several authors via various cell culture models and animal models. It is also shown to have anti-carcinogenic properties via inducing cell cycle arrest and inducing apoptotic cell death (Zeng, 2009). It is considered a pro-drug because the compound per se is not toxic unless it is metabolized by the enzymes Kynurenine aminotransferase 1 (KYAT1), Kynurenine aminotransferase 3 (KYAT3) and cystathionine γ-lyase (CTH) (Cooper et al., 2008). Of these, KYAT1 plays a vital role in MSC metabolism. KYAT1 is a PLP dependent enzyme and its main function is to cleave carbon-sulfur bonds (Cooper et al., 2008). KYAT is a bi-functional enzyme that catalyzes transamination and β-elimination reactions with single substrates. KYAT1 metabolizes MSC into β-methylselenopyruvate (MSP) and methylselenol via transamination and β-elimination catalysis, respectively. MSP, structurally resembles butyrate, which is a known HDAC inhibitor (Lee et al., 2009). HDAC inhibition results in cell cycle arrest and induction of apoptosis in cancer cells through chromatin remodeling. On the other hand, methylselenol, is a redox-active compound which redox cycles with oxygen and thiols, causing massive ROS formation and anti-proliferative and pro-apoptotic effects in several cancer cell lines (Zeng, Briske-Anderson, Idso, & Hunt, 2006). Since both metabolites possess anti-tumor and anti-proliferative properties, it is important to identify the activity of KYAT1. The following section describes protocols that are regularly used in our lab to determine the activity of KYAT1 using MSC and other substrates.

5.2 Simple spectrophotometric protocol for KYAT1 transamination activity

KYAT1 transaminates a wide variety of amino acids such as glutamine, phenylalanine, leucine, kynurenine, tryptophan, methionine, tyrosine, cysteine,

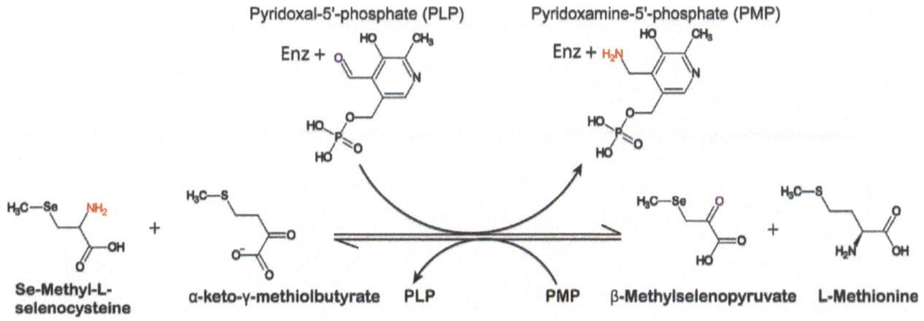

Formula 1 Transamination activity of KYAT1 with MSC in the presence of 2-ketomethiobutyrate as a co-substrate. *Reproduced from Selvam, A. K., Björnstedt, M., & Misra, S. (2018). Therapeutic potential of selenium compounds in the treatment of cancer. In B. Michalke (Ed.),* Selenium *(pp. 251–269). Cham: Springer International Publishing.*

asparagine and histidine. This enzyme is mostly catalytically active towards the L-form of amino acids. MSC is converted to β-methylselenopyruvate by transamination. Formula 1 shows the transamination activity of KYAT1 with MSC as substrate (Selvam, Björnstedt, & Misra, 2018).

5.3 Buffers and chemical preparation

5.3.1 Phosphate EDTA buffer pH 7.4

To prepare 1 M Potassium phosphate EDTA buffer, prepare two different solutions, A and B.

Solution A: Dissolve 87.09 g K_2HPO_4 (174.18 g mol^{-1}) in 0.5 L distilled water

Solution B: Dissolve 68.045 g KH_2PO_4 (136.09 g mol^{-1}) in 0.5 L distilled water

Both solutions A and B can be stored at 4 °C for longer use. To prepare 1 M Potassium phosphate EDTA buffer, add 80.2 mL of solution A, 19.8 mL of solution B and adjust the pH to 7.4 with 1 N NaOH. Finally, add 58.5 mg of EDTA disodium salt (2 mM) to the buffer and mix well to prepared 1 M Potassium phosphate EDTA buffer.

5.3.2 Ammediol-HCL buffer

Dissolve 2.1 g of 2-amino-2-methyl-1,3-propanediol in 50 mL of distilled water and adjust the pH to 9.0 with 200 mM HCl (approximately 16.7 mL) and add distilled water to a final volume of 100 mL. Buffer can be stored at 4 °C for longer use.

100 mM Se-methylselenocysteine: Dissolve 21.85 mg of MSC in 1 mL of distilled water and adjust the pH to 6.5 with 1 N NaOH. Store at −20 °C.

10 mM Phenylpyruvic acid: Dissolve 1.64 mg of phenylpyruvic acid in 1 mL of DMSO and store at −20 °C.

200 mM L-Phenylalanine: Dissolve 33.04 mg of L-Phe in 1 mL of distilled water. Store at 4 °C.

500 mM α-keto-γ-methylthiobutyric acid sodium salt: Dissolve 85.08 mg of KMB in 1 mL of distilled water. Store at −20 °C.

1 N Sodium hydroxide: Dissolve 4 g of NaOH pellets in 100 mL of distilled water.

5.3.3 Materials and equipment

1 M Potassium phosphate EDTA buffer
Whole-cell extract or pure KYAT1 enzyme
100 mM *Se*-methylselenocysteine
10 mM Phenylpyruvic acid
1 N Sodium hydroxide
0.5 mL Eppendorf tubes
37 °C Heating block
96-well UV flat-bottom plates
96-well plate spectrophotometer

5.3.4 Procedure

1. Mix the following components in a 0.5 mL Eppendorf tube.
   5 μL 1 M Potassium phosphate EDTA buffer
   2.5 μL 100 mM *Se*-methylselenocysteine
   3 μL 10 mM Phenylpyruvic acid
   39.5 μL distilled water
2. Incubate the reaction mixture at 37 °C for 5 min.
3. Add 20 μg of whole-cell extract or 100 ng of pure KYAT1 enzyme to the reaction mixture, such that the final volume of the reaction mixture is 50 μL.
4. Incubate the reaction mixture at 37 °C for 10 min.
5. Stop the reaction by adding 150 μL of 1 N NaOH.
6. Transfer the reaction mixture to 96-well UV flat-bottom plates.
7. Read the absorbance immediately at 320 nm in a 96-well plate spectrophotometer against a blank containing a complete reaction mixture without an enzyme.

5.3.5 Expected outcomes

The amount of phenylpyruvate consumed is calculated from the decrease in absorbance at 320 nm against blank. The extinction coefficient is 16,000 $M^{-1}\ cm^{-1}$.

5.3.6 Quantification

The values are calculated and denoted as mmol/min/mg of protein.

5.3.7 Advantage

This assay is simple and quick to determine the transamination activity of KYAT1 with MSC.

Note: If the spectrophotometer has an inbuilt temperature controlling facility, set the temperature to 37 °C in the spectrophotometer 10 min before the start of the experiment. Using such a spectrophotometer instead of using a heating block reduces the temperature fluctuation during the enzyme reactions.

5.4 Alternative method

5.4.1 Materials and equipment

200 mM Ammediol-HCL buffer pH 9.0
Whole-cell extract or pure KYAT1 enzyme
200 mM L-Phenylalanine
500 mM α-keto-γ-methylthiobutyric acid
1 N Sodium hydroxide
0.5 mL Eppendorf tubes
37 °C Heating block
96-Well UV flat-bottom plates
96-Well plate spectrophotometer

5.4.2 Procedure

1. Mix the following components in a 0.5 mL Eppendorf tube
   25 μL Ammediol-HCL buffer
   5 μL L-Phenylalanine
   18 μL distilled water
2. Incubate the reaction mixture at 37 °C for 2 min.
3. Add 2 μL α-keto-γ-methylthiobutyric acid and 20 μg of whole-cell extract or 100 ng of pure KYAT1 enzyme to the reaction mixture, such that the final volume of the reaction mixture is 50 μL.

4. Incubate the reaction mixture at 37 °C for 30 min.
5. Stop the reaction by adding 150 μL of 1 N NaOH.
6. Transfer the reaction mixture to 96-well UV flat-bottom plates.
7. Read the absorbance immediately at 320 nm in a 96-well plate spectrophotometer against a blank containing a complete reaction mixture without an enzyme.

5.4.3 Expected outcomes

The amount of phenylpyruvate enol formed is calculated from the increase in absorbance at 320 nm against blank. The extinction coefficient is 16,000 M^{-1} cm^{-1}.

$$\text{L} - \text{Phenylalanine} + \alpha - \text{keto} - \gamma - \text{methylthiobutyric acid}$$
$$\rightarrow \text{Phenylpyruvate enol} + \text{L} - \text{methionine}$$

5.4.4 Quantification

The values are calculated and denoted as mmol/min/mg of protein.

5.4.5 Advantage

This assay is simple, sensitive, and quick to determine the transamination activity of KYAT1 for aromatic amino acids. KYAT1 activity towards MSC can be determined indirectly from this assay by adding MSC into the assay mixture. The reaction mixtures are as follows:

(1) Blank: reaction mixture without enzyme
(2) In tube (2), a complete enzyme mixture as explained above
(3) In tube (3), a completed enzyme mixture with 5 mM of MSC (L-Phe +MSC)

The absorbance in tube 3 will be reduced compared to tube 2. The reduction in absorbance can be denoted in percentage compared to control (tube 2) and can be denoted as a competitive substrate inhibition in percentage.

Note: If the spectrophotometer has an inbuilt temperature controlling facility, set the temperature to 37 °C in the spectrophotometer 10 min before the start of the experiment. All the incubation steps should be done in the spectrophotometer. Using such a spectrophotometer instead of using a heating block reduces the temperature fluctuation during the enzyme reactions. This assay results greatly vary with temperature fluctuations and KMB incubations (KMB should always be added along with the enzyme).

5.5 Simple spectrophotometric protocol for KYAT1 β-elimination activity

KYAT1 demonstrates higher β-elimination activity towards Se-conjugate compounds over corresponding S-conjugate compounds. This is likely due to lower C—Se bond energy as compared to the C—S bond. MSC is cleaved into methylselenol by KYAT1 via β-elimination (Formula 2). Methylselenol is highly volatile and considered extremely reactive. This compound is of particular interest in pharmacological applications because of its efficient inhibition of cancer cell growth and viability. MS is highly unstable, why it is difficult to detect during the reaction. A limited number of assays have been described to detect the β-elimination reaction, with variable reliability and reproducibility. In this section, we describe two different assays to detect the β-elimination activity of KYAT1 with MSC.

5.5.1 Buffers and chemical preparation

1 M Phosphate EDTA buffer pH 7.4, 100 mM *Se*-methylselenocysteine, and 1 N Sodium hydroxide preparation as shown in the previous section.

10 mM α-keto-γ-methylthiobutyric acid sodium salt (KMB): To 20 μL of 500 mM KMB stock add 980 μL of distilled water. Store at −20 °C.

1 mM Pyridoxal 5′-phosphate hydrate (PLP): To prepare 10 mM stock, dissolve 2.47 mg of PLP in 1 mL of distilled water. Store at −20 °C. Prepare 1 mM of PLP from the above stock by diluting with 1:10 ratio V/V in distilled water.

10 mM α-Ketoglutaric acid sodium salt (α-KG): Dissolve 1.68 mg of α-KG in 1 mL of DMSO and store at −20 °C.

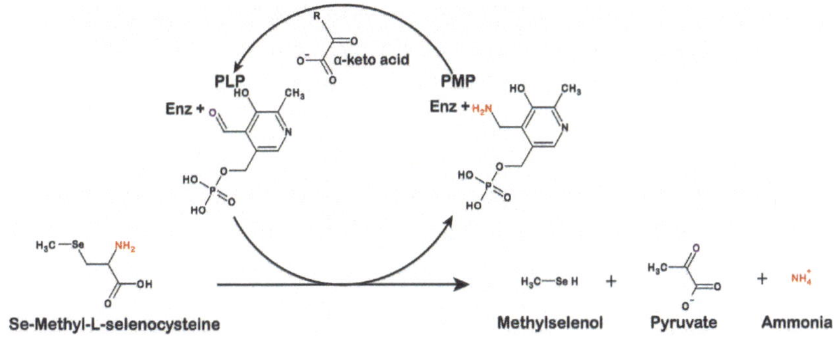

Formula 2 Beta-elimination activity of KYAT1 with MSC. *Reproduced from Selvam, A. K., Björnstedt, M., & Misra, S. (2018). Therapeutic potential of selenium compounds in the treatment of cancer. In B. Michalke (Ed.),* Selenium *(pp. 251–269). Cham: Springer International Publishing.*

5 mM 2,4-Dinitrophenylhydrazine in 2 M HCl (DNPH): Dissolve 49.6 mg of DNPH in 50 mL of 2 M HCl. Can be stored in the air-flow hood at room temperature for at least 6 months.

5.5.2 Materials and equipment

1 M Potassium phosphate EDTA buffer
Whole-cell extract or pure KYAT1 enzyme
100 mM MSC
10 mM KMB
10 mM α-KG
1 mM PLP
5 mM 2,4-Dinitrophenylhydrazine
1 N Sodium hydroxide
0.5 mL Eppendorf tubes
37 °C Heating block
96-Well flat-bottom plates
96-Well plate spectrophotometer

5.5.3 Procedure

1. Mix the following components in a 0.5 mL Eppendorf tube.
   20 μL 1 M Potassium phosphate EDTA buffer pH 7.4
   10 μL 100 mM MSC.
   2 μL 10 mM KMB.
   2 μL 10 mM α-KG.
   2 μL 1 mM PLP.
   164 μL distilled water
2. Incubate the reaction mixture at 37 °C for 2 min.
3. Add 80 μg of whole-cell extract or 800 ng of pure KYAT1 enzyme to the reaction mixture, such that the final volume of the reaction mixture is 200 μL.
4. Incubate the reaction mixture at 37 °C.
5. Take 50 μL from the enzyme reaction mixture at various time points, i.e., 0, 5, 30, and 60 min. Stop the reaction by adding 20 μL of 5 mM DNPH.
6. Incubate the reaction mixture for 10 min at room temperature.
7. Add 130 μL of 1 M NaOH to produce a reddish-brown color.
8. Transfer the reaction mixture to 96-well flat-bottom plates immediately.
9. Read the absorbance immediately at 515 nm in a 96-well plate spectrophotometer against a blank containing a complete reaction mixture without a substrate.

5.5.4 Expected outcomes

The amount of pyruvate formed is calculated from the increase in absorbance at 515 nm against blank. The extinction coefficient is 16,000 M^{-1} cm^{-1}. Pyruvate formation is measured at various time points to determine the linearity of pyruvate formation.

5.5.5 Quantification

The values are calculated and denoted as nmol of pyruvate produced/min/mg of protein.

5.5.6 Advantage

This assay is simple and quick to determine the β-elimination activity of KYAT1 with MSC.

Note: Other keto acids in the assay mixture such as KMB and α-KG will also react with DNPH to generate a reddish-brown color, which can be seen in the blank. This is the major drawback of this assay, i.e., this assay cannot differentiate between pyruvate and other keto acids. Moreover, this assay is an indirect method that measures pyruvate production and not the production of methylselenol.

5.6 Alternative and preferred method

Coupled-β-elimination KYAT1 activity assay.

5.6.1 Materials and equipment

1 M Potassium phosphate EDTA buffer
Whole-cell extract or pure KYAT1 enzyme
100 mM MSC
10 mM KMB
10 mM α-KG
1 mM PLP
0.4 mM NADPH
Mammalian thioredoxin reductase 1 (TXNRD 1)
0.5 mL Eppendorf tubes
96-Well UV flat-bottom plates
96-Well plate spectrophotometer with inbuilt heating

5.6.2 Procedure

1. Turn on the spectrophotometer and set the temperature to 37 °C.
2. Mix the following components in a 96-well UV flat-bottom plate.
   10 μL 1 M Potassium phosphate EDTA buffer pH 7.4
   5 μL 100 mM MSC

1 μL 10 mM KMB
1 μL 10 mM α-KG
1 μL 1 mM PLP
(varies) 20 μg of whole-cell extract or 200 ng of pure KYAT1 enzyme
80 μL distilled water

3. Incubate the reaction mixture at 37 °C for 5 min.
4. Add 2 μL of 0.4 mM NADPH and TXNRD1 to a final concentration of 30 nM (in 100 μL). Add distilled water to a final volume of 100 μL.
5. Continuously measure the NADPH consumption at 340 mm in the spectrophotometer from 0 to 60 min (kinetic measurement at an interval of 30 s or 1 min).
6. The assay mixture without TXNRD1 is used as a blank.

5.6.3 Expected outcomes

Reaction as follows in Formula 3: KYAT1 cleaves MSC into methylselenol (MS), MS further reacts with oxygen to generate methylseleninate. Methylseleninate will be reduced by TXNRD1 as a substrate to generate MS at the expense of 2 NADPH. We assume that 2 mol of NADPH correspond to 1 mol of MS. The extinction coefficient is 6220 M^{-1} cm^{-1}.

5.6.4 Quantification

The values are calculated at 15 and 30 min depending on the rate of the reaction and denoted as nmol of NADPH consumed/min/mg of protein.

5.6.5 Advantage

This assay is simple and quick to determine the β-elimination activity of KYAT1 with MSC and measure the formation MS rather than any other co-product.

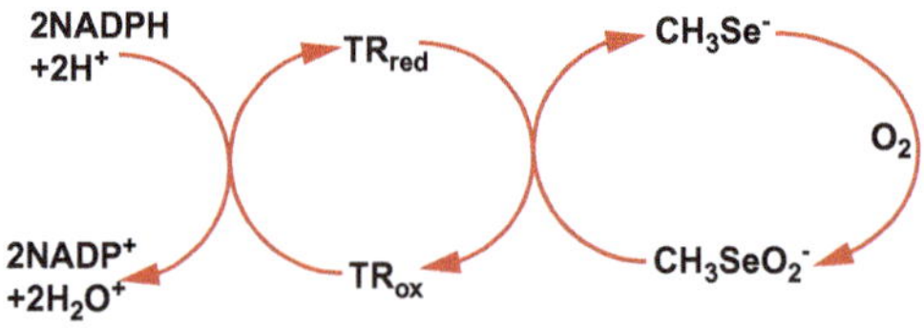

Formula 3 Proposed mechanism of TR (TXNRD) activity in reducing methylseleninate. *Modified from Selvam, A. K., & Bjornstedt, M. (2020). A novel assay method to determine the beta-elimination of Se-methylselenocysteine to monomethylselenol by kynurenine aminotransferase 1.* Antioxidants (Basel), *9(2). doi:https://doi.org/10.3390/antiox9020139.*

Note: NADPH should be prepared freshly before the start of every experiment and temperature fluctuations will greatly affect the enzyme kinetics. Thus, care should be taken by incubating the device at 37 °C at least 10 min before the start of the experiment.

6. Selenium treatment and microRNA expression

MicroRNAs are short, non-coding RNA molecules with a central role in sequence-specific post-transcriptional regulation (Lendvai et al., 2020). Although their presence has been known for decades, microRNAs are still a major research topic. Depending on the targets, microRNAs may function as oncomicroRNAs or tumor suppressor microRNAs, thereby affecting different molecular pathways in tumors (Tarek et al., 2017). Several studies suggest that due to their stability in serum, microRNAs can be used to diagnose and monitor diseases (Lendvai et al., 2020; Tarek et al., 2017).

Considering the importance of microRNAs in multiple biological mechanisms, it is worth examining how treatments may change their expression and regulation of each molecular pathway, but only a few studies have addressed microRNA levels related to selenium treatment (Lendvai et al., 2020).

Alehagen et al. shed light on the potential connection between selenium and microRNA expression. Plasma from a placebo-controlled trial, in which 443 healthy elderly were supplemented during 4 years with either placebo or a combination of selenium and Q10, was analyzed for microRNA levels. The expression of the 172 most commonly occurring microRNAs in plasma was detected using qPCR analysis. Significant differences were noted in the expression of more than 100 microRNAs with up to fourfold differences. In addition, a correlation was found between microRNA-130 and pretreatment selenium levels, and microRNA-130 was reduced after the treatment (Alehagen, Johansson, Aaseth, Alexander, & Wågsäter, 2017).

In another study, the relationship between selenoprotein P and microRNA-7 was determined by analyzing the serum of patients with hepatocellular carcinoma (HCC) and in vitro studies. Selenoprotein P is an extracellular protein produced primarily by the liver, which is responsible for transporting selenium to extrahepatic tissue and protecting against oxidative damage (Tarek et al., 2017). In this study, selenoprotein P levels measured with ELISA and microRNA-7 levels measured using qPCR analysis were also low in the serum of HCC patients compared to chronic hepatitis. Furthermore, microRNA-7 and selenoprotein P levels were associated negatively with alpha-fetoprotein, which is often used to diagnose HCC. Therefore microRNA-7 and selenoprotein P were also recommended as a potential marker of HCC in this study. In in vitro studies, sodium selenite caused altered levels of microRNA-7 and selenoprotein P in HepG2 cells. Based on these results, sodium selenite was suggested as an antitumor therapy in HCC (Tarek et al., 2017).

We have recently published a study in which the effects of selenium on the expression of several key microRNAs in liver cancer cell lines were systematically explored (Lendvai et al., 2020). The selected microRNAs were microRNA-21, -22, -24, -122, -125b, -143, -194, -199a and let-7a (normal liver); microRNA-210 (hypoxia survival); microRNA-224 (promotes proliferation).

As shown in Table 2, only a few microRNAs were significantly changed during treatment, although fold change differences were observed in almost all. However, the most pronounced change was in the expression of microRNA-122 after selenite treatment. Since this microRNA is characteristic of normal liver cells and is known to function as a tumor suppressor, this finding might reflect a beneficial effect of selenium. In contrast to selenite, MSC induced less pronounced changes. This demonstrates that different selenium species may affect the expression of microRNA differently and underline the need of measuring these effects in pharmacological experiments. In addition, microRNA antisense elements may be used to regulate protein expression in formulations of stable mRNA (Selvam et al., 2021). Therefore, in any experimental protocol using microRNA switches, the effects of concomitant selenium treatment on microRNA expression have to be monitored and considered.

6.1 Determination of microRNA expression

There are several techniques used in quantitative and qualitative microRNA analyses, including real-time qPCR, microarray hybridization and next-generation sequencing (NGS). The qPCR-based microRNA profiling is limited to analyses of known microRNA profiles but can be used on a wide variety of sample materials and is especially useful if the amount is scarce. Although NGS technology is more sensitive and provides opportunities to analyze a wide range of microRNAs including unknown, it requires a large investment and a greater degree of bioinformatic analysis. Thus, the most widely used technique is qPCR (Chugh & Dittmer, 2012).

Table 2 The effects of MSC and selenite on microRNA expression in liver cancer.

	MSC		Sodium selenite	
microRNAs studied	**Significant different**	**Fold changes different**	**Significant different**	**Fold changes different**
microRNA-122		HLE, HLF, TFK1, HuH-28	HLF	HLE, HLF, TFK1, HuH-28
Let-7a		TFK1, HuH-28		HLE, HLF, TFK1, HuH-28
microRNA-22		HLE, HuH-28	HLF	HLE, HLF, TFK1, HuH-28
microRNA-125b				HLE, TFK1, HuH-28
microRNA-143	TFK1	TFK1, HuH-28	HLF	HLE, HLF, TFK1
microRNA-21		HLE, HuH-28		HLE, HLF, TFK1, HuH-28
microRNA-24		HLE, HuH-28	HLF	HLF, TFK1, HuH-28
microRNA-194		HLE		HLF, HuH-28
microRNA-199a	HLF, TFK-1	HLE, HLF, HuH-28		HLF, TFK1, HuH-28
microRNA-224		TFK1, HuH-28	HLF	HLF, TFK1, HuH-28
microRNA-210		HLE, TFK-1	HLF, TFK-1	HLE, HLF, TFK1

MicroRNA expression in hepatocellular carcinoma (HLE and HLF) and cholangiocarcinoma cell lines (TFK-1 and HuH-28) after selenite or MSC treatment, measured using TaqMan MicroRNA Assay (Lendvai et al., 2020).

6.2 qPCR-based method for measurement of microRNA expression

The first step in analyzing microRNA expression is the extraction of total RNA. One of the most well-known methods is direct lysing with Trizol followed by RNA extraction with chloroform and isopropanol. However, Trizol-free RNA isolation kits with comparable performance are preferred due to safety concerns.

Two different methods among commercially available qPCR-based platforms can be used after total RNA isolation. One of them is the universal system which generates poly-A tails and an anchored poly-T primer for cDNA preparation. The other method uses the stem-loop primers (Taqman™ primers) which are sequence-specific thus resulting in a more sensitive system compared to linear primers (Chugh & Dittmer, 2012).

Further, the use of qPCR technology will be introduced through TaqMan MicroRNA Universal Assay.

6.3 Materials

- RNase-free sterile-filtered water
- TaqMan MicroRNA Reverse Transcription Kit (Applied Biosystems, Cat. No: 4366596)
- TaqMan MicroRNA Universal Master Mix (Applied Biosystems, Cat. No: 4440043)
- TaqMan MicroRNA primers

6.4 Equipment

- Eppendorf tubes (1.5 mL)
- Microcentrifuge, Plate centrifuge
- Gradient PCR machine (Eppendorf PCR Mastercycler)
- PCR tubes (0.5 mL)
- Vortex
- Real-Time PCR Instruments
- Optical 96-Well Thermal Cycling Plate
- Optical Adhesive Covers

6.5 Procedure

1. Isolate RNA using any methods as mentioned above.
2. Dilute total RNA samples with RNase-free water to 20 ng/μL concentration.
3. Prepare master mix on ice using the TaqMan MicroRNA Reverse Transcription Kit according to the manufacturer's recommendation.

   Note: To save the reagents, half the quantity of the recommended volume is sufficient.
4. Add 1.5 μL reverse transcription primer to the master mix and 0.5 μL of total RNA sample (20 ng/μL). Thus, the reverse transcription reaction is carried out in a final volume of 7.5 μL containing 10 ng of total RNA.

5. Mix gently and centrifuge.
6. Incubate the reaction on ice.
7. Place the PCR tubes in the Gradient PCR instrument.
8. After reverse transcription reaction, centrifuge the samples and proceed to quantify the microRNA expression or store the samples at −80 °C for later analysis.

6.6 qPCR

During qPCR, negative and positive controls should be used. The negative control contains all PCR reaction reagents except for cDNA samples. The positive control should be a known microRNA expression sample. Furthermore, it is necessary to test the microRNA housekeeping genes in addition to the selected microRNAs.

All reactions should be performed in triplicate:

1. Dilute cDNA samples on ice. cDNA samples were diluted according to the manufacturer's instructions. Use an amount of cDNA that yields a Ct value of between 15 and 35 cycles.

   Note: We recommend to use 5–10 times dilution.
2. Prepare the master mix in nuclease-free water in a 1.5 mL Eppendorf tube. Mix gently and centrifugate the 1.5 mL Eppendorf tube and aliquot 26.5 μL (for three sample sizes) into 0.5 mL Eppendorf tube.
3. Add 1.5 μL microRNA primer and 2.0 μL of cDNA sample and resuspend.

   Note: The recommended reaction volume is 20 μL but can be scaled down to 10 μL to save the reagents. In addition, 10% extra volume should be prepared to compensate for the pipetting errors.
4. Transfer 10 μL samples to each well of an Optical 96-Well Thermal Cycling Plate.
5. Seal the plate with an optical adhesive cover and briefly centrifuge the plate.
6. Load the plate in the instrument and set the program according to the manufacturer's instructions.

   Note: Change the settings of the instrument to fit the final volume of 10 μL. If the UNG enzyme is used, then incubate the reaction mixture at 50 °C for 2 min before the 95 °C step.
7. Following qPCR, determine the threshold cycles (Ct) using the real-time PCR system software and normalize the Ct value of the sample using the Ct value of endogenous controls. Compare the ΔCt of the treatment samples to ΔCt of control, determine the treatment effects on microRNA expression (ΔΔCt).

7. Selenium and the immune system

An increasing number of publications suggest that antiproliferative, chemopreventive and cytotoxic mechanisms of selenium in cancer are mainly mediated by modulation of the immune system.

7.1 Effects of selenium on antitumor immunity

The tumor microenvironment (TME) promotes tumor progression and angiogenesis by switching to M2 tumor-associated macrophages (TAMs), which release immunosuppressive and anti-inflammatory factors like interleukin (IL)-4 and IL-10. Selenium supplementation, on the other hand, causes TAMs to polarize towards antitumor proinflammatory M1 macrophages. As a result, it appears that selenium affects macrophage polarization in TME in the opposite direction of bacterial and viral infections. The TME also contains a large number of CD4+ T-lymphocytes, also known as T regulatory cells, which suppress the antitumor response. In addition, CD4+ T-lymphocyte subsets including Th2 (releasing IL-4, IL-5, and IL-13) and Th_{17} (releasing IL-17A, IL-17F, IL-21, and IL-22) have been linked to tumorigenesis and inflammation. However, selenium turns the TME's immunosuppression into proinflammatory antitumor immunity. Higher selenium intake, for example, inhibits tumor growth by suppressing antitumor immunity, which is triggered by inducing Th_1 immune responses and a decrease in myeloid-derived suppressor cells and CD4+ T-lymphocytes. Selenium supplementation, in other words, redirects CD4+ Th_2-immunity toward the Th_1 immunity that protects against cancer. Furthermore, higher selenium levels stimulate the cytotoxic activities of natural killer cells and lymphokine-activated killer cells (cytotoxic lymphocytes stimulated by IL-2 to kill tumor cells) (Razaghi et al., 2021).

It has been hypothesized that selenium may induce the proinflammatory response in the TME through two pathways. (i) selenium plays a prooxidant role inducing ROS, which activates the Akt-NF-κB pathway. (ii) selenium plays an antioxidant role through selenoprotein synthesis such as TXNRD which relocates to the nucleus and activates NF-κB, thus, resulting in activation of proinflammatory cytokine genes and leukocytes (Razaghi et al., 2021).

Thus far, research has revealed mechanistic insights into selenium's effects on immune cells and inflammatory networks; modifying these pathways could theoretically serve as a therapeutic portal. Selenium regulates redox homeostasis and inflammatory pathways in cells via selenoproteins. As a result, selenium can influence the interaction between tumor and immune cells in the TME, which may help to clarify the best uses of selenium for cancer treatment in the clinical stage (Razaghi et al., 2021).

7.2 Cytokine assays: Multiplex immunoassays

Cytokines are essential parts of innate and adaptive immunity. As key players in the TME, these molecules can act as biomarkers, i.e., systemic cytokine concentrations may serve as parameters to study inflammatory pathways in patients. However, trivial processes (e.g., sample freeze-thawing cycles and storage) can influence the reproducibility and reliability of cytokine detection. Cytokines are stable at $-80\,°C$ storage for ≤ 2 years. However, some cytokines (CXCL8, IL-1α, IL-1β, IL-10, and IL-15) are degraded $\leq 75\%$ of baseline values after 4 years. IL-6 and IL-10 are stable during several freeze-thawing cycles, while CXCL8, IFNγ, IL-4, IL-13, IL-15, IL-17 and TNFα are unstable suggesting that samples for cytokine analysis ought not to be subjected to repeated freeze-thawing cycles (de Jager, Bourcier, Rijkers, Prakken, & Seyfert-Margolis, 2009).

To assess cytokine levels, a number of protocols can be employed including bioassays, sandwich ELISA, high-performance liquid chromatography (HPLC), protein microarrays and bead-based multiplex immunoassays (MIA). Due to the interactions of cytokines, MIAs are in the vanguard of cytokine analysis measuring multiple cytokines simultaneously. However, sample standardization is required before these proteins can act as biomarkers (Keustermans, Hoeks, Meerding, Prakken, & de Jager, 2013).

In MIA, the magnetic beads are coated with specific capture primary antibodies. Subsequently, incubated with a small volume of sample. Later, labeled detection secondary antibodies bind to the bead complex making a four-member solid phase sandwich which allows the quantification of the desired compound while passing through the detection system (Keustermans et al., 2013).

In comparison to ELISA, MIAs require smaller sample volumes, are highly specific, rapid, and more sensitive showing a broader analytical range—detecting $pg\,mL^{-1}$, allowing the concurrent measurement of over 500 proteins (xMAP technology, Luminex, Texas, USA). However, MIAs, similar to the rest of antibody-based methods, are influenced by heterophilic and autoantibodies causing false positive and false negative signals. To troubleshoot this phenomenon, three methods may be undertaken as follows.

(1) Using animal serum to block heterophilic/autoantibody (ineffective if antibody titers are high).
(2) Using internal assay markers which allow the interfering heterophilic or auto-antibodies to be monitored. This allows the exclusion of unfit samples in the course of analysis.
(3) The most preferable technic is the elimination of heterophilic/autoantibodies by incubating samples with either antibodies cocktail such as HeteroBlock or protein-L (Keustermans et al., 2013).

8. Concluding remarks

Selenium compounds are emerging to become highly interesting antineoplastic compounds and pharmacological formulations based on the outstanding nucleophilic and highly reactive properties of selenium are likely future evident constituents of modern cancer therapeutics. The multimodal action of selenium is of particular interest in the current era of so-called "targeted therapies" where single mutations are targets—a strategy that has proven disappointing due to a fast accommodation and development of resistance likely explaining the failure of improving prognosis in highly resistant visceral malignancies. Selenium acts truly by a multitude of mechanisms making the development of resistance very hard if not impossible. The challenge so far has been to specifically target malignant cells with no or only minor off-target effects on benign, normal cells and tissues demonstrating the need for reliable and reproducible study models. This is demonstrated by the fact that mechanistic studies of the effects of selenium are limited by severe difficulties in accurate and reproducible determinations of the antineoplastic effects in various cell lines. Due to the multimodal action; the culture conditions, media composition and method employed to determine cell viability will greatly affect the result and must be defined for comparisons to be possible. The levels of glucose and micronutrients including minerals and vitamins in the media are very important and may alter the IC_{50} determinations by several-fold. The extracellular redox environment decides the uptake of several selenium compounds and must be defined and monitored. In this context, the expression of the x_c^- antiporter must be defined and any constituents or xenobiotics used that will modulate the expression of the x_c^- antiporter must be defined and taken into consideration when defining the IC_{50} of selenite, GS-Se-SG or selenocystine using a particular cell line. The effects of MSC must thus be interpreted in relation to information concerning the levels and relative activities of KYAT.

The leading metabolic enzyme of importance for selenite and GS-Se-SG metabolism is the selenoenzyme mammalian Thioredoxin Reductase (TXNRD), and thus the activity of TXNRD is important to monitor in studies of cytotoxicity. MicroRNAs are small RNA fragments that have several physiological functions and may determine the phenotype of specific cells. Since the expression of proteins may be regulated by microRNA antisense sequences in pharmacological and vaccine formulations, the effects of selenium treatment of microRNA levels are important for studies of complex systems in which selenium is used to modulate cell growth. In addition, some antineoplastic and growth inhibitory effects of selenium may be explained by modulation of immune function—the effects on immunological responses must be considered in mechanistic studies on selenium in cancer. We herein demonstrate that studies of selenium cytotoxicity are complex, and several factors must be considered in order to achieve reproducible and comparable results.

Acknowledgments

This work was supported by grants from Cancerfonden, Cancer och Allergifonden, Radiumhemmets forskningsfonder, Karolinska Institutet to M.B and The Wenner-Gren foundation to M.B and T.S.

Conflicts of interest

M.B. is listed as an inventor in a patent application for i.v. use of inorganic selenium in cancer patients and holds shares in SELEQ OY, a company involved in the development of Se-based formulations for prevention and treatment.

References

Alehagen, U., Johansson, P., Aaseth, J., Alexander, J., & Wågsäter, D. (2017). Significant changes in circulating microRNA by dietary supplementation of selenium and coenzyme Q10 in healthy elderly males. A subgroup analysis of a prospective randomized double-blind placebo-controlled trial among elderly Swedish citizens. *PLoS One, 12*(4), e0174880. https://doi.org/10.1371/journal.pone.0174880.

Alehagen, U., Johansson, P., Bjornstedt, M., Rosen, A., & Dahlstrom, U. (2013). Cardiovascular mortality and N-terminal-proBNP reduced after combined selenium and coenzyme Q10 supplementation: A 5-year prospective randomized double-blind placebo-controlled trial among elderly Swedish citizens. *International Journal of Cardiology, 167*(5), 1860–1866. https://doi.org/10.1016/j.ijcard.2012.04.156.

Alehagen, U., Johansson, P., Bjornstedt, M., Rosen, A., Post, C., & Aaseth, J. (2016). Relatively high mortality risk in elderly Swedish subjects with low selenium status. *European Journal of Clinical Nutrition, 70*(1), 91–96. https://doi.org/10.1038/ejcn.2015.92.

Arodin Selenius, L., Wallenberg Lundgren, M., Jawad, R., Danielsson, O., & Björnstedt, M. (2019). The cell culture medium affects growth, phenotype expression and the response to selenium cytotoxicity in A549 and HepG2 cells. *Antioxidants (Basel), 8*(5). https://doi.org/10.3390/antiox8050130.

Bjorkhem-Bergman, L., Jonsson, K., Eriksson, L. C., Olsson, J. M., Lehmann, S., Paul, C., et al. (2002). Drug-resistant human lung cancer cells are more sensitive to selenium cytotoxicity. Effects on thioredoxin reductase and glutathione reductase. *Biochemical Pharmacology*, *63*(10), 1875–1884. https://doi.org/10.1016/s0006-2952(02)00981-4.

Bjornstedt, M., & Fernandes, A. P. (2010). Selenium in the prevention of human cancers. *The EPMA Journal*, *1*(3), 389–395. https://doi.org/10.1007/s13167-010-0033-2.

Bjornstedt, M., Kumar, S., Bjorkhem, L., Spyrou, G., & Holmgren, A. (1997). Selenium and the thioredoxin and glutaredoxin systems. *Biomedical and Environmental Sciences*, *10*(2–3), 271–279.

Brodin, O., Eksborg, S., Wallenberg, M., Asker-Hagelberg, C., Larsen, E. H., Mohlkert, D., et al. (2015). Pharmacokinetics and toxicity of sodium selenite in the treatment of patients with carcinoma in a phase I clinical trial: The SECAR study. *Nutrients*, 7(6), 4978–4994. https://doi.org/10.3390/nu7064978.

Brodin, O., Hackler, J., Misra, S., Wendt, S., Sun, Q., Laaf, E., et al. (2020). Selenoprotein P as biomarker of selenium status in clinical trials with therapeutic dosages of selenite. *Nutrients*, *12*(4). https://doi.org/10.3390/nu12041067.

Chugh, P., & Dittmer, D. P. (2012). Potential pitfalls in microRNA profiling. *Wiley Interdisciplinary Reviews*, *3*(5), 601–616. https://doi.org/10.1002/wrna.1120.

Cooper, A. J., Krasnikov, B. F., Niatsetskaya, Z. V., Pinto, J. T., Callery, P. S., Villar, M. T., et al. (2011). Cysteine S-conjugate beta-lyases: Important roles in the metabolism of naturally occurring sulfur and selenium-containing compounds, xenobiotics and anticancer agents. *Amino Acids*, *41*(1), 7–27. https://doi.org/10.1007/s00726-010-0552-0.

Cooper, A. J. L., Pinto, J. T., Krasnikov, B. F., Niatsetskaya, Z. V., Han, Q., Li, J. Y., et al. (2008). Substrate specificity of human glutamine transaminase K as an aminotransferase and as a cysteine S-conjugate beta-lyase. *Archives of Biochemistry and Biophysics*, *474*(1), 72–81. https://doi.org/10.1016/j.abb.2008.02.038.

de Jager, W., Bourcier, K., Rijkers, G. T., Prakken, B. J., & Seyfert-Margolis, V. (2009). Prerequisites for cytokine measurements in clinical trials with multiplex immunoassays. *BMC Immunology*, *10*, 52. https://doi.org/10.1186/1471-2172-10-52.

Eliot, A. C., & Kirsch, J. F. (2004). Pyridoxal phosphate enzymes: Mechanistic, structural, and evolutionary considerations. *Annual Review of Biochemistry*, *73*, 383–415. https://doi.org/10.1146/annurev.biochem.73.011303.074021.

Flouda, K., Dersch, J. M., Gabel-Jensen, C., Sturup, S., Misra, S., Bjornstedt, M., et al. (2016). Quantification of low molecular weight selenium metabolites in human plasma after treatment with selenite in pharmacological doses by LC-ICP-MS. *Analytical and Bioanalytical Chemistry*, *408*(9), 2293–2301. https://doi.org/10.1007/s00216-016-9325-2.

Gabel-Jensen, C., Lunoe, K., & Gammelgaard, B. (2010). Formation of methylselenol, dimethylselenide and dimethyldiselenide in in vitro metabolism models determined by headspace GC-MS. *Metallomics*, *2*(2), 167–173. https://doi.org/10.1039/b914255j.

Han, Q., Cai, T., Tagle, D. A., & Li, J. (2010). Structure, expression, and function of kynurenine aminotransferases in human and rodent brains. *Cellular and Molecular Life Sciences*, *67*(3), 353–368. https://doi.org/10.1007/s00018-009-0166-4.

Keustermans, G. C., Hoeks, S. B., Meerding, J. M., Prakken, B. J., & de Jager, W. (2013). Cytokine assays: An assessment of the preparation and treatment of blood and tissue samples. *Methods*, *61*(1), 10–17. https://doi.org/10.1016/j.ymeth.2013.04.005.

Kumar, S., Björnstedt, M., & Holmgren, A. (1992). Selenite is a substrate for calf thymus thioredoxin reductase and thioredoxin and elicits a large non-stoichiometric oxidation of NADPH in the presence of oxygen. *European Journal of Biochemistry*, *207*(2), 435–439. https://doi.org/10.1111/j.1432-1033.1992.tb17068.x.

Lee, J. I., Nian, H., Cooper, A. J., Sinha, R., Dai, J., Bisson, W. H., et al. (2009). Alpha-keto acid metabolites of naturally occurring organoselenium compounds as inhibitors of histone deacetylase in human prostate cancer cells. *Cancer Prevention Research (Philadelphia, Pa.)*, *2*(7), 683–693. https://doi.org/10.1158/1940-6207.CAPR-09-0047.

Lendvai, G., Szekerczes, T., Kontsek, E., Selvam, A., Szakos, A., Schaff, Z., et al. (2020). The effect of methylselenocysteine and sodium selenite treatment on microRNA expression in liver cancer cell lines. *Pathology Oncology Research, 26*(4), 2669–2681. https://doi.org/10.1007/s12253-020-00870-8.

Madeja, Z., Sroka, J., Nystrom, C., Bjorkhem-Bergman, L., Nordman, T., Damdimopoulos, A., et al. (2005). The role of thioredoxin reductase activity in selenium-induced cytotoxicity. *Biochemical Pharmacology, 69*(12), 1765–1772. https://doi.org/10.1016/j.bcp.2005.02.023.

McGoldrick, T. A., Lock, E. A., Rodilla, V., & Hawksworth, G. M. (2003). Renal cysteine conjugate C-S lyase mediated toxicity of halogenated alkenes in primary cultures of human and rat proximal tubular cells. *Archives of Toxicology*, 77(7), 365–370. https://doi.org/10.1007/s00204-003-0459-6.

Misra, S., Boylan, M., Selvam, A., Spallholz, J. E., & Bjornstedt, M. (2015). Redox-active selenium compounds- -from toxicity and cell death to cancer treatment. *Nutrients*, 7(5), 3536–3556. https://doi.org/10.3390/nu7053536.

Misra, S., Selvam, A. K., Wallenberg, M., Ambati, A., Matolcsy, A., Magalhaes, I., et al. (2016). Selenite promotes all-trans retinoic acid-induced maturation of acute promyelocytic leukemia cells. *Oncotarget*, 7(46), 74686–74700. https://doi.org/10.18632/oncotarget.12531.

Nikfarjam, L., & Farzaneh, P. (2012). Prevention and detection of Mycoplasma contamination in cell culture. *Cell Journal, 13*(4), 203–212.

Nilsonne, G., Olm, E., Szulkin, A., Mundt, F., Stein, A., Kocic, B., et al. (2009). Phenotype-dependent apoptosis signalling in mesothelioma cells after selenite exposure. *Journal of Experimental & Clinical Cancer Research, 28*, 92. https://doi.org/10.1186/1756-9966-28-92.

Nilsonne, G., Sun, X., Nystrom, C., Rundlof, A. K., Potamitou Fernandes, A., Bjornstedt, M., et al. (2006). Selenite induces apoptosis in sarcomatoid malignant mesothelioma cells through oxidative stress. *Free Radical Biology & Medicine, 41*(6), 874–885. https://doi.org/10.1016/j.freeradbiomed.2006.04.031.

Olm, E., Fernandes, A. P., Hebert, C., Rundlof, A. K., Larsen, E. H., Danielsson, O., et al. (2009). Extracellular thiol-assisted selenium uptake dependent on the x(c)-cystine transporter explains the cancer-specific cytotoxicity of selenite. *Proceedings of the National Academy of Sciences of the United States of America, 106*(27), 11400–11405. https://doi.org/10.1073/pnas.0902204106.

Razaghi, A., et al. (2021). Selenium stimulates the antitumour immunity: Insights to future research. *European Journal of Cancer, 155*, 256–267. https://doi.org/10.1016/j.ejca.2021.07.013.

Riener, C. K., Kada, G., & Gruber, H. J. (2002). Quick measurement of protein sulfhydryls with Ellman's reagent and with 4,4′-dithiodipyridine. *Analytical and Bioanalytical Chemistry, 373*(4–5), 266–276. https://doi.org/10.1007/s00216-002-1347-2.

Selenius, M., Rundlöf, A. K., Olm, E., Fernandes, A. P., & Björnstedt, M. (2010). Selenium and the selenoprotein thioredoxin reductase in the prevention, treatment and diagnostics of cancer. *Antioxidants & Redox Signaling, 12*(7), 867–880. https://doi.org/10.1089/ars.2009.2884.

Selvam, A. K., & Bjornstedt, M. (2020). A novel assay method to determine the beta-elimination of se-methylselenocysteine to monomethylselenol by kynurenine aminotransferase 1. *Antioxidants (Basel), 9*(2). https://doi.org/10.3390/antiox9020139.

Selvam, A. K., Björnstedt, M., & Misra, S. (2018). Therapeutic potential of selenium compounds in the treatment of cancer. In B. Michalke (Ed.), *Selenium* (pp. 251–269). Cham: Springer International Publishing.

Selvam, A. K., Jawad, R., Gramignoli, R., Achour, A., Salter, H., & Björnstedt, M. (2021). A novel mRNA-mediated and MicroRNA-guided approach to specifically eradicate drug-resistant hepatocellular carcinoma cell lines by se-methylselenocysteine. *Antioxidants, 10*(7), 1094.

Tarek, M., Louka, M. L., Khairy, E., Ali-Labib, R., Zakaria Zaky, D., & Montasser, I. F. (2017). Role of microRNA-7 and selenoprotein P in hepatocellular carcinoma. *Tumour Biology*, *39*(5). https://doi.org/10.1177/1010428317698372. 1010428317698372.

Vadhanavikit, S., Ip, C., & Ganther, H. E. (1993). Metabolites of sodium selenite and methylated selenium compounds administered at cancer chemoprevention levels in the rat. *Xenobiotica*, *23*(7), 731–745.

Van Liefferinge, J., Bentea, E., Demuyser, T., Albertini, G., Follin-Arbelet, V., Holmseth, S., et al. (2016). Comparative analysis of antibodies to xCT (Slc7a11): Forewarned is forearmed. *The Journal of Comparative Neurology*, *524*(5), 1015–1032. https://doi.org/10.1002/cne.23889.

Wallenberg, M., Misra, S., & Bjornstedt, M. (2014). Selenium cytotoxicity in cancer. *Basic & Clinical Pharmacology & Toxicology*, *114*(5), 377–386. https://doi.org/10.1111/bcpt.12207.

Wallenberg, M., Misra, S., Wasik, A. M., Marzano, C., Bjornstedt, M., Gandin, V., et al. (2014). Selenium induces a multi-targeted cell death process in addition to ROS formation. *Journal of Cellular and Molecular Medicine*, *18*(4), 671–684. https://doi.org/10.1111/jcmm.12214.

Zeng, H. (2009). Selenium as an essential micronutrient: Roles in cell cycle and apoptosis. *Molecules*, *14*(3), 1263–1278. https://doi.org/10.3390/molecules14031263.

Zeng, H., Briske-Anderson, M., Idso, J. P., & Hunt, C. D. (2006). The selenium metabolite methylselenol inhibits the migration and invasion potential of HT1080 tumor cells. *The Journal of Nutrition*, *136*(6), 1528–1532.

CHAPTER THREE

Using selenocysteine-specific reporters to screen for efficient $tRNA^{Sec}$ variants

Christina Z. Chung[a,†], Dieter Söll[a,b,*], and Natalie Krahn[a,†]
[a]Department of Molecular Biophysics & Biochemistry, Yale University, New Haven, CT, United States
[b]Department of Chemistry, Yale University, New Haven, CT, United States
*Corresponding author: e-mail address: dieter.soll@yale.edu

Contents

Abstract

The unique properties of selenocysteine (Sec) have generated an interest in the scientific community to site-specifically incorporate Sec into a protein of choice. Current technologies have rewired the natural Sec-specific translation factor-dependent selenoprotein biosynthesis pathway by harnessing the canonical elongation factor

† These authors contributed equally.

Methods in Enzymology, Volume 662
ISSN 0076-6879
https://doi.org/10.1016/bs.mie.2021.10.005

(EF-Tu) to simplify the requirements for Sec incorporation in *Escherichia coli*. This strategy is versatile and can be applied to Sec incorporation at any position in a protein of interest. However, selenoprotein production is still limited by yield and serine misincorporation. This protocol outlines a method in *E. coli* to design and optimize tRNA libraries which can be selected and screened for by the use of Sec-specific intein-based reporters. This provides a fast and simple way to engineer tRNAs with enhanced Sec-incorporation ability.

1. Introduction

Selenium, an essential micronutrient in many organisms, including humans, is found in proteins in the form of selenocysteine (Sec) (Stadtman, 1996). Selenoproteins are found in all domains of life: eukarya, archaea, eubacteria, and even viruses (Guillin, Vindry, Ohlmann, & Chavatte, 2019). The number of selenoproteins varies with the organism, with at least 25 identified in humans. These proteins are necessary for human health, participating in redox signaling and oxidative stress response (Lobanov, Hatfield, & Gladyshev, 2009). Although Sec is a close chemical relative of cysteine (Cys) (only differing in the replacement of a sulfur atom for a selenium atom), it confers unique properties to the protein. Sec has increased nucleophilicity at neutral pH, is able to recover from oxidative inactivation (Snider, Ruggles, Khan, & Hondal, 2013), and forms strong diselenide bonds (Arai et al., 2017). The effect of Sec in selenoenzymes greatly depends on the architecture of the catalytic site. For example, *E. coli* formate dehydrogenase is 100 times more active with Sec as the active site residue than the Cys homolog (Axley, Böck, & Stadtman, 1991), while the *E. coli* Cys-containing selenophosphate synthetase is more active than its Sec-containing counterpart in *Haemophilus influenza* (Lacourciere & Stadtman, 1999). Furthermore, insertion of Sec into recombinant proteins can provide a tool for investigating mechanisms of redox enzymes (Evans et al., 2021), and modulating protein folding (Liu, Chen, & Rozovsky, 2017; Metanis & Hilvert, 2015) and stability (Arai et al., 2017). The broad applications for recombinant selenoproteins require an efficient system for protein production. Here, we discuss one of the current strategies for recombinant selenoprotein production and the associated engineered tRNAs. We then describe a method for generating a tRNA library from which we select the most efficient tRNA using Sec-specific reporters.

2. Natural selenocysteine translation

Sec incorporation occurs naturally in some bacteria, following a unique translation pathway (Fig. 1A). Like other amino acids, Sec has a dedicated tRNA (tRNA^{Sec}, encoded by the *selC* gene), but differs with the absence of a cognate aminoacyl-tRNA synthetase. Instead, it harnesses seryl-tRNA synthetase (SerRS) for aminoacylation. As such, tRNA^{Sec} is first aminoacylated with serine (Ser) to generate Ser-tRNA^{Sec}. Then selenocysteine synthase (SelA) converts Ser to Sec on the tRNA using a selenium donor generated by selenophosphate synthetase (SelD). Fully aminoacylated tRNA^{Sec} (Sec-tRNA^{Sec}) is then brought to the ribosome by a specialized elongation factor, SelB, through recognition of a UGA codon and Sec insertion sequence (SECIS) element. In bacteria, the SECIS element is in the coding region of the protein, immediately downstream of the UGA codon. Its hairpin motif and relative positioning to the UGA codon allows for SelB recognition and insertion of Sec into the growing polypeptide chain.

This multistep translation pathway is possible because of the unique structure of tRNA^{Sec}: (i) a long variable arm, (ii) increased base pairs (bp) in the D-arm, and (iii) a 13 bp acceptor domain (acceptor and T-stem combined) (Fig. 1B). For the initial step in this pathway, tRNA^{Sec} must interact with SerRS to become serylated. This requires tRNA^{Sec} to have the same identity elements as tRNA^{Ser}, namely, a long variable arm and G73 discriminator base (Giegé, Sissler, & Florentz, 1998). The sequence of the variable arm is believed to be more important in tRNA^{Sec} than tRNA^{Ser}. This is because SerRS preferentially binds 12 bp acceptor domains (as in tRNA^{Ser}) compared to 13 bp (as in tRNA^{Sec}). Therefore, residues in the variable arm of tRNA^{Sec} aid in increasing the affinity it has toward SerRS (Ohama, Yang, & Hatfield, 1994; Wu & Gross, 1993). The 13 bp acceptor domain is a distinguishing feature of tRNA^{Sec}, acting as an identity element for the Sec-specific factors (SelA and SelB).

SelA uses both the 13 bp acceptor domain and the D-arm to distinguish between tRNA^{Ser} and tRNA^{Sec}. Binding assays showed that truncating the acceptor stem of tRNA^{Sec} to create a 12 bp acceptor domain reduces affinity to SelA by three-fold (Baron & Böck, 1991). A co-crystal structure of SelA and tRNA^{Sec} revealed the direct interactions between the two molecules. It was found that the D-arm of tRNA^{Sec} has a unique shape which fits perfectly into the N-terminal domain of SelA (Itoh et al., 2013). The D-arm of

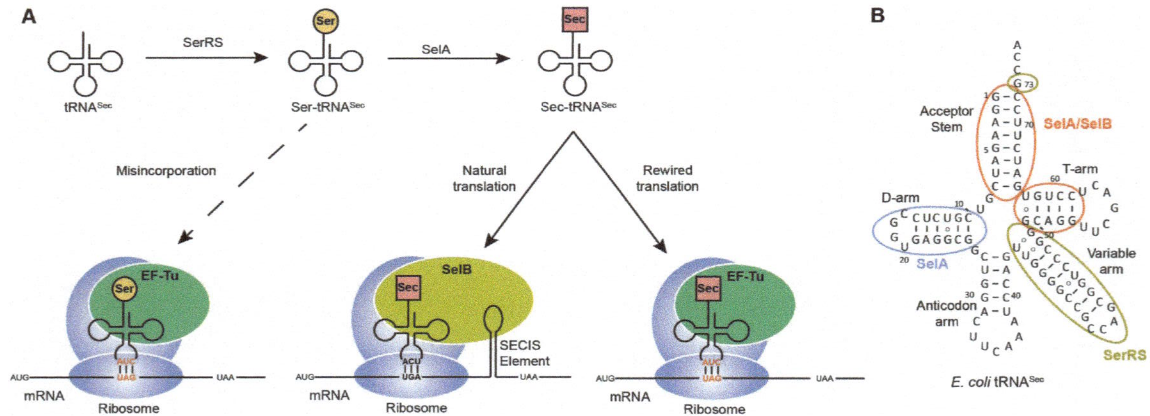

Fig. 1 (A) Selenocysteine (Sec) incorporation on the ribosome via natural and rewired translation systems. Both systems require serylation of the $tRNA^{Sec}$ by seryl-tRNA synthetase (SerRS) before the Ser residue is converted to Sec via selenocysteine synthase (SelA). Sec-$tRNA^{Sec}$ is then naturally brought to the ribosome via SelB (a specialized elongation factor) through recognition of a UGA codon and SEC Insertion Sequence (SECIS) element. In the rewired translation system, Sec-$tRNA^{Sec}$ is engineered to be recognized by EF-Tu (canonical elongation factor) and brought to the ribosome in response to a UAG codon. This system is susceptible to misincorporation by Ser as EF-Tu also recognizes Ser-$tRNA^{Sec}$. (B) Cloverleaf structure of *E. coli* $tRNA^{Sec}$ highlighting the regions of the tRNA which are recognized by the Sec incorporation machinery (SerRS, SelA and SelB). SerRS recognizes the G73 discriminator base and the long variable arm, SelA recognizes the entire D-arm and the 13 bp acceptor domain (acceptor and T-stems combined), and SelB recognizes the 13 bp acceptor domain.

tRNA^{Sec} consists of a 6 bp stem and 4 base loop (Fig. 1B), which is quite distinct from tRNA^{Ser} which has a 4 bp stem and 8–11 base loop. On the other hand, SelB has been found only to recognize the 13 bp acceptor domain, more specifically, the 8 bp acceptor stem. Deletion of 1 bp in the acceptor stem caused tRNA^{Sec} to become a substrate for the canonical elongation factor, EF-Tu, and abolished binding to SelB (Baron & Böck, 1991; Forchhammer, Boesmiller, & Böck, 1991).

3. Recombinant selenocysteine translation

Recombinant expression in *E. coli* of natural bacterial selenoproteins is possible by simply cloning their coding sequences (containing a SECIS element) into an expression plasmid, and transforming an *E. coli* strain that contains all the genes for selenoprotein synthesis (*selA, selB, selC*, and *selD*) (Mandrand-Berthelot, Wee, & Haddock, 1978; Shchedrina, Novoselov, Malinouski, & Gladyshev, 2007; Yoshizawa & Böck, 2009). Formate dehydrogenase H is the only selenoprotein in *E. coli*; its gene (*fdhF*) contains the SECIS RNA element immediately following the UGA codon to be recoded (Heider, Baron, & Böck, 1992). Since the RNA sequence of this structural element is naturally very diverse, a minimal SECIS element (17 nucleotides), containing a bulged U in the stem of the hairpin, was created for efficient recombinant Sec incorporation (Martin, Harney, & Berry, 1998). Engineering this minimal SECIS into the open reading frame (ORF) of glutathione *S*-transferase permitted site-specific insertion of Sec. Surprisingly, the additional nucleotides added into the mRNA did not alter the protein activity (Jiang et al., 2004). This result might not be true for all proteins, making it difficult to broadly apply this system. Another strategy has been to insert the SECIS element from *fdhF* into the 3′-untranslated region (UTR) of a plasmid, directly after the ORF. This was used to successfully express mammalian thioredoxin reductase (Arnér, Sarioglu, Lottspeich, Holmgren, & Böck, 1999), but again is limited in its broad application. Depending on the position of the desired Sec, the distance between the UGA codon and the SECIS element in the 3′-UTR may not be optimal to facilitate Sec incorporation. Therefore, Sec-tRNAs were created that allowed site-specific UAG programmed Sec insertion by EF-Tu, thus obviating the need for SECIS and SelB (Aldag et al., 2013; Miller et al., 2015; Mukai, Sevostyanova, Suzuki, Fu, & Söll, 2018; Thyer, Robotham, Brodbelt, & Ellington, 2015). We briefly describe their successes and limitations.

A caveat with reprogramming Sec insertion with EF-Tu, is that EF-Tu is not specific for Sec like SelB. This means that Ser-tRNA^{Sec} can also be inserted into the growing peptide resulting in Ser misincorporation (Fig. 1A) (Asahara & Uhlenbeck, 2005). In many cases Ser does not create an active enzyme, therefore small amounts are ignored (Evans et al., 2021) while large amounts can be separated (Aldag et al., 2013). One of the first attempts to engineer a Sec tRNA recognized by EF-Tu was to create a hybrid tRNA using *E. coli* tRNA^{Ser} as a scaffold. The first 7 bp in the acceptor stem of *E. coli* tRNA^{Sec} were used to replace the first 6 bp in the acceptor stem of tRNA^{Ser}. The last bp in tRNA^{Ser} was unchanged to retain the identity elements for EF-Tu. This created a tRNA with a 13 bp acceptor domain (tRNA^{UTu}) as in tRNA^{Sec}, but with identity elements from tRNA^{Ser} for recognition by EF-Tu (Aldag et al., 2013).

Due to the poor selenylation of tRNA^{UTu}, another hybrid tRNA was designed using *E. coli* tRNA^{Sec} as a scaffold. This strategy left most of tRNA^{Sec} untouched, with only the antideterminants for EF-Tu replaced, namely, the last bp in the acceptor stem and first 2 bp in the T-stem (as was left in tRNA^{UTu}). Instead of just replacing these antideterminant residues by the corresponding residues in tRNA^{Ser}, a library was created to screen for tRNAs which were accepted as a substrate by EF-Tu. This resulted in the design of tRNA^{SecUx} (Thyer et al., 2015).

A separate attempt to improve the selenylation of tRNA^{UTu} used rational engineering based on the co-crystal structure of decameric *Aquifex aeolicus* SelA protein in complex with 10 *Thermoanaerobacter tengcongensis* tRNA^{Sec} molecules (Itoh et al., 2013). The cocrystal structure provided information on the recognition of tRNA^{Sec} by SelA, and the features which allow discrimination against tRNA^{Ser}. Using tRNA^{UTu} as a scaffold, the EF-Tu recognition elements were maintained while the SelA recognition elements were varied. This resulted in the design of tRNA^{UTuX} which contained mutations in connecting regions of the tRNA as well as the D- and T-arms (Miller et al., 2015).

Since tRNA^{UTu}, tRNA^{SecUx}, and tRNA^{UTuX} all have 13 bp acceptor domains, they are ideal substrates for SelA (conversion of Ser to Sec), but not ideal for EF-Tu which prefers binding to a 12 bp domain structure (Baron & Böck, 1991). A close relative of *E. coli*, *Aeromonas salmonicida* (*As*), was found to have a SelA enzyme capable of recognizing a 12 bp acceptor domain in tRNA^{Sec} (Mukai et al., 2016). Efforts to use the *As* tRNA^{Sec} as a scaffold for EF-Tu dependent incorporation were unsuccessful (Mukai et al., 2018). From these studies, it was evident that tRNA^{Sec} has evolved specifically to be recognized by SelA and SelB and not EF-Tu, making it a poor scaffold for EF-Tu mediated Sec incorporation. Therefore, efforts were made to search for tRNA^{Sec}-like structures which may not have evolved specifically for Sec insertion. A metagenomic search identified allo-tRNAs which had many features resembling that of tRNA^{Sec}; namely, the long variable arm and unique D-arm (longer D-stem and smaller D-loop) (Mukai et al., 2017). However, unlike tRNA^{Sec}, they were all found to contain a 12 bp acceptor domain, ideal for interaction with EF-Tu. These allo-tRNAs were tested in *E. coli* and found to be capable of inserting a wide range of amino acids depending on their identity elements (Mukai et al., 2017).

One of these allo-tRNAs (tRNA^{UTu1}) was found to act as a tRNA^{Ser} in *E. coli* but contained identity elements for SelA recognition (Fig. 2). A second allo-tRNA with a different cloverleaf structure to tRNA^{UTu1} acted as a tRNA^{Ala} in *E. coli* (Mukai et al., 2017). Replacing the tRNA^{Ala} identity elements with those from tRNA^{Ser} allowed recognition by SerRS, rendering it a suitable substrate for Sec incorporation (tRNA^{UTu2}) (Fig. 2) (Mukai et al., 2018). Therefore, these tRNAs were coupled with *As* SelA and tested for their ability in site-specific Sec incorporation. tRNA^{UTu1} outperformed the aforementioned tRNAs while tRNA^{UTu2} was very poor at Sec incorporation (Chung, Krahn, Crnković, & Söll, 2021). One downfall with EF-Tu mediated Sec incorporation is the possibility for Ser to be inserted instead of Sec (Fig. 1A). Therefore, to increase the levels of Sec-tRNA, the tRNAs were engineered to increase their affinity to *As* SelA. This was done through mutation of the D-arm to match the sequence of the natural *As* tRNA^{Sec}, generating tRNA^{UTu1D} and tRNA^{UTu2D}, respectively (Mukai et al., 2018) (Fig. 2). The D-arm mutations caused a significant increase in selenocysteine incorporation, resulting in a system that is currently being used to express recombinant selenoproteins (Chung et al., 2021; Evans et al., 2021; Greene, Stubbe, & Nocera, 2019). Engineering allo-tRNAs has proven to be more fruitful in achieving increased recombinant selenoprotein yields and a recommended scaffold for future optimization.

Our lab has recently developed two Sec-specific reporters that utilize the *Synechocystis* sp. PCC6803 DnaB mini-intein M86 variant (Appleby-Tagoe et al., 2011; Chung et al., 2021). The M86 variant of the DnaB mini-intein was inserted into the kanamycin resistance (*kanR*) gene to produce KanR-M86 plasmid pAB_a02 and into superfolder GFP (sfGFP) to produce sfGFP-M86 plasmid pB_04. The first position of the intein is a UAG stop codon that can be decoded by allo-tRNA. Insertion of Sec allows the intein to complete the splicing pathway to generate a reconstituted reporter. Misincorporation of Ser generates an inactive intein which prevents the splicing pathway from occurring and generates an inactive reporter protein.

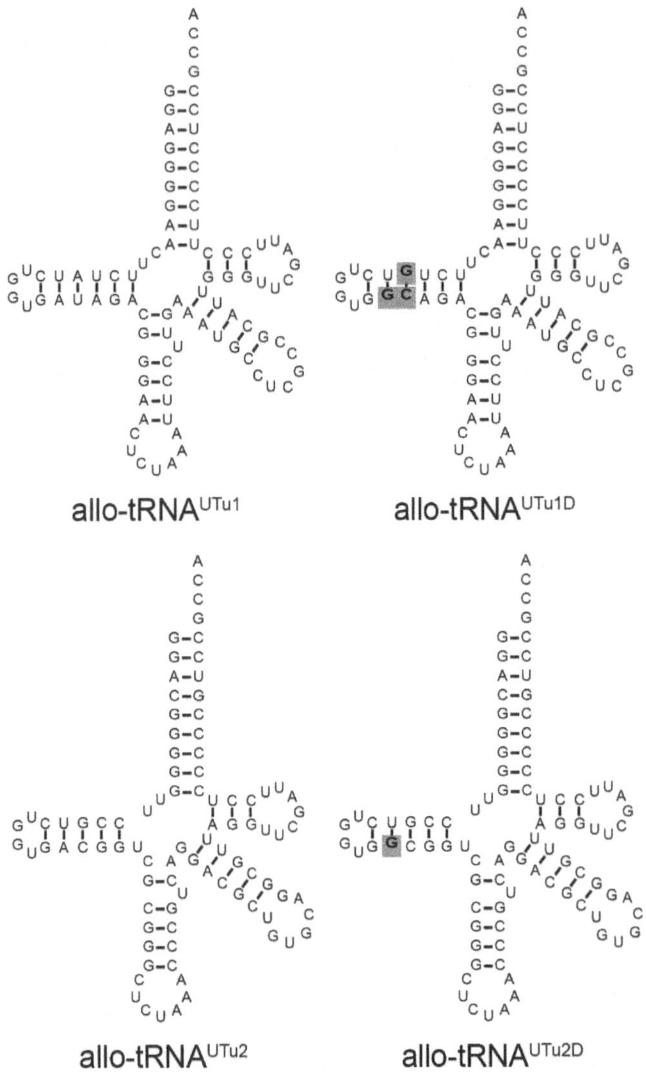

Fig. 2 Cloverleaf structures of allo-tRNAs which have been adopted for Sec incorporation using the rewired translation pathway. $tRNA^{UTu1D}$ and $tRNA^{UTu2D}$ have mutations in their D-arms (highlighted) which make them better substrates for *Aeromonas salmonicida* SelA compared to $tRNA^{UTu1}$ and $tRNA^{UTu2}$, respectively.

Moreover, translation is terminated if no readthrough occurs, leading to a truncated and nonfunctional protein (Fig. 3). The selective and fluorescent reporters can be used together to screen components of the selenoprotein synthesis system by selecting for variants that increase Sec incorporation. As mentioned above, engineering tRNA variants has led to increased Sec incorporation and the following sections will describe how the intein-based reporters can

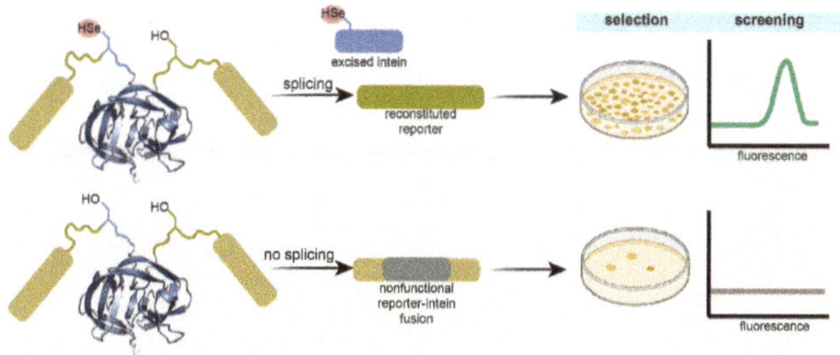

Fig. 3 Schematic of intein reporters. The M86 variant of the DnaB mini-intein is inserted within a reporter protein to disrupt activity. Insertion of Sec at the UAG codon in the first position of the intein will allow splicing of the intein from the reporter and reconstitution of a functional reporter for growth on kanamycin-containing LB agar plates (KanR-M86 plasmid) or fluorescence (sfGFP-M86 plasmid). However, misincorporation of Ser does not allow splicing to occur, resulting in little to no cell growth on kanamycin-containing LB agar plates and no fluorescence. *Figure adapted from Chung, C. Z., Krahn, N., Crnković, A., & Söll, D. (2021). Intein-based design expands diversity of selenocysteine reporters. Journal of Molecular Biology, 167199. doi: 10.1016/j.jmb.2021.167199.*

be used to screen for improved tRNA variants. These Sec-specific reporters are not limited to enhancing the tRNA but can be used to engineer any part of the selenoprotein synthesis pathway, such as SelA or EF-Tu.

4. Evolving tRNA^{Sec} variants for increased Sec incorporation

Due to the natural multistep insertion pathway for Sec, engineering an efficient recombinant system has proved difficult. Here, we describe a strategy for evolving a tRNA for increased Sec incorporation. Although using an allo-tRNA as a scaffold is recommended, this strategy can be used to optimize any starting tRNA. Since much is known about the tRNA^{Sec} features which govern aminoacylation and elongation, described above, one can focus on specific steps in the translation pathway (Fig. 1B). This protocol will describe the steps required to design a tRNA library (specifically to target one of the described interactions) and a Sec-specific screening and selection method to choose the best performing tRNA (Fig. 4).

4.1 Equipment

1. Thermocycler
2. Electrophoretic equipment to make and run agarose gels
3. Microwave (for melting agarose)
4. Gel imaging system (e.g., ChemiDoc)
5. 37 °C stationary incubator

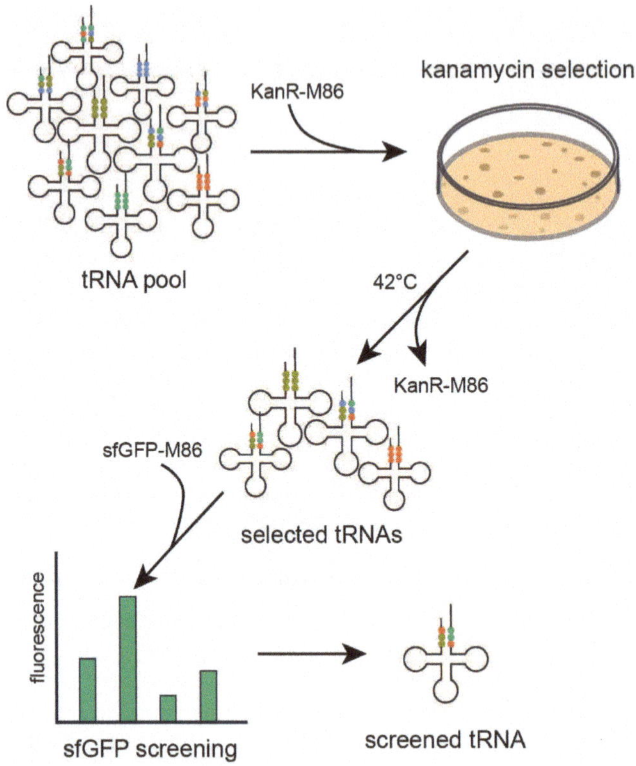

Fig. 4 Schematic of the selection protocol. Initially a library of tRNAs is created which is transformed with the KanR-M86 plasmid (Chung et al., 2021). After selection at varying concentrations of kanamycin, surviving clones are selected. The KanR-M86 plasmid is lost by incubation at 42 °C, leaving the cell with only the tRNA plasmid of interest. The sfGFP-M86 plasmid (Chung et al., 2021) is then transformed into the cells with the tRNA plasmid, to screen for the tRNA with the best Sec incorporation activity.

6. Spectrophotometer (e.g., NanoDrop)
7. Electroporator
8. Microcentrifuge
9. Incubator shaker
10. Plate reader (e.g., Synergy HTX Plate Reader)

4.2 Reagents and materials

1. tRNA plasmid (e.g., pSecUAG-Evol2 from Addgene #163148)
2. Oligonucleotide primers (see Section 5.2 for more details)
3. High-fidelity DNA polymerase (e.g., PfuUltra II Fusion)
4. 0.2 mL PCR tubes
5. 6X DNA loading dye

6. Agarose
7. 1 × TAE (Tris/Acetate/EDTA) buffer (40 mM Tris-base, 20 mM acetic acid, 1 mM EDTA)
8. DNA staining reagent (e.g., ethidium bromide or SYBR Safe)
9. DNA ladder (e.g., 1 kb or 2-Log)
10. Restriction enzymes for cleavage of insert and vector and corresponding 10 × reaction buffers (these will be specific for the plasmid containing the tRNA of interest and depend on how the library is designed)
11. *Dpn*I (e.g., NEB)
12. 1.5 mL microcentrifuge tubes
13. PCR purification kit (e.g., Macherey-Nagel or Monarch®)
14. Distilled water, autoclaved
15. Dimethyl sulfoxide (DMSO)
16. DNA ligase and corresponding reaction buffer (e.g., T4 DNA ligase)
17. Electrocompetent cloning cells (e.g., Top10, DH5α, Stellar™)
18. 0.2 cm electroporation cuvettes
19. Luria Broth (LB), autoclaved
20. Super Optimal broth with Catabolite repression (SOC) media, autoclaved (optional growth media)
21. Cell spreaders
22. 100 mg/mL ampicillin stock solution, filtered
23. 50 mg/mL kanamycin stock solution, filtered
24. Stock solutions of other necessary antibiotic solutions (e.g., for tRNA expression plasmid), filtered
25. LB agar plates with appropriate antibiotics (100 and 150 mm wide plates)
26. Plasmid miniprep kit (e.g., Macherey-Nagel or Monarch®)
27. Plasmid maxiprep kit (e.g., Qiagen®)
28. tRNA inducing agent, filtered (e.g., 20% (*w*/*v*) arabinose)
29. 20% (*w*/*v*) glucose solution, filtered
30. 100 mM sodium selenite (Na_2SeO_3) solution, filtered
31. Culture tubes (e.g., 14 mL capped round-bottom culture tubes)
32. 24-well clear plates (with plate lids)
33. 96-well black plates with clear bottoms
34. Breathable sealing membrane (e.g., Breathe-Easy® sealing membrane)
35. 1 M mM isopropyl β-d-1-thiogalactopyranoside (IPTG), filtered
36. Cells for recombinant protein production [e.g. B-95.ΔAΔ*fabR*Δ*selABC* (Mukai et al., 2015)]
37. KanR-M86 plasmid (pAB_a02) (Chung et al., 2021)
38. sfGFP-M86 plasmid (pB_04) (Chung et al., 2021)

5. Protocol

5.1 Electrocompetent cells

5.1.1 Electrocompetent cells for library preparation (e.g., Top10, DH5α, Stellar™) and library screening [e.g., B-95.ΔAΔ*fabR*Δ*selABC* (Mukai et al., 2015)] can be bought (e.g., Takara Bio or Thermo Fischer Scientific) or made in the lab [e.g., (Gonzales, Brooks, Pukatzki, & Provenzano, 2013)].

5.1.2 To prepare electrocompetent cells [e.g., B-95.ΔAΔ*fabR*Δ*selABC* (Mukai et al., 2015)] containing the KanR-M86 plasmid pAB_a02 (Chung et al., 2021), the plasmid must first be transformed into the desired electrocompetent strain. Mix 50 ng pAB_a02 into 50 μL electrocompetent cells in an electroporation cuvette and place on ice. Transform cells at 1.8 V using an electroporator. Add 1 mL growth media (e.g., LB or SOC) and allow the cells to recover at 37 °C for at least 1 h while shaking. After cell recovery, plate 100 μL of cells on a prewarmed LB agar plate (100 mm wide) containing 100 μg/mL ampicillin (amp) and evenly spread the cells using a cell spreader. Incubate the plates at 37 °C overnight.

5.1.3 The following day, pick a colony and resuspend in a culture tube containing LB with 100 μg/mL amp. Grow the preculture overnight at 37 °C while shaking. Follow any online protocol to make electrocompetent cells and include 100 μg/mL amp for all cell growth [e.g., (Gonzales et al., 2013)]. *Note: cells can be made and used right away or aliquoted and kept frozen at −80 °C until ready to use. Make sure that cells have been resuspended in 15% (v/v) glycerol before freezing.*

5.2 Primer design to generate a targeted tRNA library

Before designing primers for the tRNA library, ensure that the wildtype (WT) tRNA plasmid (e.g., pSecUAG_Evol2) contains all the necessary components required for Sec incorporation (Sawers, Heider, Zehelein, & Böck, 1991). It is important to note that the parent plasmid cannot be resistant to kanamycin (kan) or amp. Kan will be used as a selectable marker during library screening and both screening plasmids (pAB_a02 and pB_04) are resistant to amp. Alternative antibiotics to use include spectinomycin and chloramphenicol. In addition, IPTG is used to induce expression of sfGFP (using pB_04) during screening of the library; therefore, tRNA expression should have an alternate inducing agent. For example, arabinose is a common alternative provided the appropriate operon is present on the tRNA parent plasmid (*araC*) (Guzman, Belin, Carson, & Beckwith, 1995).

With regards to generating a tRNA library, this is the most difficult step in this protocol and may require considerable optimization. The first matter to consider is the design of the tRNA library. To avoid bias, the region of the tRNA which is to be randomized must be removed. This can be done using restriction enzymes (provided there are unique cut sites around the tRNA), or through amplification of the parent plasmid. We discuss in detail how to design primers and execute the latter. If restriction sites around the tRNA are already present, these steps can be omitted. The parent plasmid can be directly digested and only primers for the insert are needed (skip to step 5.2.2.2). All primers can be ordered from any oligo-producing facility with Ns in place of the bases to be randomized. The facility will insert random bases at each N in the sequence to generate a mixture of sequences.

5.2.1. Design a set of primers to amplify the parent tRNA plasmid. These need to be strategically chosen to remove the area of the tRNA that is to be randomized (see details in Fig. 5).

5.2.1.1 If a localized region is targeted, it is not necessary to remove the whole tRNA. Following the example depicted in Fig. 5A, A-Fwd1 is the primer which binds just after the randomized region in the tRNA and A-Rev1 is the primer which binds outside the tRNA. For randomizing a region near the 3′-end of the tRNA, primer A-Rev1 will bind to the tRNA just prior to the randomized area while primer A-Fwd1 will bind outside the tRNA sequence.

5.2.1.2 If multiple regions on the tRNA are to be randomized, then the strategy would be to first open the parent plasmid to remove the entire tRNA (Fig. 5B). This requires primers B-Rev1 to bind upstream of the tRNA and B-Fwd1 to bind downstream of the tRNA. Each of these primers also need to encode a unique restriction site (i.e., will not cut the plasmid and tRNA at any other site) with flanking bases for binding of the enzymes (about 6 random nucleotides). *Note: the restriction sites chosen must be unique and different from each other. If possible, choose restriction enzymes that are active in the same reaction buffer to allow for a double digest during preparation of the tRNA library. This ensures that during ligation the library is inserted in the appropriate orientation.*

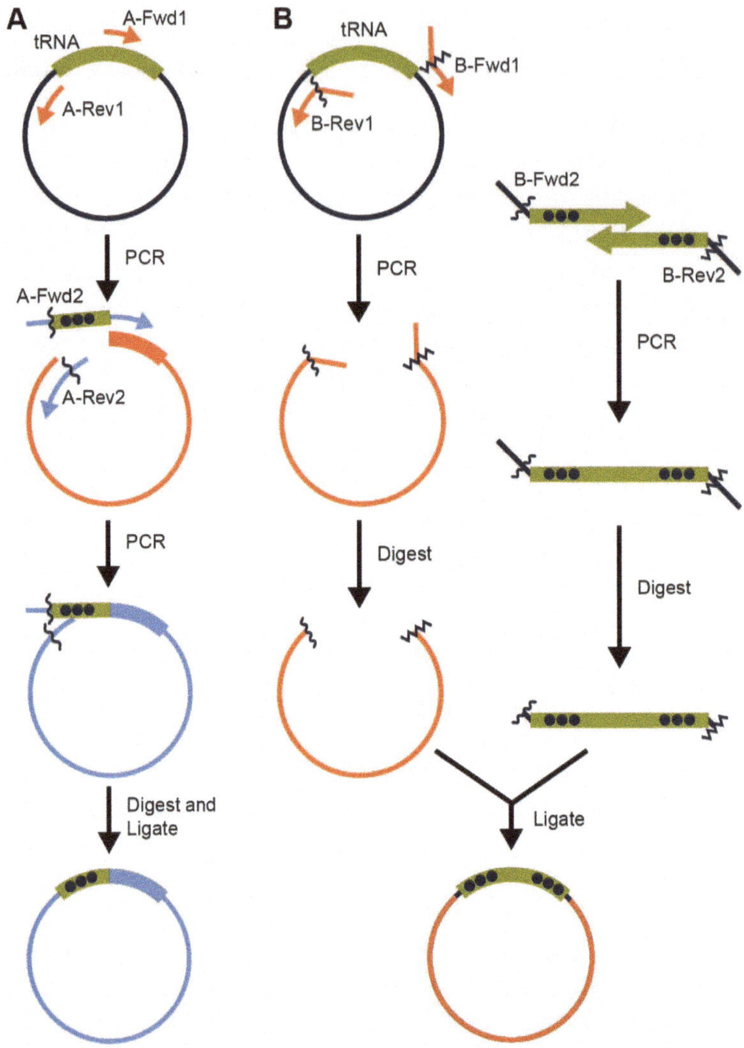

Fig. 5 Primer design and protocol to generate the tRNA library. (A) The parent plasmid is amplified around the localized tRNA region to be randomized (primers A-Fwd1 and A-Rev1). The opened plasmid is then amplified with primers containing the same restriction enzyme cut site and the randomized tRNA sequence (primers A-Fwd2 and A-Rev2). The product is digested and ligated to generate the tRNA library. (B) For multiple randomized regions, the parent plasmid is amplified around the tRNA sequence with primers that each contain a different restriction enzyme cut site (primers B-Fwd1 and B-Rev1). The tRNA insert is generated by two overlapping primers that contain the randomized regions (not present in the overlapped regions) and matching restriction sequences to the opened plasmid (primers B-Fwd2 and B-Rev2). Both products are digested and ligated together to generate the tRNA library.

Note: Each primer should start and end with a C or G, be approximately 20 nucleotides in length, have an approximate 50% GC content, and have melting temperatures (50–60°C) within a few degrees of each other. These are general guidelines for designing primers and are flexible to minor changes. Depending on the sequences in and around the tRNA, one or more of the guidelines may require some minor modifications.

5.2.2. Design another set of primers to insert the randomized tRNA sequence into the opened parent plasmid (Fig. 5). This step will differ depending on whether a single region or multiple regions are being randomized on the tRNA.

5.2.2.1 Using the example in Fig. 5A, design primer A-Rev2 to be similar in sequence to primer A-Rev1 but with a restriction enzyme cut site on the 5′-end with flanking bases (about 6 random nucleotides) for binding of the restriction enzyme. Primer A-Fwd2 will be identical in sequence to A-Fwd1 but with an additional flanking region on the 5′ end of the primer. Starting at the 5′ end, this flanking region will contain 6 random nucleotides, followed by a restriction enzyme cut site and finally the deleted section of the tRNA sequence containing the randomized positions. The restriction cut site for A-Rev2 and A-Fwd2 should be the same and unique to the parent plasmid and tRNA.

5.2.2.2 To target multiple regions in the tRNA (as in Fig. 5B), primers B-Fwd2 and B-Rev2 will create the tRNA library insert using overlap extension PCR. Each primer should encode for a specific region of randomized nucleotides as well as a restriction enzyme cut site that matches those designed in step 5.2.1.2 (B-Fwd1 with B-Rev2 and B-Rev1 with B-Fwd2). Furthermore, B-Fwd2 and B-Rev2 should overlap each other in the non-randomized section of the tRNA by 20–30 bases. *Note: overlapping the randomized regions will create bias in the tRNA library.*

5.3 Making the tRNA library

As previously mentioned, making the tRNA library can require significant optimization. It is recommended to initially follow the protocol steps on a small scale before preparing enough of the tRNA library for multiple transformations.

5.3.1 Amplify the parent plasmid using a high-fidelity DNA polymerase (e.g., PfuUltra II Fusion). For primers A-Fwd1 and A-Rev1 set up a 25 μL PCR reaction using a 0.2 mL PCR tube. For primers B-Fwd1 and B-Rev1, set up multiple 50 μL PCR reactions. Follow the details for the PCR setup in Tables 1 and 2 (adapted from the manufacturer's protocol).

5.3.2 Prepare a 1% (*w*/*v*) agarose gel while waiting for the PCR to finish. Setup the gel casting equipment with the appropriate comb. Based off the size of the gel, measure out the appropriate amount of agarose in an Erlenmeyer flask containing the correct volume of 1 × TAE buffer (e.g., 0.4 g agarose in 40 mL TAE buffer). Microwave the flask until just boiling to slowly dissolve the agarose. Add 0.25 μg/mL ethidium bromide (or another appropriate DNA staining reagent) and pour the solution into the prepared gel casting equipment. Allow the gel to fully solidify before removing the comb. Transfer the gel to the electrophoretic gel box and pour 1 × TAE buffer over the gel until the MAX fill line.

5.3.3 Once the PCR is complete, mix 5 μL of the reaction product with 1 μL of 6X DNA loading dye. Load the entire sample and an appropriate DNA ladder into the wells of the agarose gel. Run the gel at 150 V for 20–30 min depending on the size of the plasmid (larger plasmids should be run longer to verify appropriate size). Image the gel with a gel imaging system (e.g., ChemiDoc) to confirm amplification of the plasmids.

Table 1 Reaction components to amplify the parent plasmid.

Component	25 μL reaction	50 μL reaction	Final concentration
10 × PfuUltra II Reaction buffer	2.5 μL	5 μL	1 ×
10 mM dNTP mix	0.6 μL	1.2 μL	0.24 μM
10 μM Primer Fwd1	0.5 μL	1 μL	0.2 μM
10 μM Primer Rev1	0.5 μL	1 μL	0.2 μM
DNA template	X μL	X μL	5–30 ng
PfuUltra II Fusion HS DNA polymerase	0.5 μL	1 μL	
Distilled water	to 25 μL	to 50 μL	

Values have been adapted from the manufacturer's protocol.

Table 2 Thermocycler settings to amplify the parent plasmid.

Step	Temperature (°C)	Duration	Number of cycles
Denaturation	95	2 min	1
Denaturation	95	20 s	30
Annealing	50–63[a]	20 s	
Extension	72	30 s–1 min/kb[b]	
Final extension	72	3 min	1

[a]Annealing temperature will depend on the primer melting temperatures and whether DMSO is present. Consult the manufacturer's guidelines for appropriate temperatures.
[b]Extension duration may be adjusted during optimization of reaction conditions. Consult the manufacturer's guidelines for appropriate duration.

5.3.4 Add 0.5 μL *Dpn*I (10 units) to the remaining PCR product (it will be necessary to add more for multiple 50 μL reactions) and incubate at 37 °C for 1 h (or overnight) to digest the methylated parent plasmid. *Note: Using an incubator or thermocycler instead of a heat block is recommended to avoid condensation on the lid and evaporation of the sample.*

5.3.5 After digestion, PCR purify the plasmid using a PCR purification kit (follow the manufacturer's protocol) to isolate the amplified PCR product. Elute the plasmid with sterile water. Measure and record the DNA concentration using a spectrophotometer (e.g., NanoDrop).

5.3.6 The tRNA library is created using primers Fwd2 and Rev2.

5.3.6.1 For the tRNA library with a localized region randomized, use primers A-Fwd2 and A-Rev2 with the purified PCR product from step 5.3.5, and a high-fidelity DNA polymerase (e.g., PfuUltra II Fusion). Setup multiple 50 μL reactions to introduce the randomized tRNA template. Refer to Tables 2 and 3 for recommended conditions

5.3.6.2 For the tRNA library with multiple randomized regions, use primers B-Fwd2 and B-Rev2 with a high-fidelity DNA polymerase (e.g., PfuUltra II Fusion). Setup multiple 50 μL reactions to generate a tRNA library insert (Tables 4 and 5).

Note: The addition of DMSO may be required to prevent secondary structures from forming in the primer(s) containing the randomized mutations. The amount of DMSO and annealing temperature will need to be optimized to achieve a single clear band on the agarose gel. It is recommended to optimize the PCR reaction in 25 μL reactions before scaling up to multiple 50 μL reactions.

Table 3 Reaction components to introduce the localized, randomized tRNA template into the opened parent plasmid.

Component	Volume per reaction	Final concentration
10× PfuUltra II Reaction Buffer	5 μL	1×
10 mM dNTP mix	1.2 μL	0.24 μM
10 μM Primer A-Fwd2	1 μL	0.2 μM
10 μM Primer A-Rev2	1 μL	0.2 μM
DNA template	X μL	5–30 ng
DMSO	X μL	2–10%
PfuUltra II Fusion HS DNA polymerase	1 μL	
Distilled water	to 50 μL	

Values have been adapted from the manufacturer's protocol.

Table 4 Reaction components to amplify the tRNA library insert.

Component	Volume per reaction	Final concentration
10× PfuUltra II Reaction buffer	5 μL	1×
10 mM dNTP mix	1.2 μL	0.24 μM
10 μM Primer B-Fwd2	6 μL	1.2 μM
10 μM Primer B-Rev2	6 μL	1.2 μM
DMSO	X μL	2–10%
PfuUltra II Fusion HS DNA polymerase	1 μL	
Distilled water	to 50 μL	

Values have been adapted from the manufacturer's protocol.

5.3.7 Prepare a 1% (*w*/*v*) agarose gel as in step 5.3.2. Mix 5 μL of the PCR reaction product and 1 μL of 6X DNA loading dye. Load the entire sample, along with an appropriate DNA ladder, into the wells of the agarose gel. Run the gel as in step 5.3.3 to confirm amplification of the randomized plasmid (for step 5.3.6.1) or insert (for step 5.3.6.2). Visualization of bands can be achieved with ethidium bromide or any appropriate DNA staining reagent and imaged with a gel imaging system (e.g., ChemiDoc).

Table 5 Thermocycler settings to generate the tRNA library insert.

Step	Temperature (°C)	Duration	Number of cycles
Denaturation	95	2 min	1
Denaturation	95	20 s	15
Annealing	50–63[a]	20 s	
Extension	72	30 s[b]	
Final extension	72	1 min	1

[a]Annealing temperature will depend on the primer melting temperatures and if DMSO is present. Consult the manufacturer's guidelines for appropriate temperatures.

[b]Extension duration may be adjusted during optimization of reaction conditions. Consult the manufacturer's guidelines for appropriate duration.

5.3.8 PCR purify the remaining PCR product using a PCR purification kit and elute with sterile water. Measure and record the DNA concentration using a spectrophotometer (e.g., NanoDrop).

5.3.9 Digest the PCR product(s) with the appropriate restriction enzyme(s) engineered in step 5.2.2. For randomization of a localized region use the plasmid PCR product from step 5.3.8, while for multiple regions use the plasmid PCR product from step 5.3.5 and the insert PCR product from step 5.3.8. Follow the manufacturer's recommended guidelines for enzyme amount, buffer, incubation temperature, and incubation time (e.g., if primer set A has an *Eco*RI cut site, use 50 units of EcoRI-HF in a 50 μL reaction. Incubate for 1 h at 37 °C). *Note: Only one restriction enzyme will be used for one randomized region while two restriction enzymes will be used for multiple randomized regions. A double digest can be performed for the multiple randomized regions if the two restriction enzymes are active in the same reaction buffer. The amount of DNA used for the digests will depend on how many transformations are needed. Keep in mind that DNA is lost at each step.*

5.3.10 PCR purify the digested DNA using a PCR purification kit and elute with sterile water. Measure and record the DNA concentration using a spectrophotometer (e.g., NanoDrop). *Note: the restriction enzymes should not simply be heat inactivated. PCR purifying the digestion reaction transfers the cleaved DNA into water so that the ligation reaction buffer can be used in the next step. This increases the ligation efficiency of the reaction.*

5.3.11 Once the DNA has been digested to create "sticky" ends, the tRNA library plasmid must be ligated back together. With the localized randomized region this will involve just ligating the plasmid back together. However, with multiple randomized regions, the tRNA library insert will be ligated into the parent plasmid. Set up the ligation reactions in a 50 μL reaction using a DNA ligase such as T4 DNA ligase. For the library with the localized region, the only DNA present in the reaction will be the plasmid containing the randomized region. For the library with multiple randomized regions, the plasmid and tRNA insert will be present at a 1:3 ratio (plasmid:insert). Incubate at 16 °C overnight using a thermocycler. *Note: the DNA ligase and the ligation conditions may need to be optimized for efficient ligation and maximum colonies after transformation.*

5.3.12 PCR purify the plasmid using a PCR purification kit and elute with sterile water. Measure and record the DNA concentration using a spectrophotometer (e.g., NanoDrop).

5.3.13 Mix 150–200 ng DNA (containing the ligated and purified tRNA library) into 100 μL electrocompetent cells (e.g., Top10, DH5α, Stellar™). Transfer the cells into a 0.2 cm electroporation cuvette and place on ice. Electroporate the cells at 2.5 V to maximize the amount of tRNA library inserted into the cells. Add 1 mL growth media (e.g., LB or SOC) and allow the cells to recover at 37 °C for at least 1 h while shaking. Prewarm LB agar plates (100 mm wide) containing the appropriate antibiotic for the tRNA plasmid. *Note: plating at an antibiotic concentration slightly lower than the normal working concentration can enhance the number of transformants.*

5.3.14 After cell recovery, prepare 10-fold serial dilutions of the transformed cells (100 μL of each dilution). Dilutions up to 10^{-6} can be made to allow for cell counting. Plate 100 μL of each dilution onto a prewarmed LB agar plate and evenly spread the cells using a cell spreader. Incubate the plates at 37 °C overnight.

5.3.15 The next morning, count the number of cells on the two most dilute plates to approximate how many successful transformants were present in the undiluted sample. For example, if there are 20 cells on a 10^{-4} plate and 150 on a 10^{-3} plate, that would be equivalent to approximately 1.5–2 million cells in the 1 mL recovery. The number of cells needed for complete library coverage will depend on how many nucleotides are randomized. To ensure complete coverage of the library, the number of cells should be at least 5–10 times that of the actual library size (e.g., 10×4^n, where n is the number of nucleotides). Therefore, for a library with 10 sites randomized, 10^7 cells would be required. This may not be accomplished in a single transformation, and it is common for 10–20 transformations to be done to achieve this large number of cells (e.g., if you achieve 10^6 cells in a single transformation, 10 transformations would provide you with 10^7 cells, enough for complete coverage of a library with 10 sites randomized). *Optional: pick 5–10 colonies from an LB agar plate with clear individual colonies as well as 5 μL of the undiluted transformation recovery from step 5.3.14 and grow in 3–5 mL LB with the appropriate antibiotic overnight at 37 °C. Spin down the cells and prepare a plasmid miniprep following the manufacturer's protocol. Send samples for sequencing to check for library diversity before scaling up (sample from the undiluted transformation should have multiple peaks at the randomized positions to denote that each base is present).*

5.3.16 After determining the number of transformations required for appropriate library coverage, transform more of the tRNA library as previously done in step 5.3.13. After recovery, spin down the cells in a microcentrifuge (11,000 × *g*, 2 min) and remove excess growth media. Resuspend each tube of cells in 300 μL LB. Plate the entire 300 μL of cells onto a pre-warmed LB agar plate (150 mm wide) containing the appropriate antibiotic for library selection. Evenly spread the cells using a cell spreader and incubate the plates at 37 °C overnight.

5.3.17 The next morning, add 2 mL LB onto each plate and gently scrape all the cells off the plate with a cell spreader. Pipette the resuspended cells from all plates into a 50 mL Falcon tube. Spin down the cells and perform a plasmid maxiprep following the manufacturer's protocol. Resuspend the tRNA library DNA in sterile water. Aliquot and freeze the tRNA library at −20 °C or −80 °C until use.

5.4 Screening the tRNA library using Sec-intein reporters

5.4.1 Following the protocol from step 5.3.13, transform 100 ng of the library (or WT tRNA plasmid as a control) into 50 μL electrocompetent B-95.ΔAΔ*fabR*Δ*selABC* (Mukai et al., 2015) already containing the KanR-M86 plasmid pAB_a02 (Chung et al., 2021) (see step 5.1.2 for details) at 1.8 V. Add 1 mL growth media (e.g., LB or SOC) and allow the cells to recover at 37 °C for 1 h while shaking. While the cells are in recovery, prewarm LB agar plates (150 mm wide) containing 50 μg/mL amp and the appropriate antibiotic for the tRNA plasmid.

5.4.2 To plate the cells, spin the tubes in a microcentrifuge (11,000 × *g*, 2 min) and remove excess growth media. Resuspend each tube in 300 μL LB and plate the entirety of each transformation onto the prewarmed LB agar plate. Evenly spread the cells using a cell spreader and incubate the plates at 37 °C overnight. *Optional: the transformation can be performed a few times to confirm that the entirety of the library is being covered.*

5.4.3 The next morning, harvest all cells from the plate(s). To do this, add 2 mL LB onto each plate and gently scrape all cells off the plate using a cell spreader. Pipette the resuspended cells into a culture tube, adding additional LB to a final volume of 4 mL. Prepare the culture tube for tRNA expression by adding 50 μg/mL amp, the appropriate antibiotic for the tRNA plasmid, the tRNA inducing agent, and 10 μM sodium selenite (Na_2SeO_3). Incubate at 37 °C for 6–8 h while shaking.

5.4.4 Measure the OD_{600} with a spectrophotometer (e.g., NanoDrop) and dilute the cultures to an OD_{600} of 0.2 using LB. To select for promising tRNAs, plate 100 μL of the dilution onto prewarmed LB agar plates (100 mm wide) containing 50 μg/mL amp, the appropriate antibiotic for the tRNA plasmid, the tRNA inducing agent, 10 μM Na_2SeO_3, and kan (at concentrations between 0 and 400 μg/mL). Incubate the plates at 37 °C overnight. *Optional: it is recommended to do this kan selection step using the original tRNA plasmid first or concurrently with the library. This will determine the highest concentration of kan that the WT tRNA can grow on.*

5.4.5 The next day, pick 10–20 colonies that grew on plates with higher kan concentrations than the WT tRNA control grew on. Resuspend each colony in 5 mL LB containing the appropriate antibiotic for the tRNA plasmid and incubate at 42 °C overnight while shaking. The absence of amp and incubation at a higher temperature allows for the loss of the KanR-M86 plasmid. *Note: If there are no colonies on plates with increased kan concentrations compared with the WT tRNA, then colonies can be picked from the plate with the same concentration, or another transformation could be done with kan concentrations closer together* *(e.g., 125 or 175 μg/mL instead of 100, 150, and 200 μg/mL). If the opposite is true and there are hundreds of colonies on the plate with the highest concentration of kan, consider setting up another 2–3 plates with concentrations of kan 25–100 μg/mL higher to reduce the number of colonies.*

Note: As the incubation at 42°C may be stressful and prevent cell growth, cultures can be grown during the day at 37°C while shaking to stimulate growth and increased to 42°C for overnight. An alternate incubation protocol is to grow the cultures at 37°C overnight while shaking and on the following day, dilute the culture 1:1000 into 5 mL fresh LB containing the appropriate antibiotic for the tRNA plasmid and again incubate at 37°C overnight while shaking.

5.4.6 The following day, plate 100 μL from each culture onto LB agar plates (100 mm wide) containing 100 μg/mL amp and streak them out (save the remaining volume of cell culture at 4 °C). Incubate the plates at 37 °C overnight. If no colonies are observed the next day, the KanR-M86 plasmid has been lost. If colonies grow, the precultures may need to be grown longer. Repeat step 5.4.5 using 100 μL from the saved preculture described above to confirm loss of pAB_a02. *Note: an alternate way to isolate the tRNA plasmid is to run an agarose gel and gel extract the plasmid of interest. This is only possible if the tRNA plasmid is a significantly different size than the KanR-M86 plasmid (4.3 kb).*

5.4.7 Once there is no cell growth on LB agar plates containing amp, the tRNA plasmid can be extracted. Using the saved 5 mL preculture, miniprep the tRNA plasmids using a plasmid miniprep kit and elute with sterile water. Measure the DNA concentration using a spectrophotometer (e.g., NanoDrop). It is important to clearly label each selected tRNA variant to avoid confusion during the subsequent steps. *Optional: it is recommended to send the extracted tRNA plasmids for sequencing at this point to determine the diversity of the tRNA plasmids.*

5.4.8 Using the selected tRNAs from the library, cotransform 50 ng of each tRNA variant plasmid (or WT tRNA plasmid as a control) with 50 ng pB_04 (amp resistant) into electrocompetent B-95.ΔAΔ*fabR*Δ*selABC* (Chung et al., 2021; Mukai et al., 2015) as stated above in step 5.3.13 using 1.8 V in a 0.2 cm electroporation cuvette. Add 1 mL growth media (e.g., LB or SOC) and allow the cells to recover at 37 °C for 1 h while shaking. While the cells are recovering, prewarm LB agar plates (100 mm wide) containing 50 μg/mL amp and the appropriate antibiotic for the tRNA plasmid. Plate the entire transformation onto the prewarmed LB agar plates and evenly spread the cells using a cell spreader. Incubate the plates at 37 °C overnight.

5.4.9 The next day, pick four individual colonies per tRNA plasmid (four biological replicates) and resuspend each colony in 0.5 mL LB (50 μg/mL amp, the appropriate antibiotic for the tRNA plasmid, the tRNA inducing agent, 1% (*w*/*v*) glucose, and 10 μM Na_2SeO_3) using a 24-well clear plate with a lid. Incubate the cultures at 37 °C for 6–8 h while shaking.

5.4.10 Prepare a 96-well black plate with growth media (75 μL LB containing 50 μg/mL amp, the appropriate antibiotic for the tRNA plasmid, the tRNA inducing agent, 1% (*w*/*v*) glucose, and 10 μM Na_2SeO_3) for expression of sfGFP-M86. Each colony requires two wells; one with 2 mM IPTG for inducing the sfGFP-M86 plasmid and one without IPTG for background signal. Mix 75 μL of cell culture from the 24-well plate into the corresponding wells (one with and without 2 mM IPTG); the final concentration of IPTG in the 150 μL culture is 1 mM. Seal the 96-well plate with a breathable sealing membrane (e.g., Breathe-Easy® sealing membrane).

5.4.11 Place the 96-well plate into a plate reader (e.g., Synergy HTX Plate Reader). Set the plate reader to read fluorescence (e.g., 485 nm, Em. 528 nm) and OD_{600} every 15 min for 24 h. The plate should be incubated at 37 °C during the 24 h and set to shake between reads.

5.4.12 Analyze the data to determine the best tRNA candidates. Divide the fluorescence readings by the OD_{600} at 16 h and take the average of the four biological replicates for each tRNA candidate with and without IPTG. To consider the background fluorescence, subtract the value of the uninduced culture from the induced culture. Graphing these results will provide a visual representation to compare the effectiveness of each tRNA in Sec incorporation. *Note: sfGFP production can also be calculated at 8 h and 24 h timepoints and compared to 16 hr. Typically the trend is similar, however, in some cases there may be a tRNA variant which spikes up at 24 h compared to 16 hr. If there is too much variability in the four biological replicates (i.e., two are significantly different from the other two), then the transformation in step 5.4.8 and corresponding assay should be repeated.*

5.4.13 Send the tRNA plasmids (generated in step 5.4.7) which have increased sfGFP fluorescence for sequencing.

5.5 Confirm efficiency of screened tRNA for Sec incorporation

With the results obtained in step 5.4, a few or many tRNAs may have increased Sec incorporation activity with respect to the starting WT tRNA. The intein-based reporters specifically screen for the increased presence of Sec in the intein. This could result from an increase in selenoprotein production, an increase in the percent of Sec incorporated, or a mixture of the two. To illuminate the details of the results, a reporter selenoprotein should be expressed and purified (e.g., *E. coli* Grx1, mammalian GPx1, or mammalian MsrB1).

5.5.1 From the results in step 5.4, choose which screened tRNAs are worth using for selenoprotein production. Co-transform them individually with the selenoprotein expression plasmid into electrocompetent B-95.ΔAΔ*fabR*Δ*selABC* cells (or equivalent selenoprotein expression cell line). *Note: preparation of selenoprotein with the WT tRNA is advised so that direct comparisons can be made on the selenoprotein yield and amount of Sec incorporated.*

5.5.2 Follow the specific expression and purification protocol for the selected reporter selenoprotein as outlined previously [e.g., *E. coli* Grx1 and human GPx1 (Aldag et al., 2013), human MsrB1 (Kim & Gladyshev, 2004)].

5.5.3 Once the selenoprotein has been purified, quantify the protein yield obtained and confirm purity on an SDS-PAGE gel.

5.5.4 Perform the corresponding activity assays to confirm that the purified selenoprotein is active before sending for mass spectrometry (MS) (Peeler & Weerapana, 2019).

5.5.5 Following instructions from an MS facility, send the protein samples for intact MS analysis. Analyze the data to determine the percent Sec incorporation and compare results to the WT tRNA. *Note: Proteins sent for intact MS analysis should see an increase in mass of 68 Da for the Sec-containing protein compared to the Ser-containing protein. If Sec incorporation is poor, it is possible to see some glutamine misincorporation. For large proteins (>100 kDa), this small mass shift might be difficult to resolve. Therefore, MS/MS analysis should be done to look at Sec incorporation in the peptides collected from the digested protein.*

6. Expected outcomes

The aim in this protocol is to screen for an improved tRNA for recombinant Sec incorporation. Starting with a tRNA that has already been shown to site-specifically introduce Sec into a recombinant protein, this is a guide to (i) generate a tRNA library and (ii) test that library for improved insertion of Sec into a recombinant protein. After successful design and coverage of a tRNA library, top candidates will be selected for using a constitutive KanR-M86 intein reporter. These tRNAs can then be further characterized on their Sec incorporation ability through an IPTG-inducible sfGFP-M86 intein reporter. This will ideally result in a few tRNA variants that provide increased Sec incorporation when compared to results by the WT tRNA. These resulting tRNA variants can be narrowed down to a single superior tRNA variant through expression of a selenoprotein reporter. Although this step is not necessary, it elucidates the numerical difference in protein yield and amount of Sec incorporation compared to the WT tRNA.

7. Advantages

The general strategy for screening tRNA libraries for noncanonical amino acid (ncAA) incorporation requires both positive and negative selection. First, positive selection of the library is done with a reporter gene that requires readthrough of the UAG codon to confer antibiotic resistance (e.g., *cat* gene) (Wang & Schultz, 2001). This quickly eliminates any tRNAs which cannot decode UAG but fails to discriminate which amino acid is incorporated at UAG. Therefore, subsequent negative selection of the surviving clones eliminates any tRNAs which are not specific for the ncAA of interest. This is done through expression of the remaining tRNA library with a toxic gene (e.g., barnase) containing two or more UAG codons (Wang & Schultz, 2001). When this plasmid system is grown in the absence of the ncAA, only tRNAs which incorporate canonical amino acids will express the toxic gene, causing cell death; leaving the tRNAs which specifically incorporate the ncAA.

This strategy does not work for Sec incorporation since Sec is synthesized on its cognate tRNA after initial aminoacylation with Ser. Therefore, during negative selection (in the absence of the selenium precursor), Ser would be incorporated in the barnase gene for all tRNAs instead of distinguishing between Ser misincorporation and Sec incorporation. As a result, Sec-specific reporters are required. Here, we introduce intein-based Sec-specific reporters as a tool to screen tRNA libraries for Sec incorporation. These reporters have been shown to be functional when Sec is present and nonfunctional when Ser is inserted (Chung et al., 2021). Furthermore, the amount of activity inferred with a specific tRNA is directly proportional to the amount of Sec incorporated. Therefore, this protocol describes a strategy which allows screening of libraries that are specific for Sec incorporation.

8. Limitations

tRNA libraries (like all libraries) are limited by the transformation efficiency of *E. coli* (10^9). Each additional base change creates an exponential increase in the number of sequence possibilities (4^n where n is the number of bases randomized). Therefore randomizing 10–12 bases is the upper limit to ensure the recommended 10-fold coverage of the library is achieved. It is possible that even with a low number of bases randomized, continued optimization can still result in incomplete library coverage due to toxic tRNAs. Even though tRNA expression should be under an inducible promoter, leaky transcription can still occur (Siegele & Hu, 1997).

A major concern with generating libraries is introducing bias such that not all sequence possibilities are equally represented. Some of this bias is removed by excising the randomized region of the tRNA out of the parent plasmid. This prevents favoring sequences which preferentially anneal to the original tRNA sequence. Moreover, secondary structures can form in these randomized regions, preventing adequate amplification of specific sequences. The addition of DMSO into the PCR reaction reduces this bias. Furthermore, after acquiring the tRNA library, it is assumed that all tRNA variants are equally transformed into cells containing the KanR-M86 reporter. However, it is possible that this does not happen, and some candidates are removed from the tRNA library before the selection process. Some of this bias can be reduced by transforming the library multiple times for selection by the KanR-M86 reporter.

9. Optimization and troubleshooting

As mentioned previously, the hardest step in this protocol will be to generate the tRNA library. This may require significant optimization at all steps: (i) PCR, (ii) digestion and ligation, and (iii) transformation.

(i) For the PCR, adding DMSO (2–10%) when amplifying the randomized region of the tRNA is highly recommended to prevent secondary structure formation (Varadaraj & Skinner, 1994). The addition of DMSO can affect the annealing temperatures, therefore optimizing the temperature may also be required upon the addition of high concentrations of DMSO (>2%) (Chester & Marshak, 1993; Obradovic et al., 2013). If the PCR product quality is still poor after optimization, the primers may need to be modified. It is recommended for primers to be less than 60 bases and include approximately 20 nucleotides identical to the gene at both the 3′ and 5′ end (Williams, Copp, & Ackerley, 2014).

(ii) Digestion of the plasmid (and insert) should be performed with high fidelity enzymes for maximum cleavage. The ligation should be done in the recommended buffer for maximum efficiency (this requires the digested reaction to be cleaned up and eluted in water before proceeding). Ligation reactions may need to be optimized in terms of the specific enzyme used, the amount of enzyme in the reaction, duration of ligation, and ligation temperature.

(iii) Transformation of the tRNA library should be performed in electrocompetent cells. This requires the DNA to be cleaned up and eluted in water (devoid of any salts). The transformation efficiency can be increased by growing electrocompetent cells in LB with no salt, using fresh cells (i.e., before freezing), and transforming more cells (e.g., using a 100 μL aliquot of cells instead of a 50 μL aliquot).

The KanR-M86 and sfGFP-M86 reporters have been previously optimized (Chung et al., 2021) for use with B-95.ΔAΔ*fabR*Δ*selABC* cells (Mukai et al., 2015) and the pSecUAG-Evol plasmids (Mukai et al., 2018). Some minor changes or optimization (such as deletion of *selABC* in another cell line) may be required if using something different.

Acknowledgments

We would like to thank Jonathan Fischer, Jeffery Tharp, and Oscar Vargas-Rodriguez for experimental advice and helpful discussions.

Funding

This work was supported by grants from the National Institute of General Medical Sciences (R35GM122560 to D.S.) and, for the genetic studies, the Department of Energy Office of Basic Energy Sciences (DE-FG0298ER2031 to D.S.). Christina Z. Chung holds a Postdoctoral Fellowship from the Natural Sciences and Engineering Research Council of Canada (NSERC).

References

Aldag, C., Bröcker, M. J., Hohn, M. J., Prat, L., Hammond, G., Plummer, A., et al. (2013). Rewiring translation for elongation factor Tu-dependent selenocysteine incorporation. *Angewandte Chemie (International Ed. in English)*, *52*(5), 1441–1445. https://doi.org/10.1002/anie.201207567.

Appleby-Tagoe, J. H., Thiel, I. V., Wang, Y., Wang, Y., Mootz, H. D., & Liu, X. Q. (2011). Highly efficient and more general cis- and trans-splicing inteins through sequential directed evolution. *The Journal of Biological Chemistry, 286*(39), 34440–34447. https://doi.org/10.1074/jbc.M111.277350.

Arai, K., Takei, T., Okumura, M., Watanabe, S., Amagai, Y., Asahina, Y., et al. (2017). Preparation of selenoinsulin as a long-lasting insulin analogue. *Angewandte Chemie (International Ed. in English), 56*(20), 5522–5526. https://doi.org/10.1002/anie.201701654.

Arnér, E. S., Sarioglu, H., Lottspeich, F., Holmgren, A., & Böck, A. (1999). High-level expression in *Escherichia coli* of selenocysteine-containing rat thioredoxin reductase utilizing gene fusions with engineered bacterial-type SECIS elements and co-expression with the *selA*, *selB* and *selC* genes. *Journal of Molecular Biology, 292*(5), 1003–1016. https://doi.org/10.1006/jmbi.1999.3085.

Asahara, H., & Uhlenbeck, O. C. (2005). Predicting the binding affinities of misacylated tRNAs for *Thermus thermophilus* EF-Tu·GTP. *Biochemistry, 44*(33), 11254–11261. https://doi.org/10.1021/bi050204y.

Axley, M. J., Böck, A., & Stadtman, T. C. (1991). Catalytic properties of an *Escherichia coli* formate dehydrogenase mutant in which sulfur replaces selenium. *Proceedings of the National Academy of Sciences of the United States of America, 88*(19), 8450–8454. https://doi.org/10.1073/pnas.88.19.8450.

Baron, C., & Böck, A. (1991). The length of the aminoacyl-acceptor stem of the selenocysteine-specific tRNA^{Sec} of *Escherichia coli* is the determinant for binding to elongation factors SELB or Tu. *The Journal of Biological Chemistry, 266*(30), 20375–20379.

Chester, N., & Marshak, D. R. (1993). Dimethyl sulfoxide-mediated primer tm reduction: A method for analyzing the role of renaturation temperature in the polymerase chain reaction. *Analytical Biochemistry, 209*(2), 284–290. https://doi.org/10.1006/abio.1993.1121.

Chung, C. Z., Krahn, N., Crnković, A., & Söll, D. (2021). Intein-based design expands diversity of selenocysteine reporters. *Journal of Molecular Biology, 167199*. https://doi.org/10.1016/j.jmb.2021.167199.

Evans, R. M., Krahn, N., Murphy, B. J., Lee, H., Armstrong, F. A., & Söll, D. (2021). Selective cysteine-to-selenocysteine changes in a [NiFe]-hydrogenase confirm a special position for catalysis and oxygen tolerance. *Proceedings of the National Academy of Sciences of the United States of America, 118*(13). https://doi.org/10.1073/pnas.2100921118.

Forchhammer, K., Boesmiller, K., & Böck, A. (1991). The function of selenocysteine synthase and SELB in the synthesis and incorporation of selenocysteine. *Biochimie, 73*(12), 1481–1486. https://doi.org/10.1016/0300-9084(91)90181-y.

Giegé, R., Sissler, M., & Florentz, C. (1998). Universal rules and idiosyncratic features in tRNA identity. *Nucleic Acids Research, 26*(22), 5017–5035. https://doi.org/10.1093/nar/26.22.5017.

Gonzales, M. F., Brooks, T., Pukatzki, S. U., & Provenzano, D. (2013). Rapid protocol for preparation of electrocompetent *Escherichia coli* and *vibrio cholerae*. *Journal of Visualized Experiments, 80*. https://doi.org/10.3791/50684.

Greene, B. L., Stubbe, J., & Nocera, D. G. (2019). Selenocysteine substitution in a class I ribonucleotide reductase. *Biochemistry, 58*(50), 5074–5084. https://doi.org/10.1021/acs.biochem.9b00973.

Guillin, O. M., Vindry, C., Ohlmann, T., & Chavatte, L. (2019). Selenium, selenoproteins and viral infection. *Nutrients, 11*(9). https://doi.org/10.3390/nu11092101.

Guzman, L. M., Belin, D., Carson, M. J., & Beckwith, J. (1995). Tight regulation, modulation, and high-level expression by vectors containing the arabinose PBAD promoter. *Journal of Bacteriology, 177*(14), 4121–4130. https://doi.org/10.1128/jb.177.14.4121-4130.1995.

Heider, J., Baron, C., & Böck, A. (1992). Coding from a distance: Dissection of the mRNA determinants required for the incorporation of selenocysteine into protein. *The EMBO Journal, 11*(10), 3759–3766.

Itoh, Y., Bröcker, M. J., Sekine, S., Hammond, G., Suetsugu, S., Söll, D., et al. (2013). Decameric SelA*tRNA^{Sec} ring structure reveals mechanism of bacterial selenocysteine formation. *Science, 340*(6128), 75–78. https://doi.org/10.1126/science.1229521.

Jiang, Z., Arnér, E. S., Mu, Y., Johansson, L., Shi, J., Zhao, S., et al. (2004). Expression of selenocysteine-containing glutathione *S*-transferase in *Escherichia coli*. *Biochemical and Biophysical Research Communications, 321*(1), 94–101. https://doi.org/10.1016/j.bbrc.2004.06.110.

Kim, H. Y., & Gladyshev, V. N. (2004). Methionine sulfoxide reduction in mammals: Characterization of methionine-R-sulfoxide reductases. *Molecular Biology of the Cell, 15*(3), 1055–1064. https://doi.org/10.1091/mbc.e03-08-0629.

Lacourciere, G. M., & Stadtman, T. C. (1999). Catalytic properties of selenophosphate synthetases: Comparison of the selenocysteine-containing enzyme from *Haemophilus influenzae* with the corresponding cysteine-containing enzyme from *Escherichia coli*. *Proceedings of the National Academy of Sciences of the United States of America, 96*(1), 44–48. https://doi.org/10.1073/pnas.96.1.44.

Liu, J., Chen, Q., & Rozovsky, S. (2017). Utilizing selenocysteine for expressed protein ligation and bioconjugations. *Journal of the American Chemical Society, 139*(9), 3430–3437. https://doi.org/10.1021/jacs.6b10991.

Lobanov, A. V., Hatfield, D. L., & Gladyshev, V. N. (2009). Eukaryotic selenoproteins and selenoproteomes. *Biochimica et Biophysica Acta, 1790*(11), 1424–1428. https://doi.org/10.1016/j.bbagen.2009.05.014.

Mandrand-Berthelot, M.-A., Wee, M. Y. K., & Haddock, B. A. (1978). An improved method for the identification and characterization of mutants of *Escherichia coli* deficient in formate dehydrogenase activity. *FEMS Microbiology Letters, 4*(1), 37–40. https://doi.org/10.1111/j.1574-6968.1978.tb02841.x.

Martin, G. W., 3rd, Harney, J. W., & Berry, M. J. (1998). Functionality of mutations at conserved nucleotides in eukaryotic SECIS elements is determined by the identity of a single nonconserved nucleotide. *RNA, 4*(1), 65–73.

Metanis, N., & Hilvert, D. (2015). Harnessing selenocysteine reactivity for oxidative protein folding. *Chemical Science, 6*(1), 322–325. https://doi.org/10.1039/c4sc02379j.

Miller, C., Bröcker, M. J., Prat, L., Ip, K., Chirathivat, N., Feiock, A., et al. (2015). A synthetic tRNA for EF-Tu mediated selenocysteine incorporation *in vivo* and *in vitro*. *FEBS Letters, 589*(17), 2194–2199. https://doi.org/10.1016/j.febslet.2015.06.039.

Mukai, T., Englert, M., Tripp, H. J., Miller, C., Ivanova, N. N., Rubin, E. M., et al. (2016). Facile recoding of selenocysteine in nature. *Angewandte Chemie (International Ed. in English), 55*(17), 5337–5341. https://doi.org/10.1002/anie.201511657.

Mukai, T., Hoshi, H., Ohtake, K., Takahashi, M., Yamaguchi, A., Hayashi, A., et al. (2015). Highly reproductive *Escherichia coli* cells with no specific assignment to the UAG codon. *Scientific Reports, 5*, 9699. https://doi.org/10.1038/srep09699.

Mukai, T., Sevostyanova, A., Suzuki, T., Fu, X., & Söll, D. (2018). A facile method for producing selenocysteine-containing proteins. *Angewandte Chemie (International Ed. in English), 57*(24), 7215–7219. https://doi.org/10.1002/anie.201713215.

Mukai, T., Vargas-Rodriguez, O., Englert, M., Tripp, H. J., Ivanova, N. N., Rubin, E. M., et al. (2017). Transfer RNAs with novel cloverleaf structures. *Nucleic Acids Research, 45*(5), 2776–2785. https://doi.org/10.1093/nar/gkw898.

Obradovic, J., Jurisic, V., Tosic, N., Mrdjanovic, J., Perin, B., Pavlovic, S., et al. (2013). Optimization of PCR conditions for amplification of GC-rich EGFR promoter sequence. *Journal of Clinical Laboratory Analysis, 27*(6), 487–493. https://doi.org/10.1002/jcla.21632.

Ohama, T., Yang, D. C., & Hatfield, D. L. (1994). Selenocysteine tRNA and serine tRNA are aminoacylated by the same synthetase, but may manifest different identities with respect to the long extra arm. *Archives of Biochemistry and Biophysics, 315*(2), 293–301. https://doi.org/10.1006/abbi.1994.1503.

Peeler, J. C., & Weerapana, E. (2019). Chemical biology approaches to interrogate the selenoproteome. *Accounts of Chemical Research*, *52*(10), 2832–2840. https://doi.org/10.1021/acs.accounts.9b00379.

Sawers, G., Heider, J., Zehelein, E., & Böck, A. (1991). Expression and operon structure of the *sel* genes of *Escherichia coli* and identification of a third selenium-containing formate dehydrogenase isoenzyme. *Journal of Bacteriology*, *173*(16), 4983–4993. https://doi.org/10.1128/jb.173.16.4983-4993.1991.

Shchedrina, V. A., Novoselov, S. V., Malinouski, M. Y., & Gladyshev, V. N. (2007). Identification and characterization of a selenoprotein family containing a diselenide bond in a redox motif. *Proceedings of the National Academy of Sciences of the United States of America*, *104*(35), 13919–13924. https://doi.org/10.1073/pnas.0703448104.

Siegele, D. A., & Hu, J. C. (1997). Gene expression from plasmids containing the *araBAD* promoter at subsaturating inducer concentrations represents mixed populations. *Proceedings of the National Academy of Sciences of the United States of America*, *94*(15), 8168–8172. https://doi.org/10.1073/pnas.94.15.8168.

Snider, G. W., Ruggles, E., Khan, N., & Hondal, R. J. (2013). Selenocysteine confers resistance to inactivation by oxidation in thioredoxin reductase: Comparison of selenium and sulfur enzymes. *Biochemistry*, *52*(32), 5472–5481. https://doi.org/10.1021/bi400462j.

Stadtman, T. C. (1996). Selenocysteine. *Annual Review of Biochemistry*, *65*, 83–100. https://doi.org/10.1146/annurev.bi.65.070196.000503.

Thyer, R., Robotham, S. A., Brodbelt, J. S., & Ellington, A. D. (2015). Evolving tRNA^{Sec} for efficient canonical incorporation of selenocysteine. *Journal of the American Chemical Society*, *137*(1), 46–49. https://doi.org/10.1021/ja510695g.

Varadaraj, K., & Skinner, D. M. (1994). Denaturants or cosolvents improve the specificity of PCR amplification of a G + C-rich DNA using genetically engineered DNA polymerases. *Gene*, *140*(1), 1–5. https://doi.org/10.1016/0378-1119(94)90723-4.

Wang, L., & Schultz, P. G. (2001). A general approach for the generation of orthogonal tRNAs. *Chemistry & Biology*, *8*(9), 883–890. https://doi.org/10.1016/s1074-5521(01)00063-1.

Williams, E. M., Copp, J. N., & Ackerley, D. F. (2014). Site-saturation mutagenesis by overlap extension PCR. *Methods in Molecular Biology*, *1179*, 83–101. https://doi.org/10.1007/978-1-4939-1053-3_6.

Wu, X. Q., & Gross, H. J. (1993). The long extra arms of human tRNA$^{(Ser)Sec}$ and tRNA^{Ser} function as major identify elements for serylation in an orientation-dependent, but not sequence-specific manner. *Nucleic Acids Research*, *21*(24), 5589–5594. https://doi.org/10.1093/nar/21.24.5589.

Yoshizawa, S., & Böck, A. (2009). The many levels of control on bacterial selenoprotein synthesis. *Biochimica et Biophysica Acta*, *1790*(11), 1404–1414. https://doi.org/10.1016/j.bbagen.2009.03.010.

CHAPTER FOUR

Expressing recombinant selenoproteins using redefinition of a single UAG codon in an RF1-depleted *E. coli* host strain

Qing Cheng[a] and Elias S.J. Arnér[a,b,*]

[a]Division of Biochemistry, Department of Medical Biochemistry and Biophysics, Karolinska Institutet, Stockholm, Sweden

[b]Department of Selenoprotein Research, National Institute of Oncology, Budapest, Hungary

*Corresponding author: e-mail address: elias.arner@ki.se

Contents

Methods in Enzymology, Volume 662
ISSN 0076-6879
https://doi.org/10.1016/bs.mie.2021.10.004

Abstract

Selenoproteins containing the rare amino acid selenocysteine (Sec), typically being enzymes utilizing the selenium atom of Sec for promoted catalysis of redox reactions, are challenging to obtain at high amounts in pure form. The technical challenges limiting selenoprotein supply derive from intricacies in their translation, necessitating the recoding of a UGA stop codon to a sense codon for Sec. This, in turn, involves the interactions of a Sec-dedicated elongation factor, either directly or indirectly, with a structure in the selenoprotein-encoding mRNA called a SECIS element (Selenocysteine Insertion Sequence), a dedicated tRNA species for Sec with an anticodon for the UGA, and several accessory enzymes and proteins involved in the selenoprotein synthesis. Here, we describe an alternative method for recombinant selenoprotein production using UAG as the Sec codon in a specific strain of *E. coli* lacking other UAG codons and lacking the release factor RF1 that normally terminates translation at UAG. We also describe how such recombinant selenoproteins can be purified and further analyzed for final Sec contents. The methodology can be used for production of natural selenoproteins in recombinant form as well as for production of synthetic selenoproteins that may be designed to use the unique biophysical properties of Sec for diverse biotechnological applications.

1. Introduction

The human genome encompasses 25 genes encoding selenoproteins while mouse has 24 (Kryukov et al., 2003), and although many organisms and life forms completely lack selenoproteins, several thousands of selenoproteins have been found in nature in analyses of different genomes (Jiang et al., 2020; Mariotti et al., 2012; Peng et al., 2016; Rother & Quitzke, 2018). Most naturally occurring selenoproteins are likely to utilize the high chemical reactivity of Sec as a catalytic residue in the active site of enzymes catalyzing redox reactions, with its similar yet more "pronounced" biophysical features compared to the common sulfur-containing Cys analogue (Arnér, 2010; Hondal, Marino, & Gladyshev, 2013; Reich & Hondal, 2016). The price to pay for the use of Sec in an enzyme is high if considering the highly intricate, energy demanding and often inefficient translation machinery that is required for expression of a selenoprotein. There are qualitative differences between these translation machineries between the kingdoms of life, and between organisms, but all involve the use of a dedicated elongation factor specific for Sec insertion through the recoding of a UGA stop codon, achieved via interactions with a stem loop structure in the selenoprotein-encoding mRNA (a SECIS structure), use of a specific tRNA with an anticodon to the UGA that is originally serylated with the seryl moiety subsequently converted to selenocysteinyl on the tRNA, and a variety of accessory enzymes and factors supporting this translation machinery that is only used for the insertion of a Sec residue in a growing polypeptide chain at the ribosome (Allmang, Wurth, & Krol, 2009; Brocker et al., 2014; Driscoll & Copeland, 2003; Gursinsky et al., 2008; Lescure et al., 2002; Rother & Quitzke, 2018; Yoshizawa & Böck, 2009). This is thus a most complicated translation machinery, with species-specific variations, inefficiency on a cellular level compared to expression of most regular non-selenoproteins, and high complexity, further complicated by the inherent high chemical reactivity of Sec (indeed itself a possible reason why its synthesis has evolved to be so complicated, possibly as a means of controlling the reactivity of Sec that would otherwise lead to uncontrolled redox reactions in cells). As a result of this complexity in Sec insertion during translation, and due to the high inherent chemical reactivity of Sec, it has been difficult to obtain pure preparations of selenoproteins in high amounts, which is a necessity for detailed studies of their features or for using them in different biotechnological applications. In this chapter we shall provide a production protocol that was rather recently developed, which enables production of recombinant selenoproteins in *E. coli* at a higher yield and purity than what has previously been possible. We will also discuss the limiting factors that are inherent in this production system and how it may be possible to further purify the selenoproteins toward higher Sec contents. We also emphasize the need of determining the Sec contents in the final product and we provide different methods of doing so.

2. Selenoprotein production

2.1 Alternative methods to produce selenoproteins: Pros and cons

It should be noted that several alternative methods to obtain preparations of selenoproteins have been developed and the method described herein is thus not the only approach. Naturally occurring selenoproteins can of course be purified from their original sources, as has been done for many bacterial as well as eukaryotic species, including mammals. An advantage of the approach of native selenoprotein purification is that, provided that the purification is successful, the naturally occurring selenoprotein will be obtained "as is." This is often not the case with synthetic production methods, where side reactions may occur that yield a final product that is non-homogenous or, in worst case, not having the same features as the naturally occurring selenoprotein. That risk should always be considered, and in this chapter, we shall also point out the side reactions that may occur using our methodology—and how the final product must thus be characterized and/or how it can be further purified.

The methodology described herein is, to the best of our knowledge, the most successful technique yet developed for production of recombinant selenoproteins in *E. coli*. Alternatives to this approach can be recombinant production in other organisms. With regards to selenoprotein production in mammalian cells, this is also possible but yields are typically rather low compared to bacterial expression, even if measures are taken to increase the selenoprotein translation efficiency (Novoselov et al., 2007). One advantage with such approaches is, however, that even Selenoprotein P (SELENOP) could be expressed in recombinant form in mammalian cells (Tujebajeva, Harney, & Berry, 2000). Such production is likely to be difficult to achieve with the methodology that we describe herein, considering that SELENOP contains more than 10 Sec residues and is furthermore glycosylated (Tujebajeva et al., 2000). Recently human methionine sulfoxide reductase was successfully purified as a recombinant selenoprotein upon the expression in HEK cells using a photocaged Sec approach together with genetic code expansion, resulting in 12.4 μg purified MsrB1-DMNB-Sec95 obtained from 10 dishes of HEK cells (Peeler et al., 2020). It was also shown that the human glutathione peroxidase GPX4 could be expressed and purified in recombinant form using HEK cells, at an amount sufficient for solution of its crystal structure with and without a covalent inhibitor bound to its Sec residue, with the original yield being 1 mg of pure GPX4 obtained from 1 L of HEK cell culture (Moosmayer et al., 2021). These alternative methods thus yield significantly less recombinant selenoprotein compared to using *E. coli* as expression host, but should be considered as possible alternative approaches that may be superior for certain applications depending upon the intended use of the recombinant selenoprotein. Note, however, that neither *Saccharomyces cerevisiae* nor insect cells have the natural machinery for selenoprotein production, meaning that recombinant selenoprotein production in regular yeast or baculoviral production systems will be very challenging, if not impossible.

In addition to the production of recombinant selenoproteins using different live cell-based protein expression systems, there are also synthetic or semi-synthetics routes to selenoprotein production. To review such alternatives is beyond this chapter, but other chapters in this volume of *Methods in Enzymology* cover such techniques, and the interested reader may also find several good reviews on the topic (Hondal, 2009; Liu, Cheng, & Rozovsky, 2018; Metanis & Hilvert, 2014; Peeler & Weerapana, 2019).

2.2 Expressing recombinant selenoproteins in *E. coli*: Prior developments and alternative methods

The first successful production of a recombinant selenoprotein at high yields in *E. coli* was that of its endogenous formate dehydrogenase H (FDH-H, encoded by *fdhF*) produced at a yield which enabled solving its crystal structure (Boyington et al., 1997). That study thereby demonstrated that the *E. coli* selenoprotein synthesis machinery can upstand a high-level expression of selenoproteins suitable for recombinant production. Tailoring a gene encoding rat thioredoxin reductase 1 for production in *E. coli* by the engineering of a compatible SECIS element was subsequently shown to be possible, and its yield was further enhanced by the overexpression of the bacterial tRNA for Sec (selC), of the Sec-specific elongation factor (SelB) and of selenocysteine synthase (SelC) (Arnér et al., 1999). Although Sec insertion was still only approximately 15–20% in that system, mainly due to competition between SelB and termination at the UGA through the action of release factor RF-2, the production principle was much used for more than two decades as a source of thioredoxin reductase isoenzymes and several other selenoproteins (Arner, 2002; Cheng & Arner, 2018; Cheng et al., 2012). In the present protocol, we provide the protocol for a recently further developed and improved methodology for recombinant selenoprotein production in *E. coli*, utilizing insertion of Sec at a redefined UAG codon in an RF1-deficient host strain (Cheng & Arner, 2017; Cheng et al., 2021), which will be described in detail below.

It should be known that recently alternative methodologies for recombinant selenoprotein production in *E. coli* were also described. One approach furthermore involves the use of UAG recoding, like our method, but instead encompassing an altered tRNA species that becomes compatible with engineered variants of the normal EF-Tu elongation factor, instead of SelB, resulting in more flexibility in gene construct design but also somewhat lower yields and a concomitant incorporation of Ser mixed with

Sec insertion (Aldag et al., 2013; Haruna et al., 2014; Miller et al., 2015; Mukai et al., 2018). Yet another approach was recently described utilizing a photocaged Sec incorporated at a defined UAG in *E. coli*, and synthesized with a mutant *Methanosarcina mazei* pyrrolysyl-tRNA synthetase (Welegedara et al., 2018). That production gave mixtures with truncation instead of Sec insertion, because a natural *E. coli* host strain was used—perhaps the RF-1 depleted host strain used in our studies would further increase yields in that selenoprotein production method. Still, in the original study purification of the full-length selenoprotein was enabled using a C-terminal His-tag and gave final total yields of pure selenoproteins at about 1.5–4.5 mg per liter culture (Welegedara et al., 2018). Our methodology results in about 20 mg pure Sec-containing thioredoxin reductase purified per liter bacterial culture (Cheng & Arner, 2017). However, thioredoxin reductase or other selenoproteins with the Sec residue close to the C-terminus result in better purity and yield with our method, than that of selenoproteins with internal Sec residues as those result in a mixture of Sec with either Gln or Lys (Cheng & Arner, 2017). It is not known why, specifically, Gln or Lys are incorporated instead of Sec using this production technique, but it can be noted that codons for these amino acids (CAG for Gln and AAG for Lys) only differ at the first nucleotide from the UAG codon. Usually wobbling occurs at the third nucleotide position of a codon, but perhaps the highly artificial production system in absence of RF-1 facilitates this apparent "first-nucleotide-position-wobbling" phenomenon. We also found an interesting +3 frameshift at the UAG codon in our production system, resulting in a Sec-to-null variant, while we furthermore discovered that the human selenoprotein GPX4 could subsequently be enriched to full Sec content at a final yield of about 4–6 mg obtained per liter original bacterial culture (Cheng et al., 2021). The enrichment of Sec-containing GPX4 was possible due to differential affinities of Sec-containing and non-Sec-containing GPX4 variants to a bromosulfophthalein-column, but that principle for further enrichment was not possible to employ for GPX1 (Cheng et al., 2021). If also other Sec-containing selenoproteins could be purified from their corresponding non-Sec containing variants, this would naturally further improve the versatility of any type of methodology for selenoprotein production that has less than 100% Sec incorporation.

The different currently available methods for production of pure selenoproteins can clearly complement each other and should be chosen depending upon the best suitability in line with the intended use of the selenoprotein product. It is also likely that the different methodologies will improve further in the years to come, which should help to use them as tools in the further advancement of the field of selenoprotein research.

3. Using UAG redefinition as a sec codon in an RF1-depleted *E. coli* host strain

The methodology to be described herein, which was rather recently developed and hitherto used for production of selenoprotein thioredoxin reductases and glutathione peroxidases (Cheng & Arner, 2017; Cheng et al., 2021), is based upon the recoding of UAG as a redefined sense codon for Sec, using the regular bacterial tRNA for Sec (the *selC* gene product) but with a mutated anticodon for compatibility with Sec insertion at UAG. This tRNA is still compatible with the SelB elongation factor, and we thus utilize overexpression of both these factors, together with overexpression of selenocysteine synthase (SelA) for improved yields. To avoid termination of translation at the UAG, the production is performed in a unique *E. coli* strain lacking release factor RF1 that otherwise competes with SelB by terminating translation at UAG. That specific *E. coli* strain was developed by others to allow for different chemical biology applications based upon expanded use of the UAG codon (Lajoie et al., 2013). Notably, all natural UAG codons were mutated to UAA in this strain (to be compatible with release factor RF2 instead) in order to "liberate" the UAG codon for other usage (Lajoie et al., 2013). Attempting selenoprotein synthesis in that strain we subsequently discovered that in this highly unnatural cellular milieu, having a single UAG codon found in the global mRNA transcriptome and a total lack of release factor RF1 that would recognize the UAG, and using an overexpressed Sec-inserting tRNA with an anticodon for UAG together with the overexpressed SelB elongation factor, it became possible to express selenoproteins also in the absence of a SECIS element, albeit with lower efficiency than if it was maintained. If the SECIS was maintained in this setup, on the other hand, the Sec insertion became close to 100% (Cheng & Arner, 2017). This is the underlying principle for the production protocol described in this chapter, and the key features of it are summarized in Fig. 1, with the mutated tRNA species shown in further detail in Fig. 2.

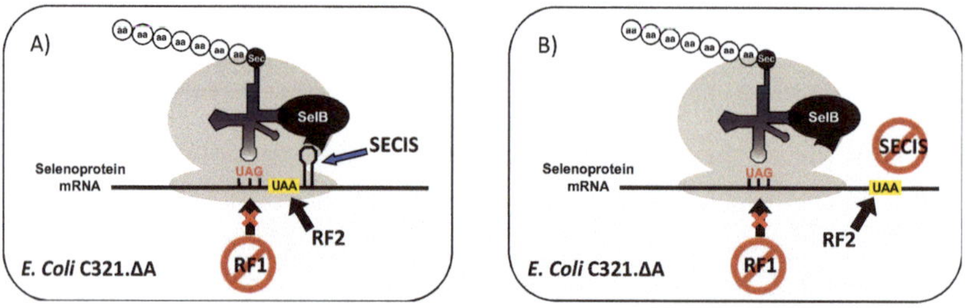

Fig. 1 *Principles of the selenoprotein production system described in this chapter.* (A) the expression of selenoproteins containing a Sec residue located close to the C-terminus, such as various thioredoxin reductase family members, is depicted. Here a bacterial SECIS element is located 11 nucleotides downstream of the Sec-encoding UAG codon, which assures maximum Sec incorporation efficiency within the RF1-depleted *E. coli* C321.ΔA strain when using a tRNA^{Sec} species with a mutated anticodon compatible with UAG (see Fig. 2). The genuine termination codon is chosen to be UAA for RF2-mediated termination and is here located between the Sec-encoding UAG codon and the SECIS element. This system results in close to 100% Sec incorporation. In (B) the production system without inclusion of a SECIS element is shown, which enables production of selenoproteins having internal Sec residues, such as glutathione peroxidases. This production results in about 20% Sec contents, which is lower than 100% mainly due to concomitant Gln- and Lys-mediated suppression of UAG.

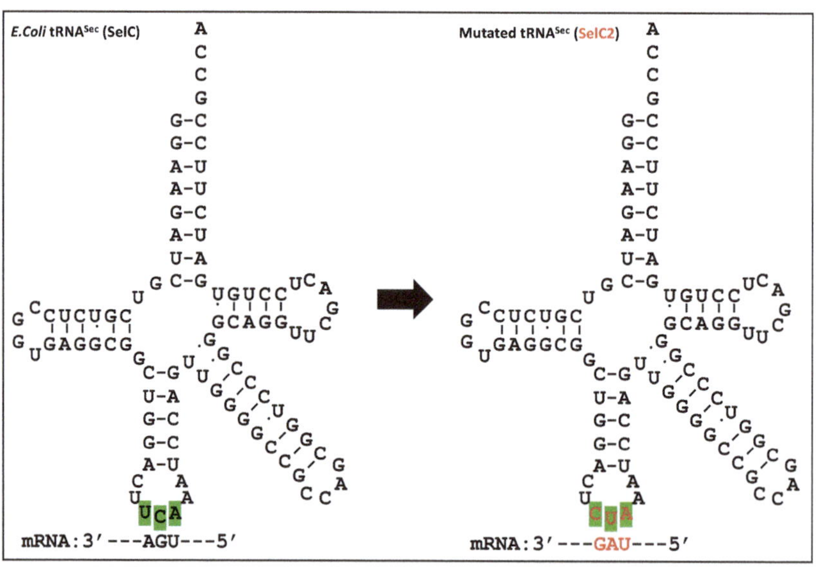

Fig. 2 *Mutation of the E. coli tRNA^{Sec} anticodon to enable Sec insertion at a dedicated UAG codon.* The method for selenoprotein production described herein utilizes SelB-mediated insertion of Sec at a dedicated UAG codon (instead of the natural UGA codon), which necessitates a corresponding mutation of the anticodon of tRNA^{Sec} (SelC, the *selC* gene product), as illustrated in this figure.

4. Materials

4.1 Equipment

A laboratory and standard equipment for routine work in biochemistry, molecular biology and microbiology will be required, including access to the following:

- PCR machine
- Variable temperature heat block or water bath
- Centrifuge, bench top and large volume (4 °C and room temperature)
- Variable temperature incubator and shaker for bacterial cultures
- Sonicator
- Nanodrop or spectrophotometer (cuvette or 96 well plate)
- Gel electrophoresis system for DNA
- Polyacrylamide electrophoresis system for protein analyses
- Dialysis tubing, Spectrum™ Spectra/Por™ 3500 and 12,000 Da MWCO
- Nickel columns with gravity flow, or FPLC system such as ÄKTAexplorer 100
- Syringes and 0.2 μm sterile filters for sterilizing solutions
- Equipment as required for analyses of selenoprotein enzyme activities (here exemplified with spectrophotometer for assessment of thioredoxin reductase or glutathione peroxidase activities)

4.2 Chemicals

The following specific chemicals, reagents and kits will typically be needed:

Common chemicals	Source	Catalog number
Tris	Sigma	GE17-1321
Hydrochloric acid (HCl)	Sigma	320331
Sodium chloride	Sigma	S9888
Methanol	Sigma	1.06009
Ethanol	VWR	20816.298
Acetic acid	Sigma	1.00063
Potassium chloride	Sigma	P3911

Continued

—cont'd

Common chemicals	Source	Catalog number
Potassium phosphate monobasic	Sigma	P5655
Potassium phosphate dibasic	Sigma	P3786
EDTA	Sigma	E9884
Sodium selenite	Sigma	S5261
Imidazole	Sigma	I2399
Glycerol	Sigma	G5516
Protein Assay Dye Reagent	Bio-Rad	#5000006
BCA Protein Assay Kit	Thermofisher	23227
Bovine Serum Albumin Standard	Thermofisher	23209
Bolt 4–12% Bis-Tris Plus gels	Thermofisher	NW04120BOX
Bolt MES running buffer (20 ×)	Thermofisher	B0002
Bolt LDS sample buffer (4 ×)	Thermofisher	B0007
Bolt sample reducing agent (10 ×)	Thermofisher	B0009

4.3 Molecular biology reagents

For the bacterial culture work and initial analyses of the expressed selenoproteins as describe in this chapter, the following will typically be needed:

Reagents	Source	Catalog number
Tryptone	Sigma	T7293
Yeast extract	Sigma	Y1625
Restriction enzymes	Thermofisher	Variable fast digest form
T4 DNA ligase	Thermofisher	EL0016
GeneJET Plasmid Miniprep Kit	Thermofisher	K0503
GeneJET Gel Extraction Kit	Thermofisher	K0692
E. coli transformation kit	Zymoresearch	T3001
Ampicillin	Fisher	10396833
Streptomycin	Sigma	S6501

—cont'd

Reagents	Source	Catalog number
IPTG	VWR	0487
DTT	VWR	443853B
L-Glutathione reduced	VWR	A2084
Hydrogen peroxide	Sigma	H1009
Cumene hydroperoxide	Sigma	247502
DTNB	Sigma	D8130
Insulin solution (about 1.6 mM)	Sigma	I9278
NADPH	Saveen Werner	A1395
Human glutathione reductase	Produced in-house, or available through diverse commercial vendors	
Human thioredoxin 1	Produced in-house, or available through diverse commercial vendors	
SUMO protease ULP1	Produced in-house, or available through diverse commercial vendors	

4.4 Plasmids

Plasmids (antibiotic resistance)	Source	Use for
pABC2a (Streptomycin)	Developed in-house (Cheng & Arner, 2017)	Selenoprotein expression
pD441a (Kanamycin)	Developed in-house (Cheng & Arner, 2017)	Non-selenoprotein expression

4.5 Sequencing primers

Name	Targets	Sequence (5′ to 3′)
441-fwd	pD441a, pABC2a	CGAAAATAATAAAGGGAAAATC
441-rev	pD441a	TGGTAGTGTGGGGACTC
ABC2-rev	pABC2a	ACCTCTGATAGTTGAGTCG

4.6 Strains

E. coli strains	Source	Catalog number
C321.ΔA	Addgene	#48998
Turbo	NEB	C2984
BL21 Star™ (DE3)	Thermofisher	C601003

4.7 Medium, solutions and buffers

1. Terrific broth (TB): Dissolve 24g yeast extract, 12g tryptone, and 5g glycerol in 900mL deionized water. Sterilize by autoclaving for 20min at 15psi (1.05kg/cm^2). Add 100mL of sterile phosphate buffer (0.17M KH_2PO_4, 0.72M K_2HPO_4) before use
2. Make various 1000× stock solutions for additions to the bacterial cultures, as outlined below. Sterilize the solutions by passing the solutions through sterile 0.22μm filters. Dilute 1000-fold directly in the bacterial cultures upon use

Reagent	Stock solution	Final concentration
Carbenicillin	50mg/mL	50μg/mL
Streptomycin	50mg/mL	50μg/mL
IPTG	0.5M	0.5mM
Sodium selenite	5mM	5μM

3. IMAC binding buffer: 50mM Tris-HCl, 100mM NaCl, 10mM imidazole, pH 7.5
4. IMAC eluting buffer: 50mM Tris-HCl, 100mM NaCl, 500mM imidazole, pH 7.5
5. Dialyzing buffer for SUMO protease ULP1: 50mM Tris-HCl, pH 8.0
6. Storage buffer for glutathione peroxidases: 50mM Tris-HCl, pH 7.5, with 100mM KCl, 5mM 2-mercaptoethanol, and 20% glycerol
7. Storage buffer for other proteins including thioredoxin reductases: 50mM Tris-HCl, pH 7.5, with 2mM EDTA, and 20% glycerol
8. Staining solution for SDS-PAGE analyses: 0.1% Coomassie blue R250 in 30% methanol and 10% acetic acid

9. Distaining solution for SDS-PAGE analyses: 30% methanol and 10% acetic acid
10. DTNB stock solution for thioredoxin reductase activity measurements: 62.5 mM 5,5′-Dithiobis(2-nitrobenzoic acid) dissolved in ethanol
11. NADPH stock solution for thioredoxin reductase or glutathione peroxidase activity measurements: 50 mM NADPH dissolved in deionized water
12. Insulin from Sigma for thioredoxin reductase activity measurements, stock solution: 10 mg/mL or 1.6 mM
13. Reduced glutathione for glutathione peroxidase activity measurements, stock solution: 10 mM dissolved in water
14. Cumene hydroperoxide for glutathione peroxidase activity measurements, stock solution: 10 mM dissolved in ethanol
15. Hydrogen peroxide for glutathione peroxidase activity measurements, stock solution: 10 mM dissolve in deionized water

5. Protocol

5.1 Plasmid construction for selenoprotein expression in an RF1-depleted *E. coli* host strain

1. For the cDNA of a certain target selenoprotein, order a synthetic DNA fragment with codon optimization for protein expression in *E. coli*. Include a His-tag and a SUMO tag at the N-terminal end of the target protein for facile protein purification. Avoid restriction sites in the synthetic DNA ORF that will overlap with those for subsequent cloning, such as *Nde*I, *Not*I and *Eco*RI.
2. The codon for selenocysteine (Sec) should be assigned to TAG, using TAA as the stop codon, and a bacterial SECIS element should be included after the TAA in case the Sec residue is close to the C-terminal end (see Fig. 1 and step 4, below).
3. The plasmid used in our work has *Nde*I (5′) and *Not*I (3′) flanking the entire ORF including the His-tag and SUMO-tag, with an *Eco*RI site inside the SUMO tag so that a shorter synthetic DNA can be utilized upon need. Detailed sequence information for the plasmid constructs can be found in Cheng et al. (2021).
4. For selenoproteins with Sec close to the C-terminal end, such as mammalian thioredoxin reductases with a C-terminal -Gly-Cys-Sec-Gly-COOH motif, include the DNA sequence for the coding of a

SECIS element (GGTTGCAGGTCTGCACC) positioned 11 nucleotides downstream of the Sec-encoding TAG (see Fig. 1A and further details in reference Cheng & Arner, 2018). However, this step is not necessary for expression of selenoproteins with internal Sec, such as glutathione peroxidases (see Fig. 1B). For further details in construct design, see the earlier publications on this method (Cheng & Arner, 2017; Cheng et al., 2021).

5. Digest the DNA fragment containing the ORF for the His- and SUMO-tagged target selenoprotein sequence using *Nde*I and *Not*I according to manufacturer's instructions.
6. Recover the fragment using a gel extraction kit according to manufacturer's instructions.
7. Ligate the fragment into the pABC2a plasmid encoding the mutated SelC for complementarity with UAG as well as SelA and SelB (Cheng & Arner, 2017), pre-cleaved at the corresponding restriction sites, using T4 DNA ligase according to manufactural instruction.
8. Transform the ligation mixture into competent *E. coli* strain Turbo and spread on an agar plate containing 50 μg/mL streptomycin for selection of transformants
9. Keep the selection plate in the incubator at 37 °C overnight.
10. Select a few colonies and perform minipreps on those clones and then sequence the plasmids using 441-fwd and ABC2-rev primers
11. Transform a plasmid with a confirmed correct sequence into competent cells of the *E. coli* stain C321.ΔA, then spreading on a selection agar plate containing 50 μg/mL carbenicillin and 50 μg/mL streptomycin subsequently kept in the incubator at 30 °C overnight (original paper describing this strain suggested growth at 34 °C (Lajoie et al., 2013)).
12. Choose one or two clones from the plate and prepare glycerol stocks for long term storage in −80 °C freezer

5.2 Culture and induction conditions

1. To 40 mL TB broth supplemented with 50 μg/mL carbenicillin and 50 μg/mL streptomycin in a 250 mL Erlenmeyer flask, inoculate the selected clone of C321.ΔA harboring the selenoprotein constructed in the pABC2a plasmid as described in Section 5.1, either from a glycerol stock or a single colony kept on a plate.
2. Shake the flask at 30 °C with 250 rpm overnight.

3. To 2L TB broth supplemented with 50 μg/mL carbenicillin and 50 μg/mL streptomycin in a 5 L Erlenmeyer flask, add 40 mL overnight culture from the previous step. This step can be scaled up using multiple bottles according to individual lab conditions and the targeted amount of final selenoprotein product.
4. Shake the flasks at 30 °C with 250 rpm for 5–6 h to let the culture reach an OD at 600 nm of approximately 1–1.5.
5. Lower the incubator temperature to 25 °C.
6. Add 5 μM selenite for selenium saturation of the selenoprotein production and 0.5 mM IPTG for induction of selenoprotein production, and continue shaking overnight.

5.3 Selenoprotein purification

1. Harvest the bacteria by centrifugation at 3000 × *g* for 20 min
2. Suspend the pelleted bacteria in IMAC binding buffer (3 mL/g wet weight)
3. Lyse the cells in the bacterial suspension by sonication according to recommended instrument settings
4. Centrifuge the sonicated lysate at 40,000 to 50,000 × *g* for 30 min, then carefully remove the supernatant from the pellet
5. Load the supernatant onto a HisPrep FF 16/10 column equipped on an ÄKTAexplorer FPLC system (Cytiva Life Sciences)
6. After loading, continue washing the column with IMAC binding buffer until the UV absorbance becomes stabilized
7. By utilizing automatic pump mixing, elute the target protein from the column with a mixture of 40% IMAC eluting buffer (pump B, about 200 mM imidazole final concentration) and 60% IMAC washing buffer (pump A). The eluent is normally 30–100 mL
8. Add 5 mM DTT and 1% His-tagged SUMO protease ULP1 to the eluent and carefully transfer the mixture into a pretreated dialysis tube
9. Seal and dialyze the tube in 5 L dialyzing buffer overnight at 4 °C or room temperature, allowing for ULP1 digestion
10. Re-apply the dialyzed mixture onto the HisPrep FF 16/10 column to separate the cleaved non-tagged target selenoprotein from its N-terminal His-tagged fusion partner as well as from the His-tagged ULP1. The target protein will be in the flowthrough fractions while other impurities will bind to the nickel column, which can subsequently be eluted again with 40% IMAC eluting buffer for further analyses, if wished for

11. The target selenoproteins should then be concentrated and buffer-exchanged to storage buffer, whereupon it can be stored in −20 °C until further analyses

6. Selenoprotein analysis and quality assessment

It is important to assess the purity and homogeneity of the selenoproteins upon expression and purification as outlined above, both considering the absence of unwanted co-purified proteins as impurities and with regards to Sec contents. Since a purification of a recombinant selenoprotein may yield a mixture of the same apparently homogenous protein but in fact containing other amino acids than Sec, it will be important not only to assess the absence of other proteins in the purified preparation, but also determine the contents of Sec in the target protein. The latter can be done by either analyses of enzyme activities, typically dependent upon the presence of Sec, or by directed selenium determinations. Here we will give examples of the quality assessments that can or should be performed with the purified selenoprotein.

6.1 Protein concentration measurements

In order to determine either Sec contents or Sec-dependent enzyme activities, first the concentration of total protein in the purified selenoprotein preparation has to be determined. This can be done by different methods, of which we here give examples of the two most often used in our own applications:

1. Being flavoproteins, the concentration of the thioredoxin reductase family of selenoproteins can be calculated by measuring the absorbance at 463 nm in a spectrophotometer and use the extinction coefficient of FAD, which is 11,300 M^{-1} cm^{-1}, to calculate the flavoprotein concentration. Thioredoxin reductase is typically a homodimer with one FAD molecule in each subunit.
2. The concentrations of other purified selenoproteins can be measured by either Bradford (Bio-Rad) or BCA (Thermofisher) protein assay kits using bovine serum albumin as standard.

The yields of finally purified recombinant selenoproteins using our methodology are for selenoproteins with C-terminal Sec residues, such as TrxRs, typically about 20–30 mg per liter original culture, while for selenoproteins with internal Sec residues, such as GPXs, typically about 5–10 mg per liter culture.

6.2 Apparent protein purity assessed by gel electrophoresis

The first and easiest test to assess the possible presence of other proteins impurities in the purified selenoprotein preparation is to analyze the product using gel electrophoresis, for which we here give a simple and straightforward approach. If evident protein bands are seen in the gel having other molecular weights than the target selenoprotein, then the expression and purification protocol, or the integrity of the plasmid construct or *E. coli* host strain, must be reevaluated.

1. Prepare up to 10 μg protein in 50 μL SDS sample buffer with reductant such as 10 mM DTT and incubate the sample at 95 °C for about 5 min.
2. Load 10–40 μL sample per well on 4–12% Bis-Tris Plus Gels and run the gel at 200 V for 24 min, or according to manufacturer's recommendations.
3. Remove the gel from the cassette and stain with commercial or homemade staining solution, shaking the gel gently in the staining solution for about 30 min.
4. Transfer the gel to destaining solution with continuous gentle shaking, wait until the background is bleached with the protein bands revealed, which normally takes from 1 h to overnight.
5. Assess the apparent protein purity on the stained gel. The purification method described above, involving two steps over a nickel column with His-tag removal before the second step, normally yields a final product with at least 95% apparent purity on the gel.

6.3 Enzyme activity measurements of sec-dependent activities of certain selenoproteins

With selenoproteins typically being Sec-dependent enzymes (as discussed above), the next quality assessment of the purified selenoprotein can be to determine its enzyme activity and to compare the specific activity (mol product/mol enzyme/min) with that of the native form of the selenoprotein (if known). Our experiences are that the percentage of turnover numbers (k_{cat}) compared to native enzymes are typically directly related to the percentage of Sec contents while K_m values are very similar to those of the native enzymes, when assessing activities with model substrates. Here we will give examples of how such activity assessments can be done for thioredoxin reductases or glutathione peroxidases, but the general principle should be applicable to any selenoprotein having a known enzymatic activity and for which an assay of activity can be established.

1. Thioredoxin reductase activity assays: the enzymatic activities of this class of selenoproteins can be measured by either a direct DTNB reduction assay or by a thioredoxin-coupled insulin reduction assay, as elsewhere described in further detail (Arnér, 2018) and shortly summarized as follows.
   a. Direct DTNB reduction assay (Fig. 3): in a microplate or cuvette, combine the components listed below in TE buffer and thereafter follow the absorbance increase over time at 412 nm. DTNB can be efficiently reduced by Sec-containing thioredoxin reductases, yielding two yellow TNB^- anions that have their maximum absorbance at 412 nm, which is the principle of the assay. Determination of mol TNB^- formed per mol (or mg) TrxR enzyme per minute will yield the specific activity of this selenoprotein. The actual number of the specific activity of this assay will differ between the species and isoforms of TrxR, and it also should be noted that some thioredoxin reductase variants lacking Sec can also have some activity in this assay

Component	Final concentration
Thioredoxin reductase	10–100 nM
DTNB	2.5 mM
NADPH	0.25 mM

   b. Thioredoxin-coupled insulin reduction assay (Fig. 4): in a microplate or cuvette, combine the below components in TE buffer and thereafter follow the absorbance decrease at 340 nm. In this assay, thioredoxin is the specific substrate that can be reduced by thioredoxin reductase. Reduced thioredoxin subsequently reduces insulin, and thereby becomes oxidized. When thioredoxin is again reduced by thioredoxin reductase

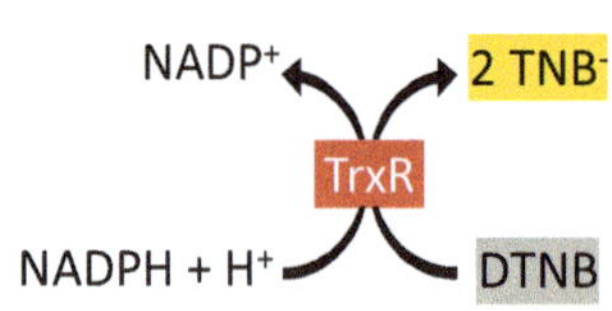

Fig. 3 *Direct DTNB reduction assay for assessment of thioredoxin reductase activity.* Selenoprotein thioredoxin reductase (TrxR) enzymes can, using NADPH, directly reduce DTNB into two molecules of TNB^-, which can be detected spectrophotometrically at 412 nm.

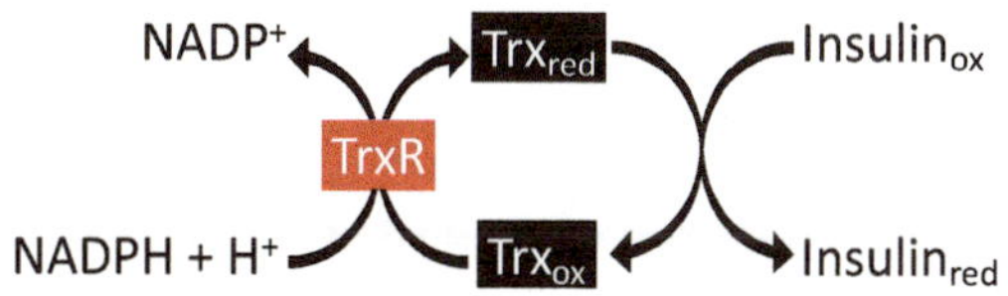

Fig. 4 *Thioredoxin-coupled insulin reduction assay for assessment of thioredoxin reductase activity.* Selenoprotein thioredoxin reductase (TrxR) enzymes will, using NADPH, maintain reduction of insulin coupled to thioredoxin. The consumption of NADPH can be detected spectrophotometrically through a time dependent decrease of absorbance at 340nm.

NADPH is consumed. This assay therefore measures NADPH consumption by following its absorbance decrease at 340nm. Unlike the direct DTNB assay, this assay is typically strictly dependent on the presence of the Sec residue in thioredoxin reductase, but again the exact specific activity will depend upon the species and isoforms of both the thioredoxin reductase and the thioredoxin used in the assay

Component	Final concentration
Thioredoxin reductase	10–100nM
Thioredoxin	1–5μM
Insulin	0.16mM
NADPH	0.25mM

2. Glutathione peroxidase activity assay: the activities of glutathione peroxidases can be measured in glutathione- and glutathione reductase (GR)-coupled assays of peroxidase activities using different peroxide substrates (Toppo et al., 2009). For this, combine in a microplate or cuvette the components listed below in TE buffer, and follow the absorbance decrease at 340nm that is due to NADPH consumption. In this assay, glutathione peroxidases will oxidize reduced glutathione, GSH, using a peroxide substrate (here hydrogen peroxide or cumene peroxide). The oxidized glutathione, GSSG, is subsequently reduced by glutathione reductase at the consumption of NADPH. This assay therefore measures NADPH consumption by following its absorbance decrease at 340nm (Fig. 5). The enzyme activities of selenoprotein glutathione peroxidases are likely to be strictly dependent on the presence of their Sec residue and the determined activity will thus be a measure of the

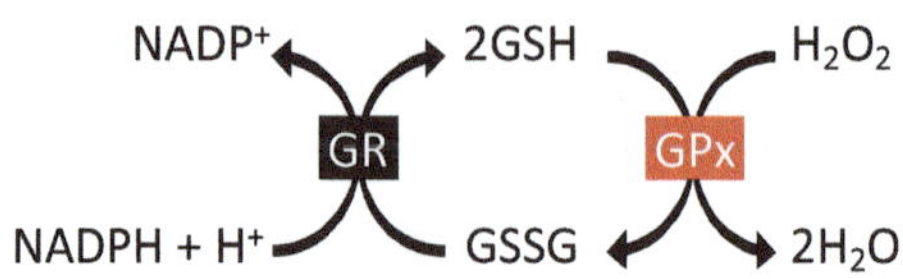

Fig. 5 *Glutathione peroxidase assay.* Selenoprotein glutathione peroxidase (GPx) enzymes will oxidize 2 molecules of reduced glutathione (GSH) to one glutathione disulfide (GSSG) during reduction of a peroxide, as here illustrated with hydrogen peroxide (H_2O_2). The GSSG will subsequently be reduced back to GSH by glutathione reductase (GR) using NADPH. The consumption of NADPH can be detected spectrophotometrically through a time dependent decrease of absorbance at 340 nm.

Sec contents (Toppo et al., 2009). However, it should be noted that some background NADPH consumption can also be observed without adding glutathione peroxidases, because the peroxide substrates can spontaneously oxidize GSH at a low speed thus triggering GR activity. Therefore, a negative control lacking the selenoprotein should also be included and the absorbance decrease by time in that control should be subtracted from that in the assay containing all components. Moreover, since glutathione peroxidases do not follow Michaelis-Menten kinetics it may be difficult to determine their exact specific activities and comparisons to the activities of native enzyme preparations are preferable if the activities shall be used as an indication of Sec contents (Cheng et al., 2021).

Component	Final concentration
Glutathione peroxidase	0.5–1 μM
Glutathione reductase	10–20 nM
H_2O_2 or cumene peroxide	0.16 mM
L-Glutathione, reduced	1 mM
NADPH	0.25 mM

6.4 Sec contents validation by direct methods

As outlined above, the purified recombinant selenoprotein should first be quality assessed in terms of apparent purity (lack of other proteins in the final product) as well as controlled for the presence of Sec. With most selenoproteins being Sec-dependent enzymes, the presence of Sec can be assessed through enzymatic activity assays performed with the purified selenoprotein, as exemplified above for thioredoxin reductases or glutathione peroxidases. However, since the translation of a recombinant selenoprotein can also result in suppression of the Sec-encoding codon with other amino acid residues than Sec (as discussed above), it will also be important to determine the exact selenium contents in the purified selenoprotein. Here we will give example of two methods by which this can be accomplished.

6.4.1 ICP-MS

Inductively coupled plasma mass spectrometry (ICP-MS) is a method suitable for the determination of the exact amounts of specific elements, such as selenium, in a given sample. Provided that the concentration of a purified selenoprotein has been well determined, the use of ICP-MS can thereby be used to determine exactly how much selenium that was present in that protein purification, which can be assumed to have been present in the form of Sec provided that reliable negative controls (recombinant proteins made not to incorporate Sec) are also included in the analysis. We also found that simultaneous determination of sulfur (S) and correlation with the expected number of S-containing amino acids in the selenoprotein (Cys and Met) can be useful as another internal control for the measurement to further solidify the determination (Cheng et al., 2021). Even if ICP-MS when correctly performed can be used to determine the amount of selenium in the purified selenoprotein, it cannot reveal what is present in place of Sec in case the selenoprotein would have less than 100% Sec contents. For that mass spectrometric analyses will instead be helpful tools.

6.4.2 LC-MS/MS

Mass spectrometry analyses and particularly liquid chromatography-tandem mass spectrometry (LC-MS/MS) will be particularly useful for the analysis of the identity of Sec, as well as other features, in the selenoprotein and the peptide(s) intended to include Sec. With the fragmentation and analytic power of LC-MS/MS it will not only be possible to conclude that Sec is present in the selenoprotein, but it will also be possible to determine the identity of alternative peptides lacking Sec. Using this methodology, it could be confirmed that the thioredoxin reductase or glutathione peroxidase 4 preparations with similar activities as the native enzymes and full selenium contents according to ICP-MS indeed contained nearly 100% Sec at the right position of the enzymes (Cheng et al., 2021). It was furthermore found, using LC-MS/MS, that the preparations of glutathione peroxidases that had only about 20% Sec also contained Gln, Lys, a small amount of Trp, and a surprising Sec-to-null variant of the enzymes (Cheng et al., 2021). Such quality assessments and insights into the nature of the recombinant selenoproteins as produced and purified would be difficult to gain with other methodologies.

7. Concluding remarks

In this chapter we have briefly discussed alternative methods to obtain pure selenoproteins, and we have given more details with regards to one methodology for the yet most effective recombinant selenoprotein production method in *E. coli* with regards to total yield and purity. We have however also discussed the factors that may lead to less than 100% Sec contents in the final product, and we have suggested what controls that need to be performed for quality assessment of the selenoprotein that has been purified. We hope that this and other methods in the field will become additionally refined and developed, but already in its present form it should likely help to further advance selenoprotein research, as the methodology outlined herein can be used as a source of selenoproteins to enable more in-depth studies of their characteristics.

Acknowledgments

The authors gratefully acknowledge funding for work in their laboratories from Karolinska Institutet, The Knut and Alice Wallenberg Foundations, The Swedish Cancer Society, The Swedish Research Council, The Cayman Biomedical Research Institute (CABRI), The Hungarian Thematic Excellence Programme (TKP2020-NKA-26), The Hungarian National Research, Development and Innovation Office (ED_18-1-2019-0025), The Hungarian National Laboratories Excellence program (under the National Tumorbiology Laboratory project) and The Hungarian Ministry of Human Capacities (ÁEEK/41872-16/2020).

Conflict of interests

The authors have shares in Selenozyme AB that is a company selling recombinant selenoproteins such as those described in this study.

References

Aldag, C., et al. (2013). Rewiring translation for elongation factor Tu-dependent selenocysteine incorporation. *Angewandte Chemie (International Ed. in English)*, *52*(5), 1441–1445.

Allmang, C., Wurth, L., & Krol, A. (2009). The selenium to selenoprotein pathway in eukaryotes: More molecular partners than anticipated. *Biochimica et Biophysica Acta*, *1790*(11), 1415–1423.

Arner, E. S. (2002). Recombinant expression of mammalian selenocysteine-containing thioredoxin reductase and other selenoproteins in Escherichia coli. *Methods in Enzymology*, *347*, 226–235.

Arnér, E. S. J. (2010). Selenoproteins—What unique properties can arise with selenocysteine in place of cysteine? *Experimental Cell Research*, *316*(8), 1296–1303.

Arnér, E. S. J. (2018). Selective evaluation of thioredoxin reductase enzymatic activities. *Methods in Molecular Biology*, *1661*, 301–309.

Arnér, E. S. J., et al. (1999). High-level expression in Escherichia coli of selenocysteine-containing rat thioredoxin reductase utilizing gene fusions with engineered bacterial-type SECIS elements and co-expression with the selA, selB and selC genes. *Journal of Molecular Biology*, *292*, 1003–1016.

Boyington, J. C., et al. (1997). Crystal structure of formate dehydrogenase H: Catalysis involving Mo, molybdopterin, selenocysteine, and an Fe4S4 cluster. *Science*, *275*(5304), 1305–1308.

Brocker, M. J., et al. (2014). Recoding the genetic code with selenocysteine. *Angewandte Chemie (International Ed. in English)*, *53*(1), 319–323.

Cheng, Q., & Arner, E. S. (2017). Selenocysteine insertion at a predefined UAG codon in a release factor 1 (RF1) depleted Escherichia coli host strain bypasses species barriers in recombinant selenoprotein translation. *The Journal of Biological Chemistry*, *292*(13), 5476–5487.

Cheng, Q., & Arner, E. S. J. (2018). Overexpression of recombinant Selenoproteins in E. coli. *Methods in Molecular Biology*, *1661*, 231–240.

Cheng, Q., et al. (2012). Site-specifically 11C-labeled Sel-tagged annexin A5 and a size-matched control for dynamic in vivo PET imaging of protein distribution in tissues prior to and after induced cell death. *Biochimica et Biophysica Acta*, *1830*(3), 2562–2573.

Cheng, Q., et al. (2021). Production and purification of homogenous recombinant human selenoproteins reveals a unique codon skipping event in E. coli and GPX4-specific affinity to bromosulfophthalein. *Redox Biology*, *46*, 102070.

Driscoll, D. M., & Copeland, P. R. (2003). Mechanism and regulation of selenoprotein synthesis. *Annual Review of Nutrition*, *23*, 17–40.

Gursinsky, T., et al. (2008). Factors and selenocysteine insertion sequence requirements for the synthesis of selenoproteins from a gram-positive anaerobe in Escherichia coli. *Applied and Environmental Microbiology*, *74*(5), 1385–1393.

Haruna, K., et al. (2014). Engineering the elongation factor Tu for efficient selenoprotein synthesis. *Nucleic Acids Research*, *42*(15), 9976–9983.

Hondal, R. J. (2009). Using chemical approaches to study selenoproteins-focus on thioredoxin reductases. *Biochimica et Biophysica Acta*, *1790*(11), 1501–1512.

Hondal, R. J., Marino, S. M., & Gladyshev, V. N. (2013). Selenocysteine in thiol/disulfide-like exchange reactions. *Antioxidants & Redox Signaling*, *18*(13), 1675–1689.

Jiang, L., et al. (2020). The algal selenoproteomes. *BMC Genomics*, *21*(1), 699.

Kryukov, G. V., et al. (2003). Characterization of mammalian selenoproteomes. *Science*, *300*(5624), 1439–1443.

Lajoie, M. J., et al. (2013). Genomically recoded organisms expand biological functions. *Science*, *342*(6156), 357–360.

Lescure, A., et al. (2002). Protein factors mediating selenoprotein synthesis. *Current Protein & Peptide Science*, *3*(1), 143–151.

Liu, J., Cheng, R., & Rozovsky, S. (2018). Synthesis and semisynthesis of selenopeptides and selenoproteins. *Current Opinion in Chemical Biology*, *46*, 41–47.

Mariotti, M., et al. (2012). Composition and evolution of the vertebrate and mammalian selenoproteomes. *PLoS One*, 7(3), e33066.

Metanis, N., & Hilvert, D. (2014). Natural and synthetic selenoproteins. *Current Opinion in Chemical Biology*, *22*, 27–34.

Miller, C., et al. (2015). A synthetic tRNA for EF-Tu mediated selenocysteine incorporation in vivo and in vitro. *FEBS Letters*, *589*(17), 2194–2199.

Moosmayer, D., et al. (2021). Crystal structures of the selenoprotein glutathione peroxidase 4 in its apo form and in complex with the covalently bound inhibitor ML162. *Acta Crystallographica Section D: Structural Biology*, 77(Pt. 2), 237–248.

Mukai, T., et al. (2018). A facile method for producing selenocysteine-containing proteins. *Angewandte Chemie (International Ed. in English)*, *57*(24), 7215–7219.

Novoselov, S. V., et al. (2007). A highly efficient form of the selenocysteine insertion sequence element in protozoan parasites and its use in mammalian cells. *Proceedings of the National Academy of Sciences of the United States of America*, *104*(19), 7857–7862.

Peeler, J. C., & Weerapana, E. (2019). Chemical biology approaches to interrogate the selenoproteome. *Accounts of Chemical Research*, *52*(10), 2832–2840.

Peeler, J. C., et al. (2020). Generation of recombinant mammalian selenoproteins through genetic code expansion with photocaged selenocysteine. *ACS Chemical Biology*, *15*(6), 1535–1540.

Peng, T., et al. (2016). Comparative genomics reveals new evolutionary and ecological patterns of selenium utilization in bacteria. *The ISME Journal*, *10*(8), 2048–2059.

Reich, H. J., & Hondal, R. J. (2016). Why nature chose selenium. *ACS Chemical Biology*, *11*(4), 821–841.

Rother, M., & Quitzke, V. (2018). Selenoprotein synthesis and regulation in Archaea. *Biochimica et Biophysica Acta - General Subjects*, *1862*(11), 2451–2462.

Toppo, S., et al. (2009). Catalytic mechanisms and specificities of glutathione peroxidases: Variations of a basic scheme. *Biochimica et Biophysica Acta*, *1790*(11), 1486–1500.

Tujebajeva, R. M., Harney, J. W., & Berry, M. J. (2000). Selenoprotein P expression, purification, and immunochemical characterization. *The Journal of Biological Chemistry*, *275*(9), 6288–6294.

Welegedara, A. P., et al. (2018). Site-specific incorporation of selenocysteine by genetic encoding as a photocaged unnatural amino acid. *Bioconjugate Chemistry*, *29*(7), 2257–2264.

Yoshizawa, S., & Böck, A. (2009). The many levels of control on bacterial selenoprotein synthesis. *Biochimica et Biophysica Acta*, *1790*(11), 1404–1414.

CHAPTER FIVE

Selenocysteine substitutions in thiyl radical enzymes

Juan Carlos Cáceres[a], Clara A. Bailey[a], Kenichi Yokoyama[b], and Brandon L. Greene[a,*]
[a]Department of Chemistry and Biochemistry, University of California, Santa Barbara, CA, United States
[b]Department of Biochemistry, Duke University, Durham, NC, United States
*Corresponding author: e-mail address: greene@chem.ucsb.edu

Contents

Abstract

Cysteine thiyl radicals are implicated as cofactors in a variety of enzymatic transformations, as well as transient byproducts of oxidative stress, yet their reactivity has undermined their detailed study. Selenocysteine exhibits a lower corresponding selenyl radical reduction potential, thus taming this radical reactivity without significant steric perturbation, potentially affording a glimpse into otherwise fleeting events in thiyl radical catalysis. In this chapter, we describe a suite of fusion protein constructs for general and efficient production of site-specifically incorporated selenoproteins by a recently developed nonsense suppression technology. As a proof of concept, we produced *NikJ*, a member of the radical S-adenosyl methionine enzyme family involved in the biosynthesis of peptidyl nucleoside antibiotics. We place emphasis throughout the plasmid

Methods in Enzymology, Volume 662
ISSN 0076-6879
https://doi.org/10.1016/bs.mie.2021.10.014

assembly, protein expression, and selenium quantitation on accommodating the structural and functional diversity of thiyl radical enzymes. The protocol produces *NikJ* with near quantitative selenocysteine insertion, 50% nonsense read-through, and facile protein purification.

1. Introduction

Amino acid radicals are essential cofactors in a number of enzymes and have been the subject of extensive study (Frey, 2001; Stubbe & van der Donk, 1998). Most elusive among the amino acid radicals are cysteine thiyl radicals, catalyzing remarkably diverse chemical transformations acting as H-atom acceptors, the product of H-atom donation, or as radical nucleophiles. When generated outside of their respective active site, free cysteine thiyl radicals rapidly and nonspecifically activate C—H bonds in peptides (Nauser, Casi, Koppenol, & Schöneich, 2008; Schöneich, 2011), reactivity that could potentially degrade the enzyme scaffold unless properly controlled. Therefore, enzymatic thiyl radical generation and catalytic outcomes must occur with high fidelity to maintain function over multiple turnovers. Targeting enzyme thiyl radicals continues to be an attractive therapeutic strategy (Greene et al., 2020), but rational design of mechanism-based inhibitors requires knowledge of the radical generation and active site chemistry, the details of which remain poorly understood for most enzymes. Thus far, the class II ribonucleotide reductase (RNR) from *Lactobacillus leichmannii* is the only example of an experimentally observed kinetically and chemically competent thiyl radical (Licht, Gerfen, & Stubbe, 1996).

Selenocysteine, a naturally occurring variant of cysteine where the sulfur atom is replaced with selenium, is a minimally invasive analog with opportunistic properties for experimental investigation of thiyl radical enzymes. The reduction potential of the selenocysteinyl radical is estimated to be 500 mV lower than that of the cysteinyl radical, corresponding to a 55 kJ/mol lower bond dissociation energy (Nauser, Dockheer, Kissner, & Koppenol, 2006), suggesting that selenocysteine may arrest radical chemistry long enough for direct experimental observation, analogous to other amino acid surrogates acting as radical traps (Minnihan, Seyedsayamdost, Uhlin, & Stubbe, 2011). The native translational machinery for site-selective selenocysteine incorporation into translating proteins (selenoproteins) is complex and difficult to repurpose for recombinant protein expression on the scale necessary for mechanistic studies (Serrão et al., 2018). Several approaches have been developed for recombinant selenoprotein expression that bypass the native machinery and are reviewed elsewhere

(Metanis & Hilvert, 2014; Wang, Liu, Chang, Xu, & Wang, 2021). We have previously applied the method of Söll (Mukai, Sevostyanova, Suzuki, Fu, & Söll, 2018) for the recombinant expression of the *Escherichia coli* (*E. coli*) class Ia RNR α_2 subunit (Greene, Stubbe, & Nocera, 2019). This method employs an engineered allo-tRNA^{sec} that efficiently interacts with both *E. coli* serine aminoacyl tRNA synthetase and the elongation factor Tu and is cognate for the amber stop codon. Co-expression of *SelA* and *SelD* catalyzes the conversion of serine acylated allo-tRNA^{sec} to selenocysteine for amber codon suppression, and the tRNA, *SelA* and *SelD* are all encoded in the pSecUAG-Evol2 plasmid. RNR cloned into pET-14b and co-expressed with pSecUAG-Evol2 in ME6 cells generated selenocysteine substituted *E. coli* RNR at the thiyl radical cysteine site (C_{439}) with a yield of >2 mg protein per gram cell paste and 80–100% selenocysteine loading, but the expression and purification strategy was not designed for general application to diverse thiyl radical enzymes.

In this chapter, we describe the development of several "plug-and-play" fusion protein plasmids for the recombinant expression of site-specifically labeled selenoproteins. The constructs developed harbor C-terminal poly-histidine tags for immobilized metal ion affinity chromatography (IMAC), as well as removable fusion proteins, namely, the small ubiquitin-like modifier protein (SUMO, *Saccharomyces cerevisiae* SMT3) or maltose binding protein (MBP). All constructs can generate near native protein following purification and protease digestion with proteases that can also be recombinantly expressed targeting SUMO (Ulp1) (Guerrero, Ciragan, & Iwaï, 2015) or tobacco etch virus (TEV) protease recognition sites (van den Berg, Löfdahl, Härd, & Berglund, 2006). Lastly, the antibiotic selection markers have been altered for improved plasmid retention during scale-up, a persistent problem with prior constructions (Greene et al., 2019). Detailed instructions for Gibson assembly of arbitrary proteins of interest into the developed fusion protein expression vectors are detailed for ease of implementation. The radical S-adenosyl methionine (SAM) enzyme *NikJ* (Lilla & Yokoyama, 2016) serves as a test case, exemplifying several common obstacles to recombinant protein expression and non-sense suppression, including low overall expression and solubility, post-translational modifications (a 4Fe-4S cluster), oxygen sensitivity, and similar properties of the full-length and truncated protein products. We hope that these vector constructs and the methods detailed herein will expand the potential of selenocysteine incorporation in thiyl radical enzymes as a mechanistic probe, and contribute more broadly to nonsense suppression-based methods for unnatural amino acid incorporation in recombinantly produced proteins.

2. Recombinant expression of selenoproteins by non-sense suppression

2.1 Notes on choosing the right vector

Thiyl radical enzymes vary widely in size, tertiary and quaternary structure, solubility, post-translational modifications, cofactor composition, and in the role of dynamics in catalysis. This diversity of structure and function necessitates a nuanced vector design for expression and isolation of selenocysteine substituted proteins for mechanistic study. The ideal expression vector imbues the protein of interest (POI) with features that assist in soluble expression and purification, while not interfering with downstream enzymatic function, either by ease of removal or innocuity. To increase the solubility of recombinantly expressed proteins, tight control of expression or fusion to a soluble carrier protein, or both, have been shown to be effective (Guerrero et al., 2015; Raran-Kurussi, Keefe, & Waugh, 2015). Vector designs that improve purification efficiency generally rely on fusion of the POI to an affinity tag domain, either a polyhistidine tag (His-tag) for IMAC or fusion to a small molecule binding protein whose ligand can be functionalized onto a stationary phase. These protein or peptide fusion constructs are appended to the N- or C-terminus of the POI, which can interfere with function or protein–protein interactions. Additionally, nonsense suppression methods such as those described herein, generate both full-length and truncated protein products, thus N-terminal affinity tags may not be effective in discriminating between truncated and full-length protein.

No one expression vector solution is viable for all POI, and thus consideration of the functional properties of the POI is essential. The *E. coli* class Ia RNR α_2 subunit, for example, has been studied by selenocysteine substitution at the thiyl radical forming C_{439} by the method of Söll (Greene et al., 2019; Mukai et al., 2018). Using a C-terminal affinity tag may be useful in the purification of the full-length selenocysteine incorporated enzyme from the truncated impurity, but the C-terminal tail of the RNR α subunit is dynamic and contains two cysteines essential for the re-reduction of active site during turnover (Åberg et al., 1989; Booker, Licht, Broderick, & Stubbe, 1994; Greene et al., 2020). Fortunately, the central location of the target cysteine C_{439} in the overall protein sequence (761 residues total) destabilized the prematurely truncated 438-residue protein, and this impurity precipitates with the cell debris upon lysis, making an N-terminal tag more appropriate (Greene et al., 2019). Nevertheless, it has been shown that an N-terminal His-tag on RNR α_2 decreases its activity for unknown reasons (Minnihan et al., 2011). Thus, there is a need for modular strategies for N- and C-terminal fusion proteins that can be readily cleaved to produce the native or near-native POI.

The expression vector of choice must also be maintained and replicated by the host in concert with the pSecUAG-EVOL2 plasmid containing the selenocysteine biosynthetic machinery and UAG anti-codon allo-$tRNA^{UTu}$ for successful expression of the selenocysteine loaded POI, and the two plasmids should be under orthogonal induction control as the selenocysteine-charged allo-$tRNA^{UTu}$ must be present prior to POI mRNA translation. The pSecUAG-EVOL2 plasmid is induced by L-arabinose (*araBAD* promotor) and harbors a kanamycin (Kan) resistance cassette *KanR*. Previously pET and pUC vectors have been chosen for their relatively high expression levels, orthogonal origin of replication, and ampicillin selection marker (Evans et al., 2021; Greene et al., 2019; Mukai et al., 2018). We have previously described plasmid instability in scale-up expression due to seed culture β-lactamase activity and culture acidification during growth, resulting in loss of the pET vector (Greene et al., 2019). This can be partially ameliorated by washing the saturated inoculum cultures to remove excess β-lactamase, introducing buffer into the culture medium, and replacement of ampicillin by carbenicillin, but we have found more stable plasmid continuity with vectors that confer chloramphenicol resistance.

In light of the multitude of challenges and considerations necessary to produce recombinant selenoproteins by non-sense suppression methods at the scale necessary for *in vitro* study, we have designed a suite of vectors for efficient selenoprotein expression and purification designated pCm 1–6, shown in Fig. 1A. A user guide for selecting an appropriate vector for a given application is also provided in Fig. 1B. The library contains six plasmids with C-terminal 8 × His-tags that are removable by proteases for applications where metal affinity tags may interfere with function or reconstitution. Plasmids pCm2, pCm3 and pCm5 contain the SMT3 SUMO fusion fragment, which acts as a low molecular weight soluble fusion protein and as a high-fidelity protease recognition site for Ubl-specific protease 1 (Ulp1), cleaving SUMO at its C-terminal GG*S motif. Plasmids pCm4–6 contain a maltose binding protein (MBP) domain, which may be more effective in maintaining POI solubility for troublesome proteins, and serves as an orthogonal affinity tag for purification on amylose resin (Duong-Ly & Gabelli, 2015; Raran-Kurussi et al., 2015). For all plasmids the entire C-terminal fused domain can be removed by TEV protease

A

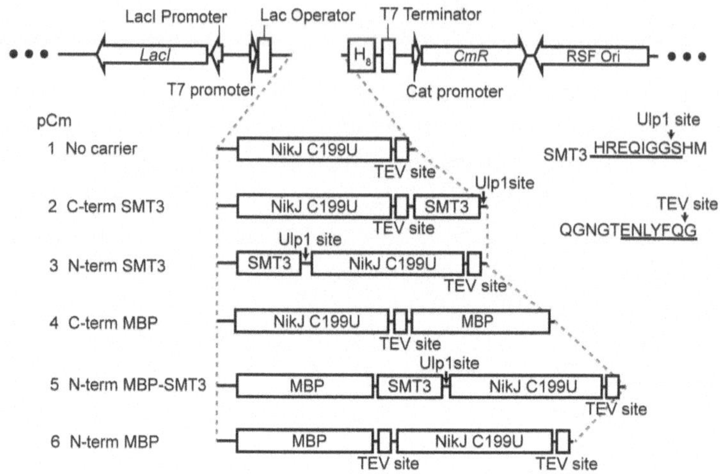

B

Criteria	Comments	Recommended plasmids N-term His-Tag	Recommended plasmids C-term His-Tag
Will C-term modification affect function?	Yes: Avoid C-term tags or use a C-term cleavable tag. No: Use C-terminal tag to separate full-length and truncated.	pET 14-16, pET19, pET32	pET-20-22, pET32, pCm1-6
Will a His-tag interfere with function?	Yes: Ensure protease cleavable or avoid all together. No: Use any C-term His-tag from row 1.	pET 14-16, pET19, pET32	pCm1-6
Is expression low or inconsistent?	Yes: Use a stronger selection marker. No: Use any plasmid from row 2.		pCm1-6
Is solubility an issue?	Yes: Use MBP or SMT3 fusion proteins. No: Use any plasmid from row 3.		pCm2-6
Is IMAC insufficient?	Yes: Use MBP protein fusion and amylose affinity chromatography. No: Use any plasmid from row 4.		pCm4-6

Fig. 1 Selenoprotein expression and purification strategy. (A) Schematic representation of pCm plasmids for selenoproteins expression and purification. Top: the common scaffold for all pCm plasmids with open variable region. Arrowed blocks represent coding sequences for: *LacI*, lactose operon repressor; H_8, octahistidine tag; *CmR*, chloramphenicol resistance gene. Bottom: the variable region of the pCm plasmids for fusion protein expression. SMT3, *Saccharomyces cerevisiae* SUMO homolog SMT3; MBP, maltose binding protein; TEV, tobacco etch Virus protease site; Ulp1, Ubl-specific protease 1; POI, protein of interest. A detailed view of the Ulp1 and TEV protease recognition sites is also provided. (B) User guide for selecting an appropriate plasmid for the expression and purification of a given selenocysteine mutant POI.

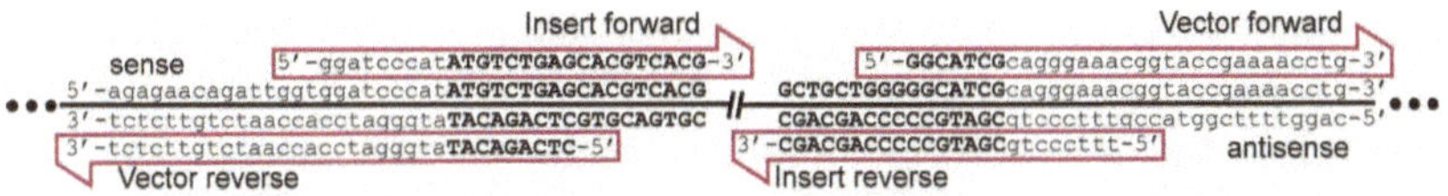

Plasmid	Primer Sense	Vector	Insert
pCm1	forward	5′-**GGCATCG**cagggaaacggtaccgaaaacctg-3′	5′-gagatatacc**ATGTCTGAGCACGTCACGC**-3′
	reverse	5′-**GCTCAGACAT**ggtatatctccttattaaagt taaacaaaattatttctacaggg-3′	5′-tttccctg**CGATGCCCCCAGCAGC**-3′
pCm2	forward	5′-**GGCATCG**cagggaaacggtaccgaaaacctg-3′	5′-gagatatacc**ATGTCTGAGCACGTCACGC**-3′
	reverse	5′-**GCTCAGACAT**ggtatatctccttattaaagt taaacaaaattatttctacaggg-3′	5′-gtttccctg**CGATGCCCCCAGCAGC**-3′
pCm3	forward	5′-**GGCATCG**cagggaaacggtaccgaaaacctg-3′	5′-ggatcccat**ATGTCTGAGCACGTCACG**-3′
	reverse	5′-**CTCAGACAT**atgggatccaccaatctgttctct-3′	5′-tttccctg**CGATGCCCCCAGCAGC**-3′
pCm4	forward	5′-**GGCATCG**cagggaaacggtaccgaaaacctg-3′	5′-gagatatacc**ATGTCTGAGCACGTCACGC**-3′
	reverse	5′-**GCTCAGACAT**ggtatatctccttattaaagt taaacaaaattatttctacaggg-3′	5′-tttccctg**CGATGCCCCCAGCAGC**-3′
pCm5	forward	5′-**GGCATCG**cagggaaacggtaccgaaaacctg-3′	5′-ggatcccat**ATGTCTGAGCACGTCACG**-3′
	reverse	5′-**CTCAGACAT**atgggatccaccaatctgttctct-3′	5′-tttccctg**CGATGCCCCCAGCAGC**-3′
pCm6	forward	5′-**GGCATCG**cagggaaacggtaccgaaaacctg-3′	5′-ttcagggg**ATGTCTGAGCACGTCACG**-3′
	reverse	5′-**CTCAGACAT**cccctgaaaatacaggttttcgaa-3′	5′-tttccctg**CGATGCCCCCAGCAGC**-3′

Fig. 2 Gibson assembly design for the cloning of a *NikJ* homolog into pCm family of plasmids. Top: schematic of the designed primers annealed with the Gibson assembly product DNA. Bold uppercase bases correspond to NikJ-coding DNA and the lowercase bases to the vector DNA, primers are shown as arrows. Bottom: list of primers designed for cloning a *NikJ* homolog into pCm1–6. To design custom primers for the cloning of a POI cDNA replace the bold bases for the corresponding cDNA of the POI gene.

treatment. In total, these six plasmids should provide broad-spectrum support for selenoprotein expression and purification as demonstrated below. The pCm plasmids also show improved continuity in scale-up due to the introduced chloramphenicol resistance (*CmR*). A guide for Gibson assembly of these plasmids with arbitrary POI genes is also provided in Fig. 2. The vectors have been deposited in the Addgene database for accessibility, and both Ulp1- and TEV protease-expressing plasmids and protocols are also available.

3. Assembling pCm-based vectors and optimizing selenoprotein expression

The following protocol describes how to clone cDNA of a POI into the pCm plasmid family and test the expression and solubility of the resulting selenoprotein. We use as an example *Streptomyces tendae NikJ*, which catalyzes the SAM-dependent formation of octosyl acid 5′-phosphate (OAP) from 3′-enolpyruvyl uridine 5′-phosphate by a radical mechanism involving enantioselective substrate radical quenching by an H-atom transfer from

C_{199}. The *NikJ* gene had previously been mutated by site-directed mutagenesis to convert the former cysteine coding TCG codon to the amber TAG codon for directed selenocysteine insertion ($C_{199}U$). Despite the low expression levels of $C_{199}U$ *NikJ*, its oxygen intolerance in both apo and holo form, and poor solubility during purification, the protein was expressed and recovered at a scale commiserate with detailed *in vitro* study. For a general thiyl radical enzyme or other POI the TGT/TGC → TAG mutation in the desired cysteine-to-selenocysteine mutation can be generated by site-directed mutagenesis before or after Gibson assembly. A detailed protocol for this site-directed mutagenesis is provided in the section "Mutate codon in desired position to TAG" of Chung et al. (Chung, Miller, Söll, & Krahn, 2021).

3.1 Materials

- Chloramphenicol (Cm)
- Kanamycin (Kan)
- Agar
- Agarose
- L-Arabinose (ara)
- Sodium selenite (Na_2SeO_3)
- Isopropyl β-D-1-thiogalactopyranoside (IPTG)
- 2 × Laemmli Buffer
- S.O.C. medium
- Luria-Bertani (LB) medium
- Terrific broth (TB) medium
- *Dpn*I restriction enzyme (New England Biolabs)
- Phusion® High-Fidelity DNA Polymerase (New England Biolabs, or another high-fidelity polymerase)
- NEBuilder® HiFi DNA Assembly Master Mix (New England Biolabs, or another Gibson assembly kit)
- InVision™ His-Tag In-Gel Stain (Thermo Scientific)
- Imperial™ Protein Stain (Thermo Scientific, or another protein stain)
- BugBuster® Protein Extraction Reagent (EMD Millipore)
- QIAprep® Spin Miniprep Kit (Qiagen, or another plasmid DNA isolation kit)
- 0.1 cm MicroPulser Electroporation Cuvettes
- Electrocompetent *E. coli* DH5α cells (Thermo Scientific)
- Electrocompetent *E. coli* BL21-AI™ cells (Thermo Scientific)

- pSecUAG-Evol2 (Addgene plasmid # 163148)
- pCm plasmid of choice (See Fig. 2, Addgene plasmids #174210, #174361, #174362, #174363, #174364, #174365)
- DNA of the protein of interest (POI) with an introduced TAG mutation for cysteine (TGT/TGC)
- Custom oligonucleotide primers (see below)
- Thermocycler
- Temperature controlled incubator shaker capable of lower than 30 °C temperatures
- MicroPulser Electroporator (Biorad)
- UV–Vis Spectrophotometer
- (Optional) SnapGene pro 5.3

3.2 Gibson assembly (2–3 days)

3.2.1 Design oligonucleotides to PCR amplify the POI gene (insert primers, Fig. 2) and the appropriately selected pCm plasmid (vector primers, Fig. 2) for Gibson assembly into the pCm vector of interest (Fig. 1). The primers listed in Fig. 2 can serve as a template, where the bold upper-case bases indicate the bases of the gene to be cloned, in this example *NikJ*. The number of overlapping base pairs necessary for proper annealing will depend on the GC content of the gene, but ~10 bp with 50% GC content has proven successful in our preparations. One primer pair should be obtained for the vector and another for the insert for each pCm-POI assembly

Note: The tool NEBuilder® HiFi DNA Assembly from the software SnapGene Pro can be used to design appropriate primers to clone the protein of interest in the open reading frame of the different pCm plasmids.

3.2.2 Amplify the pCm vector of choice and the gene of interest by PCR using the designed primers from step 3.2.1 and the Phusion® High-Fidelity DNA Polymerase, or other high-fidelity polymerase, according to the manufacturer's instructions. Resolve 3 μL of the PCR product by gel electrophoresis in an agarose gel to confirm the amplification of a single DNA fragment of the appropriate base pair composition

3.2.3 To avoid the transformation with template DNA, gel purify the pCm PCR amplicon or add 1 μL of *Dpn*I to digest any residual template plasmid in the samples according to the manufacturer's instructions

3.2.4 Assemble the vector and inserts using the NEBuilder® HiFi DNA Assembly Master Mix following the manufacturer's instructions. Transform 4 µL of the assembled product in 50 µL of electrocompetent *E. coli* DH5α cells by electroporation and outgrow cells in 1 mL of preheated S.O.C. media at 37 °C for 1 h. Plate the competent cells in LB-agar plates supplemented with 30 µg/mL Cm and grow at 37 °C until colonies appear

Note: Chemically competent cells can also be used with lower transformation efficiency.

3.2.5 Purify the assembled plasmid by Qiagen Spin Miniprep Kit according to the manufacturer's instructions from liquid culture growth of transformed colonies and confirm the correct gene insertion by sequencing the purified plasmid

3.3 Expression and amber codon suppression optimization (6 days)

3.3.1 Once the cDNA has been correctly cloned and its sequence verified, co-transform >200 ng of the pCm plasmid and >200 ng of pSecUAG-Evol2 plasmid in electrocompetent *E. coli* BL21-AI™ cells by electroporation. Outgrow cells in preheated S.O.C. media at 37 °C for 2 h and plate the transformed cells on LB-agar plates supplemented with 30 µg/mL Cm and 50 µg/mL Kan. Incubate the plates at 37 °C overnight or until colonies appear

Note: The selection of a suitable *E. coli* strain for the expression of Sec containing enzymes prove to be a challenge as pSecUAG-Evol2 requires ara for induction and most commercial expression strains can metabolize ara, compromising its function in pSecUAG-EVOL2 induction. We have expressed *NikJ* in BL21, C41, C43 and BL21-AI *E. coli* cell lines, and only the BL21-AI strain, an *araB* knockout mutant unable to consume ara, produced the full-length protein at detectable levels. The T7 RNA polymerase of the BL21-AI strain is under the control of the *araBAD* promoter, providing more control over the expression of the POI in pCm vectors. Alternative strains of *E. coli* have also been used successfully (Evans et al., 2021; Mukai et al., 2018).

Note: We have observed low co-transformation efficiency for these plasmids; to improve the efficiency use high DNA concentration and recuperate the cells at 37 °C with prewarmed S.O.C. medium for ≥2 h. Total DNA for transformation should not exceed 10% of total cell volume.

3.3.2 Test POI expression and amber codon suppression by inoculating four 5 mL cultures of TB medium, supplemented with 30 μg/mL Cm and 50 μg/mL Kan with a single colony each from step 3.3.1 and incubate overnight at 37 °C and 200 rpm

Note: For *NikJ* the cultures were also supplemented with 1 mg/L $FeCl_3$ and 24.28 mg/L of L-cysteine, which did not affect expression or amber codon suppression.

3.3.3 Inoculate a 50 mL culture of TB medium supplemented with 30 μM of $NaSeO_3$, 30 μg/mL Cm and 50 μg/mL Kan with the overnight pre-culture from step 3.3.2, using enough cells to reach an OD_{600} of 0.01, and continue incubating the cultures at 37 °C at 200 rpm

3.3.4 To test the expression of both the pSecUAG-EVOL2 and pCm plasmids induce the four cultures with the following ara and IPTG concentrations at $OD_{600} = 0.5$ and 1.0, respectively: no ara no IPTG (−ara,−IPTG), 1% *w/v* ara no IPTG (+ara,−IPTG), no ara 1.0 mM IPTG (−ara,+IPTG), and 1% w/v ara 1 mM IPTG (+ara, +IPTG). Collect 1 mL samples from each culture at $OD_{600} = 1.0$ prior to IPTG addition, spin down the cells by centrifugation, and flash freeze for storage at −80 °C. Repeat the sample collection procedure ≥24 h after IPTG induction

3.3.5 Cell pellets collected before and ≥24 h after IPTG induction should be analyzed by SDS-PAGE to confirm expression. Resuspend the pre- and postinduction cell pellets to equivalent cell density; we routinely dilute the pellets from 1 mL samples in 100 μL of 1 × Laemmli buffer per 1 unit of OD_{600}, and load 10 μL in a 15 μL well for 1 mm SDS-PAGE. Resolve the gel by electrophoresis, stain with InVision™ His-Tag In-Gel Stain according to the manufacturer's instructions, and image by UV transillumination. Next, stain the gel with Imperial™ Protein Stain or equivalent protein stain and image again. Compare the ratio of expression of full-length/truncated protein in the gel. Neither full-length or truncated POI should be expressed in the −ara,−IPTG or +ara,−IPTG cultures, only truncated protein should be observed in the −ara,+IPTG cultures, and an ideal ratio of ~1:1 full-length to truncated POI should be observed in the +ara,+IPTG. Fig. 3A shows a representative His-Tag Stain of *NikJ* expression levels in each of the 6 pCm vectors

Note: Specific His-tag detection is necessary as expression of the full-length protein might be low and difficult to observe in a Coomassie stained gel, as is the case for *NikJ*. We use InVision™ His-Tag In-Gel Stain for

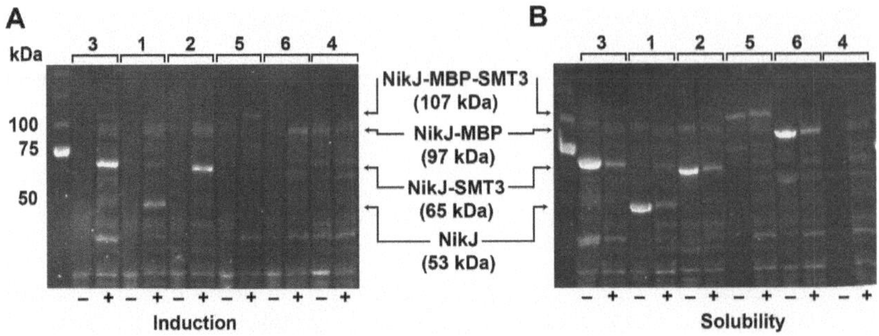

Fig. 3 Expression and solubility of $C_{199}U$ *NikJ* from pCm1–6 vectors. (A) His-tag stained SDS-PAGE gel of co-transformed BL21-AI *E. coli* cells with pSecUAG-Evol2 and pCm plasmids 1–6. Lanes were loaded with culture samples pre- and 24h post-IPTG induction. Numbers above lanes corresponds to pCm vectors according to Fig. 1A. (B) His-tag stained SDS-PAGE of protein extraction from samples from induced cultures in A, lane numbering corresponds to plasmids numbers on A and "–" sign indicates insoluble cell debris fraction and "+" sign soluble protein fraction.

convenience; however, Western blotting can provide higher sensitivity for low expression proteins.

3.3.6 If full-length protein expression levels are low, repeat steps 3.3.1–3.3.5 varying the temperature, shaking speed, ara, and IPTG levels from 4 to 48h after induction. Repeat these steps until an optimized full-length expression yield is realized or try cloning the POI in another plasmid

3.4 Cell lysis and solubility analysis (6days)

3.4.1 Once the appropriate induction conditions have been determined, the relative solubility of the expressed fusion protein should be evaluated to improve the subsequent protein purification yield and long-term stability. Repeat protocol 3.3, inducing both pSecUAG-Evol2 and pCm-POI using the optimized conditions developed therein, and collect all of the final culture volume by centrifugation in a pre-weighed collection centrifuge tube. Measure the wet cell weight

3.4.2 To analyze the fusion protein solubility, resuspend 200mg of the wet cell pellet from step 3.4.1 in 1mL of BugBuster® Protein Extraction Reagent and follow the manufacturer's instructions to obtain soluble and insoluble protein preparations. Quantify the protein concentration in both soluble and insoluble fractions and dilute as necessary to achieve 2mg/mL total protein concentration. Load equal amounts of samples from the soluble and insoluble proteins in an SDS-PAGE, resolve by electrophoresis, and stain the gel with InVision™ His-Tag In-Gel Stain. Image the His-tag stained gel for full-length protein distribution between soluble and insoluble fractions. Next, stain the total proteins with Imperial™ Protein Stain and after destaining, image total protein distribution. Compare the solubility of the different fusion proteins and choose the one with the highest full-length expression and solubility for scaling up the expression and to purify the full-length protein. Fig. 3B shows a representative solubility analysis for *NikJ* expressed from pCm1–6 vectors

4. Milligram scale selenoprotein preparation

Most fusion proteins expressed in pCm plasmids are designed to be purified in 1–2 chromatographic steps. The first chromatographic step involves IMAC to separate the POI from other cellular proteins, yielding the full-length fusion protein free of truncated impurities. In a second step, the POI can be separated from the His-tag or additional fusion proteins by protease digestion, followed by a second chromatographic step that depends on the desired product. A His-tagged TEV protease will remove the C-terminal His-tag in all pCm constructs, and the SMT3 or MBP domain for pCm2 and pCm4/6, respectively. The C-terminal His-tag on a pCm2-expressed POI can also be removed by Ulp1 while retaining the solubilizing fusion protein SMT3. The His-tag-less POI can be isolated in a second IMAC step, this time as the flow through, removing the remaining tagged impurity and Ulp1 or TEV protease. Alternatively, solubilizing fusion proteins SMT3 and MBP-SMT3 can be removed by Ulp1 in pCm3 and pCm5, respectively, and the POI can be isolated by IMAC with a retained His-tag. Here, the Ulp1 is also retained on the IMAC column, and would require a subsequent purification step for removal (e.g., TEV protease followed by a third IMAC step). Additional purification steps may be necessary based on the properties of the protein, and the desired purity level. For plasmids pCm4, pCm5 and pCm6 amylose resin can be used to further purify the fusion protein.

Selenocysteine is prone to oxidation, forming diselenides, mixed selenosulfides, or selenic acids. To maintain the selenocysteine in the selenol form, it is best to remove oxygen altogether by performing purification under an inert atmosphere such as in a glove box. If a glove box is not available, excess reductant should be included in all steps along the purification process and in storage. Furthermore, both Ulp1 and TEV protease are cysteine proteases and require supplementing the protease buffer with reductant. Appropriate reductants include tris(2-carboxyethyl)phosphine (TCEP), dithiothreitol (DTT) or β-mercaptoethanol (βME), the concentrations of which will depend on the purification step (see below).

Due to the uniqueness of the selenium atom in selenocysteine relative to the remaining 20 canonical amino acids, selenium quantitation by elemental analysis is a powerful technique for the analytical determination of selenocysteine content in the POI. Selenium quantitation is important as the biosynthesis of selenocysteine occurs on the serine acylated Ser-allo-$tRNA^{Sec}$, being catalyzed by *SelA* producing the Selenocysteine acylated Sec-allo-$tRNA^{Sec}$ using as substrate selenophosphate produced by *SelD*. If the activity of SelA and SelD cannot keep up with the POI expression, amber codon suppression can occur with insertion of serine rather than selenocysteine. For some thiyl radical enzymes, this serine substituted protein may not be of significant concern. In the *E. coli* class Ia RNR the thiyl radical that initiates nucleotide reduction is generated by radical transfer from a stable diiron-tyrosyl radical cofactor in a second β_2 subunit. Neither radical transfer nor substrate reduction occur in the $C_{439}S$ α_2 mutant, and thus this impurity is experimentally "silent" both in activity assays and in radical trapping experiments (Greene et al., 2019). Conversely, in *PolH*, a functional homolog of *NikJ*, the $C_{209}A$ mutant retains ~40% of wild type activity, and both $C_{209}A$ and $C_{209}S$ mutants lose enantioselectivity as a result of the loss of the H-atom donor C_{209} (Lilla & Yokoyama, 2016). Thus, determining the effect of the *NikJ* $C_{199}U$ mutation on enzyme activity requires precise quantification of the contaminating $C_{199}S$ content.

Mass spectrometry is also a powerful tool for analyzing selenoproteins and determining semi-quantitatively the misloaded serine contamination. Intact or fragmented mass spectrometry and liquid chromatography-tandem mass spectrometry (LC MS/MS) can be used to detect the selenocysteine loaded protein by total mass or selenocysteine peptide detection. Importantly, it is not necessarily valid to assume that digested selenocysteine and serine peptide fragments will ionize similarly, thus the ratio of these peaks may not reflect the actual U/S ratio. This can be accounted for by authentic synthesis of the respective peptides and constructing a calibration curve (Lee et al., 2018).

Below, we describe a protocol for selenoprotein purification under an inert atmosphere and selenium quantitation by inductively coupled plasma optical emission spectroscopy (ICP-OES), again using $C_{199}U$ *NikJ* as an example. The purification and fusion tag removal can be accomplished in

1 day, minimizing the opportunities for selenocysteine oxidation or protein degradation. The purification yielded 0.4 mg/g cell paste. Following purification and ICP-OES selenium quantitation, the 4Fe-4S cluster of *NikJ* was reconstituted and the enzyme was determined to be active and enantioselective for the H-atom donation to the OAP radical intermediate, implying the $C_{199}U$ mutant is chemically competent and useful for subsequent investigations into the fate of thiyl (selenyl) radicals in this enzyme.

4.1 Materials

- Chloramphenicol (Cm)
- Kanamycin (Kan)
- Agar
- L-arabinose (ara)
- Sodium selenite (Na_2SeO_3)
- Isopropyl β-D-1-thiogalactopyranoside (IPTG)
- DL-Dithiothreitol (DTT)
- Streptomycin sulfate
- Lysozyme (chicken egg white, Sigma-Aldrich)
- Ribonuclease A (bovine pancreas, Sigma-Aldrich)
- 2× Laemmli Sample Buffer
- S.O.C. medium
- LB medium
- TB medium
- Buffer X (50 mM Tris pH 7.6, 300 mM sodium chloride, 5% *w/v* Glycerol, 1 mM DTT)
- Lysis buffer (Buffer X, 1 mM PMSF)
- Buffer A (Equilibration/Wash buffer: Buffer X, 20 mM Imidazole)
- Buffer B (Elution buffer: Buffer X, 400 mM Imidazole)
- Imidazole
- InVision™ His-Tag In-Gel Stain (Thermo Scientific)
- Imperial™ Protein Stain (Thermo Scientific, or other Coomassie stain)
- Guanidinium chloride
- Trypsin (bovine pancreas, Sigma-Aldrich)
- α-Chymotrypsin (bovine pancreas, Sigma-Aldrich)
- Calcium chloride ($CaCl_2$)
- Trace-metal grade nitric acid
- Selenium Standard (TraceCERT, Sigma-Aldrich)
- Electrocompetent *E. coli* BL21-AI™ cells (Thermo Scientific)

- pSecUAG-Evol2 (Addgene plasmid # 163148)
- pCm plasmid from Gibson assembly
- Amicon® Ultra-15 Centrifugal Filter Units (EMD Millipore)
- Cytiva® HiTrap Desalting 5 mL columns with Sephadex G-25 resin
- HisPur™ Cobalt Resin column
- 0.1 cm MicroPulser Electroporation Cuvettes
- Temperature controlled incubator shaker capable of lower than 30 °C temperatures
- Biorad® MicroPulser Electroporator
- UV–Vis Spectrophotometer
- Emulsiflex C3 French press or other cell lysis tool
- iCAP 6300 ICP-OES (Thermo Fisher Scientific)

Optional

- Argon or Nitrogen gas
- Schlenk line
- Coy Lab's Vinyl Anaerobic Chamber (<20 ppm oxygen)
- Polyhistidine-tagged Ulp1
- Polyhistidine-tagged TEV protease

4.2 1.5 L protein expression (2–3 days)

4.2.1 Co-transform electrocompetent *E. coli* BL21-AI cells with >200 ng of the optimal Gibson assembled pCm plasmid and >200 ng of pSecUAG-Evol2 by electroporation. Outgrow cells in preheated S.O.C. media at 37 °C for 2 h and plate the transformed cells in LB-Agar supplemented with 30 μg/mL Cm and 50 μg/mL Kan. Incubate the plates at 37 °C until colonies appear.

4.2.2 Inoculate a 50 mL preculture of TB medium, supplemented with 30 μg/mL Cm and 50 μg/mL Kan with a single co-transformed colony and incubate overnight at 37 °C

4.2.3 Inoculate a pre-warmed 1.5 L culture of TB medium in a 6 L flask, supplemented with 30 μM of $NaSeO_3$, 30 μg/mL Cm and 50 μg/mL Kan, with the overnight pre-culture from 4.2.2, using enough cells to reach an OD_{600} of 0.01

Note: $FeCl_3$ and L-cysteine are included for *in vivo* assembly of the *NikJ* 4Fe-4S cluster. These may be omitted for non-iron-sulfur proteins, but other supplements may be necessary for different selenoproteins of interest (*e.g.* metals, cofactors, or chelators).

4.2.4 Express the protein using the optimized conditions developed in step 3.3 on the 1.5 L scale. For *NikJ*, the expression was carried out at

37 °C and 200 rpm shaking speed until the culture reached an OD_{600} of 0.5 (2–3 h after inoculation), then add 1% *w*/*v* of ara to induce the expression of the selenocysteine incorporation machinery by pSecUAG-Evol2 and the T7 RNA polymerase of the BL21-AI cells. Once an OD_{600} of 1.0 is reached (4–5 h after inoculation) induce with 1 mM IPTG, and lower the temperature to 30 °C for 24 h with continued shaking, reaching a final OD_{600} of 8–10

4.2.5 Collect the cells by centrifugation and purify the protein immediately or flash freeze the pellet in liquid nitrogen and store at −80 °C. Typical cell yields vary between 10 and 20 g wet cell weight per liter of TB medium

4.3 Protein purification (1 day)

4.3.1 Homogeneously resuspend the cell pellet in Lysis Buffer at 2.5–5 mL/g cell paste, then lyse the cells using an Emulsiflex C3 French Press under Argon as described by Schleicher et al. (Schleicher, Fritz, Seifert, & Steuber, 2020) by three passes at 14,000 psi. Centrifuge the lysate and centrifuge at 30,000 × *g* for 30 min. Collect the supernatant (38–150 mL).

Note: Initial protein purifications can be carried out in Buffer X, but optimal buffer conditions should be determined once pure protein is obtained. For *NikJ*, all buffers were supplemented to a final concentration of 500 mM NaCl and 10% glycerol.

Note: All steps were carried out anaerobically, but the protocol described herein is more general. A reducing agent in the buffer is critical to keep the selenocysteine and native cysteines reduced in either case. An effective reductant system may include 0.5–2 mM TCEP, 1–10 mM DTT, or 5–20 mM βME. Reductants can poison metal affinity chromatography resins. Consult the resin manual to determine the appropriate reductant concentration.

Note: Alternative cell lysis protocols may be more appropriate for individual POIs and laboratory capabilities. We have observed similar results with BugBuster® Protein Extraction Reagent and ultrasonification, both of which can be adapted inside of a glove box to avoid oxygen exposure.

4.3.2 While stirring the supernatant from step 4.3.1, add a 1.5% *w*/*v* of streptomycin sulfate dropwise from a 6% w/v stock solution (0.2 volume equivalents), and incubate for ≥10 min. Centrifuge the solution at 30,000 × *g* for 30 min to remove DNA

4.3.3 Apply the clarified supernatant from step 4.3.2 to a Buffer A pre-equilibrated HisPur™ cobalt resin column at a flow rate of 1 mL/min to maximize resin adsorption of the His-tagged protein. Wash the column with >10 column volumes of Buffer A and then elute the protein with 3 column volumes of Buffer B. Pool fractions that contain protein as determined by the Bradford assay or absorbance at 280 nm and supplement with DTT to a final concentration of 10 mM

4.3.4 Concentrate the protein-containing solution in an Amicon® Ultra-15 centrifugal filter of appropriate molecular weight cutoff to ≤1.5 mL and buffer exchange the protein in a Buffer X pre-equilibrated HiTrap desalting 5 mL column. Quantify the protein concentration and use immediately, remove affinity tags and fusion proteins, or flash freeze in liquid nitrogen and store at −80 °C. Fig. 4 shows a representative anaerobic purification of pCm2 expressed $C_{199}U$ *NikJ*

Note: For many proteins, purification to this point may be sufficient.

4.4 (Optional) fusion protein or His-tag removal (1 − 3 h)

Note: We produce and purified TEV protease by recombinant expression using the plasmid Addgene #125194 (van den Berg et al., 2006) and

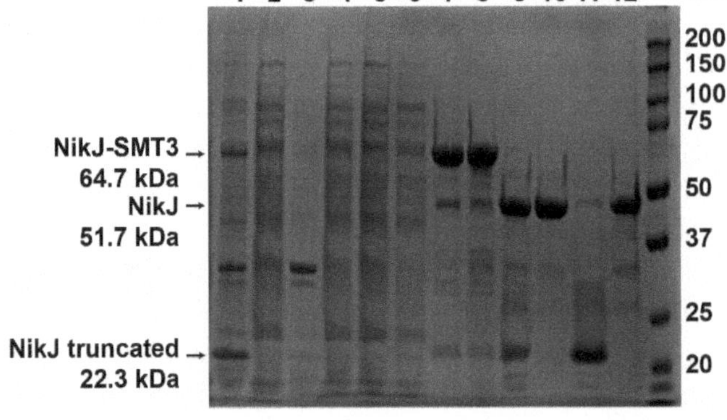

Fig. 4 Selenoprotein purification and protease cleavage. A SDS-PAGE analysis of *NikJ*-SMT3 $C_{199}U$ purification. Lanes: 1, cell lysate debris; 2, cell lysate supernatant; 3, DNA precipitate; 4, DNA precipitation supernatant; 5, HisPur™ Cobalt Resin column flow through at 20 mM imidazole; 6, wash with 20 mM imidazole, 7, eluted *NikJ*-SMT3 $C_{199}U$ (400 mM imidazole); 8, 1:10 TEV digestion at time 0; 9, 1:10 TEV digestion at time 30 min; 10, TEV digested protein in the flow through of a HisPur™ Cobalt Resin column; 11, eluted TEV and SMT3 (400 mM imidazole), 12, concentrated and desalted purified *NikJ*.

Ulp1 protease using the plasmid pFGET19_Ulp1 (Addgene Plasmid #64697) (Guerrero et al., 2015) according to the authors instructions. Both are commercially available in His-tagged form. Protease reaction times from other sources might differ from those reported herein.

4.4.1 For TEV protease digestion, add 1 mg of TEV protease per 10 mg of protein and incubate at room temperature for 30 min in Buffer X

Note: The POI sample for digestion must be previously desalted in Buffer X and free of imidazole.

4.4.2 Load the digested POI sample onto a HisPur™ Cobalt Resin column pre-equilibrated with Buffer A, wash with 2 CV of Buffer A, and collect all POI containing flow through. Follow the protocol in 4.3.4 to concentrate and buffer exchange the protein for storage

4.4.3 For Ulp1 digestion add 1 mg of Ulp1 per 100 mg of protein and incubate at room temperature for 30 min in Buffer X

4.4.4 If digesting pCm2 load the digested POI sample onto a HisPur™ Cobalt Resin column pre-equilibrated with Buffer A, wash with 2 CV of Buffer A, and collect all POI containing flow through. Follow the protocol in 4.3.4 to concentrate and buffer exchange the protein for storage

Note: If digesting pCm3 or pCm5 the POI retains the C-terminal His-tag, and thus must be eluted from the IMAC column at high imidazole concentrations. The POI will likely co-elute with the His-tagged Ulp1, potentially requiring subsequent purification steps.

4.4.5 Analyze samples before and after digestion and the IMAC column flow through and high imidazole elution by SDS-PAGE to assess the purity of the produced protein. Fig. 5 shows a representative protease digestion for pCm2 and pCm5 by Ulp1 or TEV protease

4.5 Selenium quantification by ICP-OES (2 days)

Selenium quantitation for thiyl radical protein preparations by the aforementioned methods have typically yielded the $C_{199}U$ mutant *NikJ* with 0.4 mg/g of cell paste and 73% selenium incorporation. The protein was purified from cells expressing the pCm2 vector and digested with TEV protease to yield a >90% pure near-native protein as determined by SDS-PAGE (Fig. 4). Note: The limit of detection for the quantification of selenium using our ICP-OES method is 0.03 ppm, which corresponds to 2.7 nmol Se. In the following protocol, we supply enough protein to yield a theoretical concentration of 0.1 ppm Se, and our calibration curve spans 0–0.5 ppm.

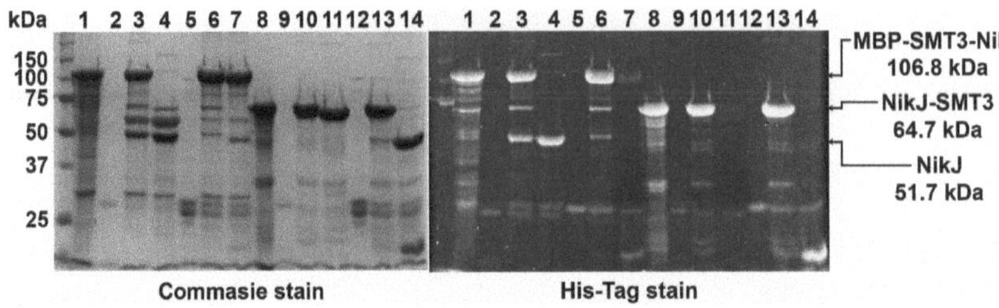

Fig. 5 SDS-PAGE analysis of protease cleavage of partially purified MBP-SMT3-*NikJ*C_{199}U and *NikJ*C_{199}U-SMT3 stained with Coomassie stain and also His-Tag stain. Lanes: 1, MBP-SMT3-*NikJ*C_{199}U; 2, Ulp1 protease; 3, MBP-SMT3-*NikJ*C_{199}U + ULP1 protease 0 min incubation; 4, MBP-SMT3-*NikJ*C_{199}U + Ulp1 protease 30 min incubation; 5, TEV protease; 6, MBP-SMT3-*NikJ*C_{199}U + TEV protease 0 min incubation; 7, MBP-SMT3-*NikJ*C_{199}U + TEV protease 30 min incubation, 8, *NikJ*C_{199}U-SMT3; 9, Ulp1 protease; 10, *NikJ*C_{199}U-SMT3 + Ulp1 protease 0 min incubation; 11, *NikJ*C_{199}U-SMT3 + Ulp1 protease 30 min incubation; 12, TEV protease; 13, *NikJ*C_{199}U-SMT3 + TEV protease 0 min incubation; 14, *NikJ*C_{199}U-SMT3 + TEV protease 30 min incubation.

We require 7 mL of final solution to obtain method triplicates. Samples and standards should be prepared by weight. We routinely achieve >0.7 selenium atoms per protein monomer (>70%). The remaining <30% is assumed to be C_{199}S.

4.5.1 Dilute 8.9 nmol of purified protein in 132 μL of concentrated trace-metal grade nitric acid (to yield 2% nitric acid upon dilution) and allow to react at room temperature overnight. Heat samples to 90°C for 2 h and allow the tubes to cool to room temperature. Add 70 μL of trace-metal grade hydrogen peroxide and heat the sample at 90°C for an hour, until the solution is clear and there is no further bubbling.

4.5.2 Dilute the samples to 7 g final solution weight with water

4.5.3 Prepare method blank samples, one without protein and one with a wild-type protein, using the above digestion procedure

4.5.4 Prepare selenium standards (TraceCERT, Sigma-Aldrich) by diluting to a relevant range for the calibration curve in 2% nitric acid

4.5.5 Analyze prepared samples by ICP-OES according to the instrument standard operating procedure

4.5.6 Calculate the Selenium per protein monomer (Se/POI) by using the equation shown below

$$\frac{Se}{POI} = \frac{(ppm\ Se) \times (g\ total\ solution)}{(10^6) \times \left(78.96 \frac{g}{mol\ Se}\right) \times (mol\ protein\ in\ sample)}.$$

5. Evaluating the effect of cysteine-to-selenocysteine substitution

After successful expression and purification of a selenocysteine-substituted thiyl radical enzyme, the selenocysteine effect can be assessed in several ways including activity assays and spectroscopic analysis. In the case of the example *NikJ*, selenocysteine was introduced in the position of the C_{199} which is proposed to rapidly quench the substrate OAP radical enantioselectively (Lilla & Yokoyama, 2016). The pure $C_{199}U$ *NikJ* was isolated with 0.8 Fe/monomer and reconstituted to 2.3 Fe/monomer, slightly below the wild type value of 3.0. The 4Fe-4S cluster reconstituted *NikJ* was determined to be catalytically active in the presence of SAM (0.5 mM), EP-UMP (0.5 mM), sodium dithionite (2 mM) and DTT (2 mM), producing enantiopure U-OAP from EP-UMP. Fig. 6 shows U-OAP formation by *NikJ* and $C_{199}U$ *NikJ*, the mutant exhibits no detectable formation of

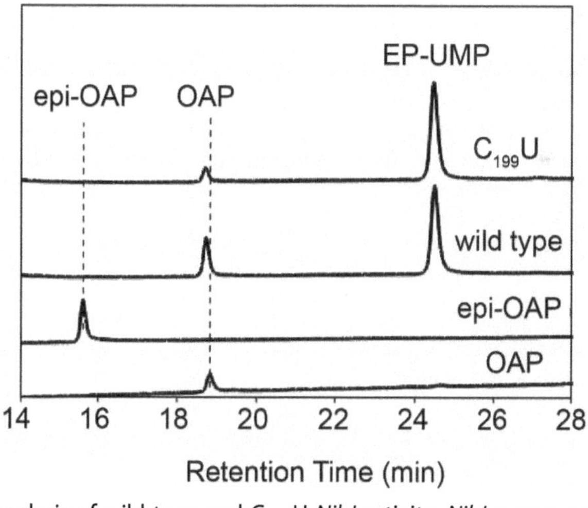

Fig. 6 HPLC analysis of wild type and $C_{199}U$ *NikJ* activity. *NikJ* assays were performed with 10 μM of *NikJ* in the presence of SAM (0.5 mM), EP-UMP (0.5 mM), sodium dithionite (2 mM), DTT (2 mM), and 40 mM NaCl in 50 mM HEPES pH 7.0 or 50 mM MES pH 6.0. The reactions were initiated by the addition of the enzyme and incubated at 28 °C for 1320 min. At each time point, 15 μL of the assay mix was mixed with equal volume of ethanol and stored at −20 °C. After removal of the protein precipitation by centrifugation, the supernatant was diluted fourfolds in 10 mM ammonium acetate pH 6.0, and analyzed by HPLC (Dionex IC5000+, Thermo) equipped with a DNAPac PA-100 column (Thermo). The HPLC analysis was performed with a linear gradient of 10–300 mM ammonium acetate pH 6.0 over 40 min.

epi-OPA (<5% epi-OAP), suggesting the selenocysteine may donate an H-atom from the selenol (R-SeH) form as opposed to electron transfer from the selenolate (R-Se$^-$, pK_a in solution of 5.2). Further investigations will shed light on the stability of the putative selenyl radical and the role of H-atom or electron transfer in $C_{199}U$ *NikJ*.

Acknowledgments

This work was supported, in part, by a National Institute of General Medical Sciences R01 GM115729 (to K.Y.). The UCSB MRL Shared Experimental Facilities housing the ICP-OES are supported by the MRSEC Program of the NSF under Award No. DMR 1720256; a member of the NSF-funded Materials Research Facilities Network (www.mrfn.org). The plasmids pFGET19_Ulp1 and pHYRSF53 were a gift from Hideo Iwai (Addgene plasmid # 64697 and # 64696). pET41-MBP-t-A3A-t-His was a gift from Rahul Kohli (Addgene plasmid # 109231). pSecUAG-Evol2 was a gift from Dieter Söll (Addgene plasmid # 163148). TEVSH was a gift from Helena Berglund (Addgene plasmid # 125194).

References

Åberg, A., Hahne, S., Karlsson, M., Larsson, Å., Ormö, M., Åhgren, A., et al. (1989). Evidence for two different classes of redox-active cysteines in ribonucleotide reductase of Escherichia coli. *Journal of Biological Chemistry*, *264*, 12249–12252. https://doi.org/10.1016/S0021-9258(18)63849-9.

Booker, S., Licht, S., Broderick, J., & Stubbe, J. (1994). Coenzyme B12-dependent ribonucleotide reductase: Evidence for the participation of five cysteine residues in ribonucleotide reduction. *Biochemistry*, *33*, 12676–12685. https://doi.org/10.1021/bi00208a019.

Chung, C. Z., Miller, C., Söll, D., & Krahn, N. (2021). Introducing Selenocysteine into recombinant proteins in *Escherichia coli*. *Current Protocols*, *1*. https://doi.org/10.1002/cpz1.54.

Duong-Ly, K. C., & Gabelli, S. B. (2015). Affinity purification of a recombinant protein expressed as a fusion with the maltose-binding protein (MBP). *Tag. Methods in Enzymology*, *559*, 17–26. https://doi.org/10.1016/bs.mie.2014.11.004.

Evans, R. M., Krahn, N., Murphy, B. J., Lee, H., Armstrong, F. A., & Söll, D. (2021). Selective cysteine-to-selenocysteine changes in a [NiFe]-hydrogenase confirm a special position for catalysis and oxygen tolerance. *Proceedings of the National Academy of Sciences*, *118*. https://doi.org/10.1073/pnas.2100921118.

Frey, P. A. (2001). Radical mechanisms of enzymatic catalysis. *Annual Review of Biochemistry*, *70*, 121–148. https://doi.org/10.1146/annurev.biochem.70.1.121.

Greene, B. L., Kang, G., Cui, C., Bennati, M., Nocera, D. G., Drennan, C. L., et al. (2020). Ribonucleotide reductases: Structure, chemistry, and metabolism suggest new therapeutic targets. *Annual Review of Biochemistry*, *89*, 45–75. https://doi.org/10.1146/annurev-biochem-013118-111843.

Greene, B. L., Stubbe, J., & Nocera, D. G. (2019). Selenocysteine substitution in a class I ribonucleotide reductase. *Biochemistry*, *58*, 5074–5084. https://doi.org/10.1021/acs.biochem.9b00973.

Guerrero, F., Ciragan, A., & Iwaï, H. (2015). Tandem SUMO fusion vectors for improving soluble protein expression and purification. *Protein Expression and Purification*, *116*, 42–49. https://doi.org/10.1016/j.pep.2015.08.019.

Lee, W., Kasanmascheff, M., Huynh, M., Quartararo, A., Costentin, C., Bejenke, I., et al. (2018). Properties of site-specifically incorporated 3-Aminotyrosine in proteins to study redox-active Tyrosines: *Escherichia coli* ribonucleotide reductase as a paradigm. *Biochemistry*, *57*, 3402–3415. https://doi.org/10.1021/acs.biochem.8b00160.

Licht, S., Gerfen, G. J., & Stubbe, J. (1996). Thiyl radicals in ribonucleotide reductases. *Science*, *271*, 477–481. https://doi.org/10.1126/science.271.5248.477.

Lilla, E. A., & Yokoyama, K. (2016). Carbon extension in peptidylnucleoside biosynthesis by radical SAM enzymes. *Nature Chemical Biology*, *12*, 905–907. https://doi.org/10.1038/nchembio.2187.

Metanis, N., & Hilvert, D. (2014). Natural and synthetic selenoproteins. *Current Opinions in Chemical Biology*, *22*, 27–34. https://doi.org/10.1016/j.cbpa.2014.09.010.

Minnihan, E. C., Seyedsayamdost, M. R., Uhlin, U., & Stubbe, J. (2011). Kinetics of radical intermediate formation and deoxynucleotide production in 3-Aminotyrosine-substituted Escherichia coli ribonucleotide reductases. *Journal of the American Chemical Society*, *133*, 9430–9440. https://doi.org/10.1021/ja201640n.

Mukai, T., Sevostyanova, A., Suzuki, T., Fu, X., & Söll, D. (2018). A facile method for producing Selenocysteine-containing proteins. *Angewandte Chemie International Edition*, *57*, 7215–7219. https://doi.org/10.1002/anie.201713215.

Nauser, T., Casi, G., Koppenol, W. H., & Schöneich, C. (2008). Reversible intramolecular hydrogen transfer between cysteine Thiyl radicals and glycine and alanine in model peptides: Absolute rate constants derived from pulse radiolysis and laser flash photolysis. *The Journal of Physical Chemistry B*, *112*, 15034–15044. https://doi.org/10.1021/jp805133u.

Nauser, T., Dockheer, S., Kissner, R., & Koppenol, W. H. (2006). Catalysis of electron transfer by Selenocysteine. *Biochemistry*, *45*, 6038–6043. https://doi.org/10.1021/bi0602260.

Raran-Kurussi, S., Keefe, K., & Waugh, D. S. (2015). Positional effects of fusion partners on the yield and solubility of MBP fusion proteins. *Protein Expression and Purification*, *110*, 159–164. https://doi.org/10.1016/j.pep.2015.03.004.

Schleicher, L., Fritz, G., Seifert, J., & Steuber, J. (2020). Anoxic cell rupture of Prevotella bryantii by high-pressure homogenization protects the Na+−translocating NADH: Quinone oxidoreductase from oxidative damage. *Archives of Microbiology*, *202*, 1263–1266. https://doi.org/10.1007/s00203-019-01805-x.

Schöneich, C. (2011). Cysteine residues as catalysts for covalent peptide and protein modification: A role for thiyl radicals? *Biochemical Society Transactions*, *39*, 1254–1259. https://doi.org/10.1042/BST0391254.

Serrão, V. H. B., Silva, I. R., da Silva, M. T. A., Scortecci, J. F., de Freitas Fernandes, A., & Thiemann, O. H. (2018). The unique tRNASec and its role in selenocysteine biosynthesis. *Amino Acids*, *50*, 1145–1167. https://doi.org/10.1007/s00726-018-2595-6.

Stubbe, J., & van der Donk, W. A. (1998). Protein radicals in enzyme catalysis. *Chemical Reviews*, *98*, 705–762. https://doi.org/10.1021/cr9400875.

van den Berg, S., Löfdahl, P.-Å., Härd, T., & Berglund, H. (2006). Improved solubility of TEV protease by directed evolution. *Journal of Biotechnology*, *121*, 291–298. https://doi.org/10.1016/j.jbiotec.2005.08.006.

Wang, Y., Liu, P., Chang, J., Xu, Y., & Wang, J. (2021). Site-specific Selenocysteine incorporation into proteins by genetic engineering. *Chembiochem*, *22*, 1–8. https://doi.org/10.1002/cbic.202100124.

CHAPTER SIX

Expression of selenoproteins via genetic code expansion in mammalian cells

Jennifer C. Peeler[a,*] and Eranthie Weerapana[b]
[a]Department of Chemistry, Colgate University, Hamilton, NY, United States
[b]Department of Chemistry, Boston College, Chestnut Hill, MA, United States
*Corresponding author: e-mail address: jpeeler@colgate.edu

Contents

Abstract

Selenoproteins, which contain the 21st amino acid selenocysteine, play roles in maintaining cellular redox homeostasis. Many open questions remain in the field of selenoprotein biology, including the functions of a number of uncharacterized human selenoproteins, and the properties of selenocysteine compared to its analogous amino acid cysteine. The mechanism of selenocysteine incorporation involves an intricate machinery that deviates from the mechanism of incorporation for the canonical 20 amino acids. As a result, recombinant expression of selenoproteins has been historically challenging, and has hindered a deeper evaluation of selenoprotein biology. Genetic code expansion methods, which incorporate protected analogs of selenocysteine, allow the endogenous selenocysteine incorporation mechanism to be bypassed entirely to facilitate selenoprotein expression. Here we present a method for incorporating a photocaged selenocysteine amino acid (DMNB-Sec) into human selenoproteins directly in mammalian cells. This approach offers the opportunity to study human selenoproteins in their native cellular environment and should advance our understanding of selenoprotein biology.

Methods in Enzymology, Volume 662
ISSN 0076-6879
https://doi.org/10.1016/bs.mie.2021.10.015

1. Introduction

Selenocysteine (Sec) is the naturally-occurring 21st proteinogenic amino acid. Sec is an analog of cysteine (Cys) with selenium in the place of sulfur. Proteins that contain Sec are named selenoproteins. Humans have 25 known selenoproteins. Of the human selenoproteins that have been characterized, many perform biological roles vital to maintaining redox homeostasis. Examples of such well-characterized selenoproteins include a number of thioredoxin reductases (Trxnrd1-3), glutathione peroxidases (Gpx1-4,6), and methionine sulfoxide reductase B1 (MsrB1) (Labunskyy, Hatfield, & Gladyshev, 2014).

Many selenoproteins have homologs that are capable of catalyzing the same reactions but utilize Cys instead of Sec. These homologs appear in organisms like yeast, which lack selenoproteins. Additionally, homologous proteins may appear in the same organism; for example, human MsrB2 and MsrB3 also reduce oxidized methionine residues like MsrB1, but have different subcellular localizations. The presence of these Cys homologs invites the question of why Sec is utilized. One current hypothesis is that Sec is protected from the irreversible overoxidation to which Cys is susceptible (Reich & Hondal, 2016). A study of the Sec residue of Gpx4 in a model of ferroptotic cell death supports this hypothesis. The mutation of Sec to Cys led to irreversible overoxidation, loss of enzymatic activity, and increased cell death compared to the wild type Sec-containing enzyme (Ingold et al., 2018). It remains to be seen whether this is also true in other selenoproteins.

The endogenous mechanism of Sec incorporation differs from the mechanism for incorporation of the 20 canonical amino acids (Fig. 1). First, Sec is biosynthesized on its own tRNA. The tRNA is initially charged with the amino acid serine (Ser). In eukaryotic systems, the Ser side chain is then phosphorylated. This phosphorylated Ser tRNA molecule is then the substrate for the enzyme SerS, which catalyzes the incorporation of a selenium atom into the amino acid side chain (Xu et al., 2007). This noncanonical process is likely employed in part so that a native aminoacyl tRNA synthetase (RS) does not need to distinguish between Sec and Cys. Second, incorporation of the Sec residue into the growing polypeptide chain requires a number of cis- and trans-acting factors. The codon encoding Sec is UGA, which is canonically a stop codon (Low & Berry, 1996). Successful recoding of UGA requires a selenocysteine insertion sequence (SECIS)

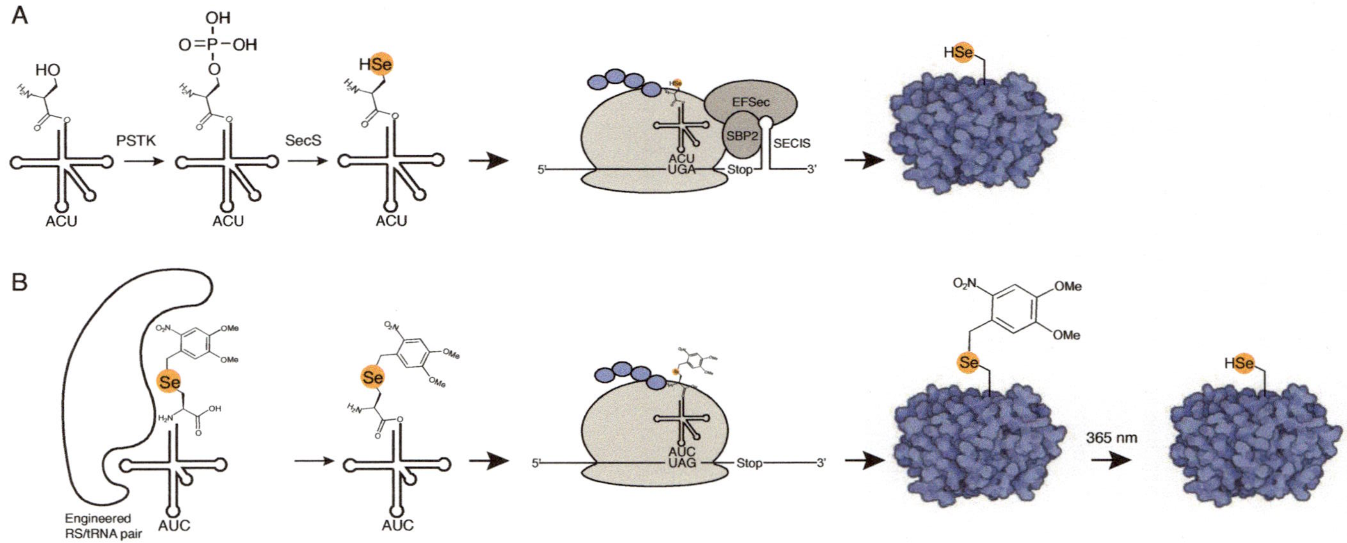

Fig. 1 Mechanisms of selenoprotein expression in mammalian cells. (A) Endogenous mechanism of Sec incorporation in mammalian cells. (B) Genetic code expansion method, utilizing an engineered RS/tRNA pair and the DMNB-Sec amino acid, for selenoprotein production in mammalian cells. Protein structures were illustrated using Illustrate (Goodsell, Autin, & Olson, 2019).

element; in the mammalian system the SECIS element is a stem-loop structure that appears in the 3′ UTR of the mRNA (Labunskyy et al., 2014). Additionally, a number of accessory proteins are required for successful Sec incorporation. The mammalian system minimally requires the SECIS binding protein 2 (SBP2) and the Sec-specific translation elongation factor (EFSec) (Donovan & Copeland, 2010). This noncanonical mechanism for selenoprotein translation means that traditional approaches to recombinant protein expression are challenging, and in some cases, impossible with selenoproteins. The challenges associated with the expression of native selenoproteins and their mutants have contributed to our limited understanding of the functions of selenoproteins, as well as the role of Sec compared to Cys.

In an effort to better study selenoprotein biology, a number of different approaches have been developed to generate Sec-containing proteins. This topic has recently been reviewed (Liu, Cheng, & Rozovsky, 2018; Peeler & Weerapana, 2019). *In vitro* methods include native chemical ligation and expressed protein ligation (Liu, Cheng, et al., 2018). Proteins produced by these methods are limited to *in vitro* assays. Additionally, endogenous prokaryotic and eukaryotic Sec incorporation mechanisms have been reengineered for optimized recombinant expression. A number of prokaryotic expression systems have been developed that make use of reengineered tRNAs, reengineered accessory proteins, and codon reprograming (Fu, Söll, & Sevostyanova, 2018). Again, human selenoproteins that are produced in these optimized prokaryotic systems are limited to *in vitro* assays of their function.

Optimization of the endogenous selenoprotein expression system in mammalian cells include the use of nonnative SECIS elements to increase recombinant selenoprotein translation efficiency (Novoselov et al., 2007). Instead of optimizing the endogenous Sec incorporation mechanism, genetic code expansion (GCE) can be used to bypass this mechanism entirely. GCE utilizes an engineered, orthogonal RS and tRNA pair to incorporate a noncanonical amino acid (ncAA) in response to a designated codon. Hundreds of ncAAs have been incorporated through GCE in a variety of model organisms using a number of engineered RS/tRNA pairs (Chin, 2017). GCE has been used to produce selenoproteins in *E. coli,* yeast, and mammalian cells. All of the published GCE strategies to express selenoproteins make use of ncAAs that are protected (caged) variants of Sec. Engineered, orthogonal RS/tRNA pairs are utilized to incorporate the caged Sec ncAA and a subsequent uncaging step is utilized to generate the Sec-containing proteins. These caging groups provide two advantages. Firstly, evolving a RS that could differentiate between uncaged Sec and endogenous Cys would be challenging if not impossible. Therefore, the structurally distinct photocaging group allows for the directed evolution of an orthogonal RS to incorporate the ncAA. Secondly, the process of uncaging provides the ability to temporally control Sec reactivity and presumably selenoprotein function. Photocaged-Sec ncAAs have been used in yeast and *E. coli* in order to incorporate Sec into model proteins like GFP (Rakauskaitė et al., 2015; Welegedara, Adams, Huber, Graham, & Otting, 2018). These systems have not been used to produce native selenoproteins; furthermore, yeast do not naturally contain selenoproteins, so presumably any mammalian selenoproteins produced by such methods would also be limited to *in vitro* assays. Two recent approaches, however, have been used to generate native selenoproteins and are compatible with mammalian expressions systems, which should enable the study of human selenoproteins in their native cellular environment. An engineered pyrrolysine RS/tRNA pair has been used to incorporate an allyl-protected Sec variant which can be deprotected using palladium. This system has been used to express the human selenoprotein Gpx1 in *E. coli* and a Sec-containing GFP variant in human cells (Liu, Zheng, et al., 2018). An engineered leucyl RS/tRNA pair has been used to incorporate a photocaged Sec variant (4,5-dimethoxy-2-nitrobenzyl Sec or DMNB-Sec) which can be deprotected using 365 nm light (Fig. 1) (Peeler et al., 2020). This system has been used to express the human selenoprotein MsrB1 in human cells. This protocol provides details of how to utilize the leucyl RS/tRNA pair to incorporate DMNB-Sec into a selenoprotein of interest (SelenoPOI) in response to an amber stop codon (UAG) in human cells, as well as how to uncage the amino acid, generating a native selenoprotein.

2. Methods

2.1 Before you begin

Here we present a general method for the expression of native selenoproteins in mammalian cells via GCE (Fig. 2). This process involves (1) the incorporation of the photocaged Sec variant (DMNB-Sec) into a full-length protein in response to the codon TAG and (2) the subsequent uncaging of this amino acid to generate the selenoprotein. Theoretically, this approach can be utilized to express any native human selenoproteins

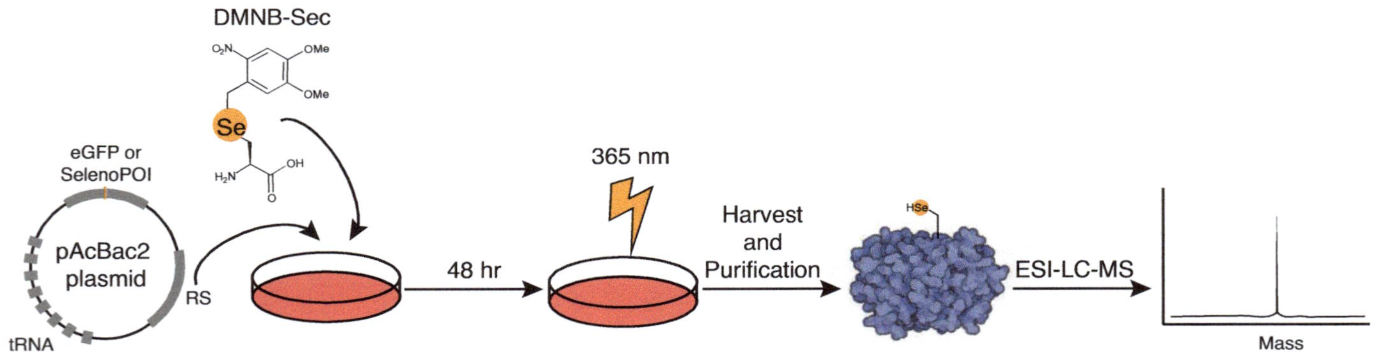

Fig. 2 Workflow of Sec incorporation in mammalian cells. Protein structures were illustrated using Illustrate (Goodsell et al., 2019).

(though expression of SelenoP, which contains 10 Sec residues, would be highly inefficient) or to incorporate Sec into a protein that does not naturally harbor the residue.

We recommend utilizing various controls in order to test the effectiveness of incorporation, uncaging, purification, and detection steps in the production of a SelenoPOI (Fig. 3). First, eGFP with a TAG codon at residue 39 (eGFP-39TAG) is a useful control for evaluating and optimizing GCE-based incorporation of DMNB-Sec. eGFP will only fluoresce if the TAG codon is successfully suppressed; truncated protein will not be fluorescent. Therefore, expressing eGFP-39TAG in the presence of DMNB-Sec and monitoring lysate fluorescence is recommended to confirm that DMNB-Sec is effectively incorporated. Expressing eGFP-39TAG in the absence of DMNB-Sec is a helpful negative control for evaluating background incorporation of other amino acids at the TAG codon. eGFP-39TAG grown in the presence of DMNB-Sec can be further utilized to test the effectiveness of uncaging. Exposing eGFP-DMNB-Sec to 365 nm light should yield the uncaged variant of the protein, eGFP-Sec. Using ESI-LC-MS to compare the mass of irradiated and nonirradiated purified protein samples is recommended to confirm that photo-uncaging of DMNB-Sec is effective.

In addition to the eGFP controls, we recommend controls specific to your SelenoPOI. First, expression of a variant where Sec is replaced with Cys bypasses both the endogenous Sec incorporation mechanism and the GCE-based incorporation mechanism of DMNB-Sec. This mutant can be a useful positive control for SelenoPOI-specific purification and detection protocols. This mutant could be produced through traditional recombinant expression, or theoretically could be produced using genetic code

Plasmid	DMNB-Sec	365 nm light	Resulting Protein
pAcBac2- EcLeuRS-BH5 T252A eGFP-39-TAG	-		Truncated eGFP or eGFP-Leu/Ile-39
pAcBac2- EcLeuRS-BH5 T252A eGFP-39-TAG	+	-	eGFP-DMNB-Sec-39
pAcBac2- EcLeuRS-BH5 T252A eGFP-39-TAG	+	+	eGFP-Sec-39
pAcBac2- EcLeuRS-BH5 T252A SelenoPOI-Cys	-		SelenoPOI-Cys
pAcBac2- EcLeuRS-BH5 T252A SelenoPOI-TAG	-		Truncated SelenoPOI or SeleonPOI-Leu/Ile
pAcBac2- EcLeuRS-BH5 T252A SelenoPOI-TAG	+	-	SelenoPOI-DMNB-Sec
pAcBac2- EcLeuRS-BH5 T252A SelenoPOI-TAG	+	+	SelenoPOI-Sec (native selenoprotein)

Fig. 3 Recommended controls for GCE-based selenoprotein expression. Note that while truncated eGFP or SelenoPOI would be generated in the absence of DMNB-Sec, truncated protein products would not be purified using a C-terminal epitope tag.

expansion with DMNB-Cys. The RS/tRNA pair presented in this work should be capable of incorporating DMNB-Cys and this control would account for any changes in protein expression or folding that may be caused by the presence of the photocaging group (Lemke, Summerer, Geierstanger, Brittain, & Schultz, 2007; Rakauskaitė et al., 2015; Wu, Deiters, Cropp, King, & Schultz, 2004). Similar to the eGFP controls, expressing SelenoPOI-TAG (the SelenoPOI open reading frame with the codon TAG at the residue normally encoding Sec) in the presence and absence of DMNB-Sec can help to determine the efficiency of DMNB-Sec incorporation. Comparing the masses of irradiated and nonirradiated SelenoPOI expressed in the presence of DMNB-Sec is also recommended in order to test the efficiency of DMNB-Sec photo-uncaging in the SelenoPOI. Downstream assays of SelenoPOI function can also benefit from the use of SelenoPOI-DMNB-Sec as a negative control. The photocaged residue should be nonreactive, and therefore any Sec-dependent functions of the protein should be lost when the photocaging group remains intact. Additionally, the photocaging group should allow for temporal and spatial control over Sec reactivity and therefore SelenoPOI activity during *in vivo* assays of SelenoPOI function. The general method below can be utilized to express any and all of the above control proteins.

2.2 Key resources table

Reagent or resource	Source	Identifier
Chemicals, peptides, and recombinant proteins		
4,5-dimethoxy-2-nitrobenzyl-Sec (DMNB-Sec)	Synthesis in Rakauskaitė et al. (2015)	
Experimental models: cell lines		
HEK293	ATCC	CRL-1573
Recombinant DNA		
pAcBac2-EcLeuRS-BH5 T252A eGFP-39-TAG	Published in Peeler et al., (2020)	

2.3 Materials and equipment

Equipment

1. Tissue culture equipment including biosafety cabinet and incubator
2. Centrifuge

3. 365 nm lamp
4. UV/Vis spectrometer
5. Fluorimeter
6. Electrospray ionization–liquid chromatography–mass spectrometer (ESI-LC-MS)

Reagents

1. pAcBac2-EcLeuRS-BH5 T252A eGFP-39-TAG control plasmid. This ampicillin-resistant plasmid encodes the engineered *E. coli* leucyl RS/tRNA pair that has previously been utilized to incorporate caged amino acids in eukaryotic cells (Lemke et al., 2007; Rakauskaitė et al., 2015; Wu et al., 2004). Additionally, this plasmid contains the open reading frame for eGFP with the stop codon TAG at residue 39. eGFP truncated at residue 39 will not be fluorescent, while successful suppression of the TAG codon will lead to generation of full-length, fluorescent protein. This eGFP expression system serves as a helpful positive control that can be easily used to optimize conditions for DMNB-Sec incorporation and/or uncaging. Finally, this plasmid is capable of being utilized to generate baculovirus for infection of the target cell line instead of transfection, though this capability has not been utilized in this protocol.
2. pAcBac2-EcLeuRS-BH5 T252A *SelenoPOI*-TAG plasmid. The eGFP open reading frame from pAcBac2-EcLeuRS-BH5 T252A eGFP-39-TAG plasmid can be removed using SfiI cut sites on the 5′ and 3′ end of the open reading frame and replaced with the open reading frame for a selenoprotein of interest (SelenoPOI) via ligation or Gibson assembly. The SelenoPOI open reading frame should include: (1) a TAG codon mutation at the Sec residue (this will require mutation from the native Sec codon, TGA) and (2) an appropriate C-terminal epitope tag for purification of the protein. The use of a N-terminal epitope tag for purification will result in the purification of truncated gene product in addition to the full-length protein, while the use of a C-terminal tag will allow you to purify full-length protein. We have used a 6xHis epitope tag, though other tags can be used as needed.
3. pAcBac2-EcLeuRS-BH5 T252A *SelenoPOI*-Cys plasmid. We have also found it helpful to include a control where your SelenoPOI is expressed with a Cys residue in the place of Sec. Expression of this variant will bypass both the endogenous Sec insertion mechanism and the GCE strategy used for incorporation of DMNB-Sec. This protein variant can serve as a helpful positive control for purification and ESI-LC-MS-based detection of the SelenoPOI.

4. Cell line for expression. We recommend HEK293 cells for their ease of transfection and general success with GCE.
5. Transfection reagents. We have used polyethylenimine (PEI) MAX (Polysciences, 24765-1) prepared as a 1 mg/mL stock in water and stored at −80 °C, though other transfection reagents can also be used.
6. 4,5-Dimethoxy-2-nitrobenzyl-Sec (DMNB-Sec) (Rakauskaitė et al., 2015).
7. Sterile water.
8. 5 M NaOH.
9. PBS.
10. Cell solubilization reagents. We utilize CelLyticM (Millipore Sigma, C2978).
11. Protein purification reagents, based on the chosen epitope tag. We utilize Ni-NTA Resin (Thermo Scientific, 88221) in order to purify proteins utilizing the C-terminal 6xHis tag on eGFP and the selenoprotein MsrB1.
12. Reagents to measure protein concentration. We utilize Coomassie Protein Assay Reagent (Thermo Scientific, 1856209), though other reagents can also be used.

Alternatives: In addition to choosing your specific SelenoPOI, you should utilize the cell line, transfection reagents, and protein purification reagents that are best suited for your applications.

2.4 Step-by-step method details

2.4.1 Expression of SelenoPOI or eGFP containing DMNB-Sec

Timing: 72 h

1. Transfect target cell line with appropriate plasmid(s). The scale of transfection is dependent upon expression levels of the SelenoPOI and required amounts of protein for downstream assays. We transfect HEK293T cells using PEI MAX. One day before transfection, split cells into 10 cm tissue culture-treated dishes so that they will be 90% confluent on the day of transfection. On the day of transfection, prepare a transfection mixture of 210 μL serum-free DMEM, 12 μg plasmid DNA, and 50 μL of MAX PEI (from 1 mg/mL stock) per 10 cm plate. Incubate the reaction mixture for 20 min at room temperature. Gently add the transfection mixture to the cell media, trying to not disrupt the cells.
2. Add DMNB-Sec to growth media. DMNB-Sec should be stored as a solid, protected from light, and stock solutions should be prepared only immediately before addition to the growth media. Prepare a stock

solution of DMNB-Sec in sterile water and add 5 M NaOH dropwise until the solid is fully in solution. Gently add an appropriate volume of stock solution to the cell media after adding the transfection mixture. For a final concentration of 12.5 μM in a 10 cm dish containing 10 mL of media, add 1.25 μL of a 100 mM stock solution.

Note: We have utilized DMNB-Sec at final concentrations from 12.5 to 100 μM. Lower concentrations may decrease protein yields while higher concentrations may exhibit cell toxicity. Testing a dose curve of DMNB-Sec concentrations may be appropriate, and utilizing the eGFP-39TAG construct for such a dose curve allows for fluorescence-based assaying of expression level compared to DMNB-Sec concentration.

3. Grow cells for 48 h.

   Note: The length of expression may need to be optimized depending on transfection conditions and the specific SelenoPOI.

2.4.2 Uncaging of DMNB-Sec and harvesting of cells

Timing: 30 min

4. Expose cells to 365 nm light. After expression of the protein containing DMNB-Sec, expose cells to 365 nm light lamp for 10 min, keeping cells on ice to prevent overheating.

   Note: We have observed successful uncaging both by exposing cells to 365 nm light before purification as well as exposing purified protein to 365 nm light. In both cases we have observed complete uncaging after 10 min.

5. Harvest and solubilize cells. 48 h after transfection, and after uncaging, harvest cells by scraping. Wash cells with PBS and pellet by centrifugation.

   Pause point: The pellet can be frozen and stored at −80 °C for several weeks at this point.

2.4.3 Evaluation of incorporation and uncaging

Timing: Variable

Quick, simple, and quantifiable evaluation of successful GCE suppression of the TAG codon can be conducted utilizing eGFP fluorescence measurements of cell lysate. This approach is recommended for the first attempt at GCE as well as attempts at optimization (for example, using a new cell line or a new transfection reagent). In order to confirm that the TAG suppression was due to incorporation of DMNB-Sec and not one of the 20 canonical amino acids and furthermore to confirm successful uncaging of DMNB-Sec, eGFP or your SelenoPOI should be purified, and the intact protein mass should be assayed via ESI-LC-MS. General methods for each of these experiments is provided below.

6. Generate cell lysates by solubilizing the cell. While any solubilization strategy can be used, we use CelLyticM, according to manufacturer's instructions.
7. If using eGFP, monitor lysate fluorescence. Especially for the first attempt at GCE with DMNB-Sec, eGFP fluorescence in the cell lysate can be utilized to assay successful suppression of the TAG codon. Fluorescence can be measured with your preferred format and instrument, with excitation at 488 nm and emission at 509 nm. We find that expression in a single well of a 12-well or 6-well plate is sufficient to produce measurable quantities of eGFP. Literature reports have shown a fivefold increase in the fluorescence levels of lysates for cells grown in the presence of DMNB-Sec vs those grown in the absence of DMNB-Sec (Peeler et al., 2020; Rakauskaitė et al., 2015). After establishing or optimizing the protocol with the eGFP control, this step can be bypassed, and protein (eGFP or SelenoPOI) can be purified from lysates to evaluate via ESI-LC-MS.

   Note: Lysate protein concentration may also be monitored at this step, though we find that information on the net yield of eGFP is more relevant than measures of the ratio of eGFP to total protein. Overall yield of eGFP is a factor both of translational efficiency (which would theoretically increase with DMNB-Sec concentration) at cell survival (which would theoretically decrease with DMNB-Sec concentration if concentration reached the point of toxicity).
8. Purify protein and measure protein concentration. Purify SelenoPOI or the eGFP control protein from the cell lysates using a purification strategy compatible with the chosen C-terminal epitope tag. We utilize Ni-NTA resin according to the manufacturer's instructions. After purification, determine protein concentration. We utilize Coomassie Protein Assay Reagent according to manufacturer's instructions. Based on the protein concentration, determine the appropriate volume of protein to analyze by ESI-LC-MS. We have found that approximately 200 ng of protein is appropriate for loading, and this can be produced from five to ten 100 mm diameter plates of cells.

   Note: An SDS-PAGE gel can be run at this point to confirm purity of the sample if desired.

   Optional: Purified protein can be uncaged at this point following the methods detailed in Section 2.4, step 4.

9. Analyze intact purified proteins by ESI-LC-MS. To confirm both successful incorporation of DMNB-Sec and successful photo-uncaging to generate a Sec-containing protein, determine the mass of the intact purified protein sample via ESI-LC-MS. For the eGFP-39TAG-6xHis control, whole protein mass predictions should take into account chromophore formation, loss of N-terminal methionine, and acetylation. Therefore, the predicted mass for the caged eGFP-DMNB-Sec39 is 29,865.32 Da, and the predicted mass for the uncaged eGFP-Sec39 is 29,669.24 Da. For cells expressing eGFP-39TAG grown in the absence of DMNB-Sec, we expect very little protein expression since the majority of the gene product is truncated at residue 39. However, we have successfully purified small amounts of full-length protein from these samples via the C-terminal 6xHis epitope tag and analyzed the protein via ESI-LC-MS. The mass of this protein product matches the expected mass of eGFP with leucine (Leu) or isoleucine (Ile) incorporation at residue 39 (29,632.35 Da), which is expected since the substrate of the parent *E. coli* RS/tRNA pair is Leu. It should be noted that this mass is not observed in ESI-LC-MS analysis of eGFP-39TAG grown in the presence of DMNB-Sec, suggesting that the DMNB-Sec substrate successfully outcompetes Leu and that misincorporation of Leu or Ile is not a concern when DMNB-Sec is present in the growth media.

Note: When analyzing SelenoPOI products, the difference between DMNB-Sec-containing proteins and uncaged Sec-containing proteins should be 196.08 Da and the difference between Sec-containing proteins and Leu/Ile-containing proteins should be 36.89 Da.

Note: Previous reports have observed both the formation of diselenide-based dimers as well as dehydroalanine formation after DMNB-Sec uncaging in the yeast-based GCE system (Rakauskaitė et al., 2015). We have not observed the formation of these products for eGFP or MsrB1 in the human cell-based GCE system (Peeler et al., 2020). However, if any masses are detected that do not match the anticipated masses for eGFP or SelenoPOI containing DMNB-Sec or Sec, these transformations should be considered.

Critical: The use of MALDI instead of ESI for sample ionization will likely result in uncaging of the DMNB-Sec and therefore should be avoided.

10. If possible, assay protein function. If a biochemical function has been characterized for the SelenoPOI and an *in vitro* enzymatic assay has been established, testing the activity of the SelenoPOI after uncaging is recommended to ensure that incorporation of the DMNB-Sec residue has not irreversibly altered protein folding and therefore protein function.

2.5 Applications

The GCE-based approach described above allows for the expression of selenoproteins. After purification and confirmation of successful DMNB-Sec incorporation and uncaging, the purified protein can be used in *in vitro* assays of selenoprotein function. However, a great advantage of this specific method is that human selenoproteins can be expressed and assayed in their native cellular environments. Therefore, after confirming that the GCE approach is effective at producing a given SelenoPOI, assays of selenoprotein function can be performed in the cell culture model system (presumably with SelenoPOI knockout) by performing the transfection and uncaging steps presented above and leaving the cells intact. This approach should enable the study of the cell biological consequences of selenoprotein function, including comparison of the activity of mutants of the SelenoPOI.

Additionally, the caging group on DMNB-Sec and the ability to uncage the protein in cells via exposure to 365 nm light provides additional experimental functionality. First, assaying of the caged variant (which has not been exposed to 365 nm light) should provide a negative control for studies on the consequences of Sec reactivity and Sec-dependent protein functions in a cell biological assay. Furthermore, uncaging via 365 nm light allows for temporal as well as spatial control over Sec reactivity and Sec-dependent protein functions, which could be of great experimental value.

3. Conclusions and outlooks

Selenoprotein biology plays an important role in cellular redox homeostasis. However, the endogenous selenoprotein translation system makes experimental expression of selenoproteins challenging, and therefore our understanding of selenoprotein biology has been limited. In fact, nearly half of the 25 known human selenoproteins still do not have known functions. Here, we present a method for expressing human selenoproteins via GCE in human tissue culture cell lines. Utilization of this approach should promote the study of human selenoproteins, especially their cell biological functions. We anticipate that this technique will help to enable the study of selenoproteins and lead to better understanding of both selenoprotein biological function and the use of Sec vs its analogous amino acid Cys.

References

Chin, J. W. (2017). Expanding and reprogramming the genetic code. *Nature, 550*(7674), 53–60. https://doi.org/10.1038/nature24031.

Donovan, J., & Copeland, P. R. (2010). Threading the needle: Getting selenocysteine into proteins. *Antioxidants & Redox Signaling, 12*(7), 881–892. https://doi.org/10.1089/ars.2009.2878.

Fu, X., Söll, D., & Sevostyanova, A. (2018). Challenges of site-specific selenocysteine incorporation into proteins by Escherichia coli. *RNA Biology, 15*(4–5), 461–470. https://doi.org/10.1080/15476286.2018.1440876.

Goodsell, D. S., Autin, L., & Olson, A. J. (2019). Illustrate: Software for biomolecular illustration. *Structure (London, England: 1993), 27*(11). 1716–1720.e1 https://doi.org/10.1016/j.str.2019.08.011.

Ingold, I., Berndt, C., Schmitt, S., Doll, S., Poschmann, G., Buday, K., et al. (2018). Selenium utilization by GPX4 is required to prevent hydroperoxide-induced ferroptosis. *Cell, 172*(3), 409–412. https://doi.org/10.1016/j.cell.2017.11.048.

Labunskyy, V. M., Hatfield, D. L., & Gladyshev, V. N. (2014). Selenoproteins: Molecular pathways and physiological roles. *Physiological Reviews, 94*(3), 739–777. https://doi.org/10.1152/physrev.00039.2013.

Lemke, E. A., Summerer, D., Geierstanger, B. H., Brittain, S. M., & Schultz, P. G. (2007). Control of protein phosphorylation with a genetically encoded photocaged amino acid. *Nature Chemical Biology, 3*(12), 769–772. https://doi.org/10.1038/nchembio.2007.44.

Liu, J., Cheng, R., & Rozovsky, S. (2018). Synthesis and semisynthesis of selenopeptides and selenoproteins. *Current Opinion in Chemical Biology, 46*, 41–47. https://doi.org/10.1016/j.cbpa.2018.04.008.

Liu, J., Zheng, F., Cheng, R., Li, S., Rozovsky, S., Wang, Q., et al. (2018). Site-specific incorporation of selenocysteine using an expanded genetic code and palladium-mediated chemical deprotection. *Journal of the American Chemical Society, 140*(28), 8807–8816. https://doi.org/10.1021/jacs.8b04603.

Low, S. C., & Berry, M. J. (1996). Knowing when not to stop: Selenocysteine incorporation in eukaryotes. *Trends in Biochemical Sciences, 21*(6), 203–208. https://doi.org/10.1016/S0968-0004(96)80016-8.

Novoselov, S. V., Lobanov, A. V., Hua, D., Kasaikina, M. V., Hatfield, D. L., & Gladyshev, V. N. (2007). A highly efficient form of the selenocysteine insertion sequence element in protozoan parasites and its use in mammalian cells. *Proceedings of the National Academy of Sciences of the United States of America, 104*(19), 7857–7862. https://doi.org/10.1073/pnas.0610683104.

Peeler, J. C., & Weerapana, E. (2019). Chemical biology approaches to interrogate the selenoproteome. *Accounts of Chemical Research, 52*(10), 2832–2840. https://doi.org/10.1021/acs.accounts.9b00379.

Peeler, J. C., Falco, J. A., Kelemen, R. E., Abo, M., Chartier, B. V., Edinger, L. C., et al. (2020). Generation of recombinant mammalian selenoproteins through genetic code expansion with photocaged selenocysteine. *ACS Chemical Biology, 15*(6), 1535–1540. https://doi.org/10.1021/acschembio.0c00147.

Rakauskaitė, R., Urbanavičiūtė, G., Rukšėnaitė, A., Liutkevičiūtė, Z., Juškėnas, R., Masevičius, V., et al. (2015). Biosynthetic selenoproteins with genetically-encoded photocaged selenocysteines. *Chemical Communications, 51*(39), 8245–8248. https://doi.org/10.1039/C4CC07910H.

Reich, H. J., & Hondal, R. J. (2016). Why nature chose selenium. *ACS Chemical Biology, 11*(4), 821–841. https://doi.org/10.1021/acschembio.6b00031.

Welegedara, A. P., Adams, L. A., Huber, T., Graham, B., & Otting, G. (2018). Site-specific incorporation of selenocysteine by genetic encoding as a photocaged unnatural amino acid. *Bioconjugate Chemistry, 29*(7), 2257–2264. https://doi.org/10.1021/acs.bioconjchem.8b00254.

Wu, N., Deiters, A., Cropp, T. A., King, D., & Schultz, P. G. (2004). A genetically encoded photocaged amino acid. *Journal of the American Chemical Society, 126*(44), 14306–14307. https://doi.org/10.1021/ja040175z.

Xu, X.-M., Carlson, B. A., Mix, H., Zhang, Y., Saira, K., Glass, R. S., et al. (2007). Biosynthesis of selenocysteine on its tRNA in eukaryotes. *PLoS Biology*. https://doi.org/10.1371/journal.pbio.0050004.

CHAPTER SEVEN

Applying selenocysteine-mediated expressed protein ligation to prepare the membrane enzyme selenoprotein S

Rujin Cheng[a], Jun Liu[a,b], Vidyadhar Daithankar[a,c], and Sharon Rozovsky[a,*]

[a]Department of Chemistry and Biochemistry, University of Delaware, Newark, DE, United States
[b]Department of Pharmaceutical Chemistry, University of California, San Francisco, CA, United States
[c]Syva formulation, Siemens Healthineers, Newark, DE, United States
*Corresponding author: e-mail address: rozovsky@udel.edu

Contents

Methods in Enzymology, Volume 662
ISSN 0076-6879
https://doi.org/10.1016/bs.mie.2021.10.023

Abstract

The intrinsically disordered membrane-bound selenoprotein s (selenos) takes part in the protein quality control pathway, vesicle trafficking, and NF-kB signaling. The reactive selenocysteine (Sec) at the penultimate position is responsible for its enzymatic activity. We report the preparation of the soluble segment as well as the full-length selenos using expressed protein ligation. This chapter discusses the practical considerations of expressed protein ligation using selenopeptides and describes our optimized procedure for the semi-synthesis of selenos.

1. Introduction

Selenos (previously known as SelS, VIMP, and SEPS1 (Gladyshev et al., 2016)) is implicated in several human diseases, such as cardiovascular diseases and diabetes mellitus. Its best-understood function is in the ER-associated degradation (ERAD) pathway (Lilley & Ploegh, 2004; Ye, Shibata, Yun, Ron, & Rapoport, 2004), but it also has cellular roles related to vesicle trafficking and the maintenance of protein complexes (Turanov et al., 2014). In addition, it contributes to NF-kB signaling, and reduced expression of selenos leads to increased production of inflammatory cytokines (Capelle et al., 2021; Liu & Rozovsky, 2015).

Selenos contains the rare amino acid selenocysteine (Sec, U) in a disordered segment at position 188 out of 189. We have shown that selenos can act as a reductase (Liu, Li, & Rozovsky, 2013; Liu & Rozovsky, 2013), however, it is challenging to prepare selenos because selenoproteins' biosynthesis necessitates the co-translational insertion of Sec by recoding the UGA codon (Cheng, Liu, Wang, Forstner, & Rozovsky, 2021). During heterologous expression, the presence of the UGA codon can lead to premature termination of translation, resulting in truncated proteins without Sec and any succeeding amino acids. In the case of selenos, separating the full-length and truncated proteins is demanding, as they differ by only two amino acids. An alternative approach that circumvents these challenges is the semi-synthesis of selenos. Here, we describe the preparation of selenos with Sec at high yield and purity using expressed protein ligation (EPL). EPL and the closely related native chemical ligation (NCL) are methods used

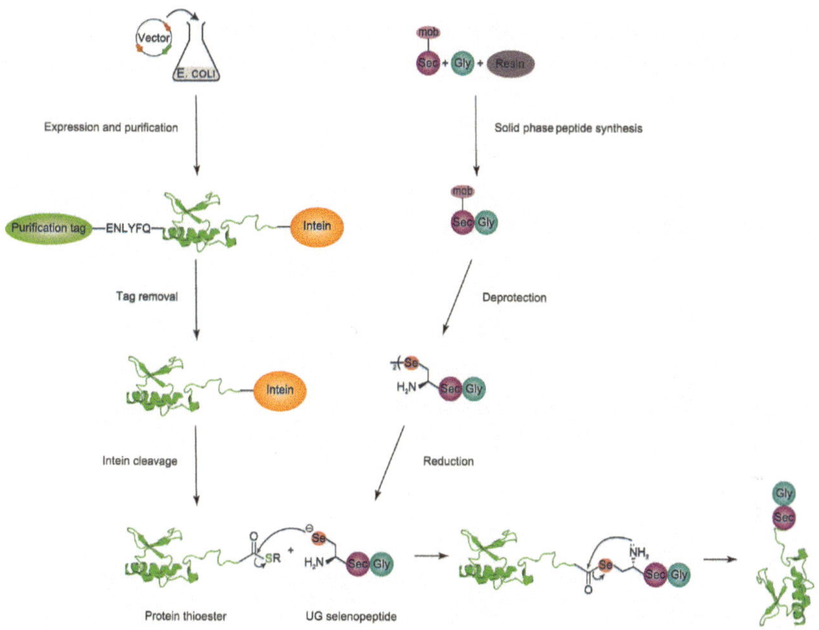

Fig. 1 Illustration of preparation of selenos by Sec mediated EPL using a selenos fragment with a C-terminus thioester and a selenopeptide UG with Sec at its N-terminal.

to prepare proteins via a chemical reaction to introduce amide bonds between respective protein fragments (Dawson & Kent, 2000; Dawson, Muir, Clark-Lewis, & Kent, 1994; Thompson & Muir, 2020) (Fig. 1).

Chemical ligations that exploit selenocysteine (Sec-NCL and Sec-EPL) have recently been reviewed (Conibear, Watson, Payne, & Becker, 2018; Liu, Cheng, & Rozovsky, 2018). Such ligations were previously used to generate different selenoproteins (Dery et al., 2017; Eckenroth et al., 2006; Eckenroth, Lacey, Lothrop, Harris, & Hondal, 2007; Liu, Chen, & Rozovsky, 2017; Mitchell et al., 2017). In the case of selenos, depending on the specific experiment, it is possible to employ its cytoplasmic, exposed soluble segment composed of residues 52-189 (abbreviated as cselenos), or the full-length enzyme (selenos residues 1-189). To prepare cselenos or selenos, a short UG peptide (selenos residues 188-189) needs to be ligated to a cselenos fragment (residues 52-187) or a selenos fragment (residues 1-187) with a C-terminus thioester. In the following, we describe the preparation of the selenopeptide UG, cselenos- and selenos-thioesters, and finally their ligation to produce cselenos and selenos.

2. Considerations for Sec-mediated EPL using selenopeptides

In the procedure we describe here, the N-terminal protein thioester is ligated with a peptide that carries an N-terminal Sec (Fig. 1). The most common side reaction is the hydrolysis of the protein-thioester, which will lead to the presence of unligated proteins. Hence, the reaction conditions are poised to minimize side reactions, such as hydrolysis of the thioester or elimination of selenium from Sec. By exploiting the high reactivity of Sec, it is possible to achieve high ligation efficiency with little hydrolysis—hence high purity (Liu et al., 2018). We discuss the procedure, potential issues that may arise in each step, and provide insights on how to address them best.

2.1 Selenopeptide preparation

Conventional solid phase peptide synthesis (SPPS) uses amine and side chain protected amino acids as building blocks (Amblard, Fehrentz, Martinez, & Subra, 2006). For Sec incorporation, SPPS requires both reagents to suppress Sec racemization and extensive protection of its side chain during coupling reactions. While there are multiple approaches to protecting Sec for SPPS (Fig. 2), Fmoc-(S)-4-methoxybenzyl selenocysteine is commercially available and thus frequently selected. When using Fmoc-(S)-4-methoxybenzyl selenocysteine, 2,2'-dithiobis (5-nitropyridine) (DTNP) is commonly used during removal of the p-methoxybenzyl protecting group, and the amount of DNTP and the duration of cleavage need to be carefully calibrated (Schroll, Hondal, & Flemer, 2012).

As an alternative to Fmoc-(S)-4-methoxybenzyl selenocysteine, it is possible to use selenocystine for selenopeptide synthesis. This approach bypasses Sec side-chain protection and deprotection during synthesis altogether. Selenocystine is commercially available, inexpensive, and can be used to generate bis-N-Fmoc selenocystine by a single reaction (Flemer, 2015). Synthesis using Bis-N-Fmoc-selenocystine does not involve extensive Sec-related protection and deprotection. Therefore, it does not require DTNP and has fewer side products after peptide cleavage. Fmoc-chloroalanine can also be used as the precursor of Sec (Aravindhan, Singh, Zeller, & Butcher, 2019). The selenium is introduced into the peptide after it is synthesized by converting chloroalanine to Sec using Na_2Se_2. The three most employed SPPS approaches are shown in Fig. 2. Regardless of the selected synthetic route, purification of the peptide before ligation is essential. It removes side products and enables accurate determination of the selenopeptides' solubility and quantity.

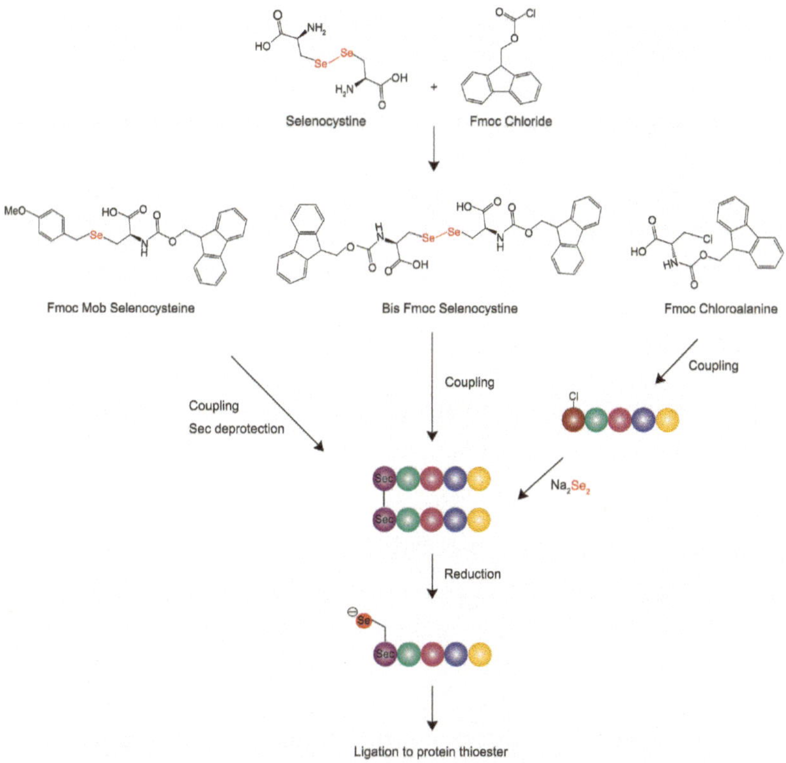

Fig. 2 Three approaches, using different building blocks for introducing Sec, for synthesizing selenopeptides. Left: Fmoc-(S)-4-methoxybenzyl selenocysteine, Sec is introduced with side chain protection. Middle: Bis-Fmoc-selenocystine, Sec is introduced without side chain protection. Right: Fmoc-chloroalanine, Sec is introduced by replacing a chloroalanine in the synthesized peptide. In this chapter, we describe SPPS with Fmoc-Mob-selenocysteine (left).

2.2 Preparation of cselenos- and selenos-thioester

In EPL, the protein thioester is prepared by fusing the desired protein fragment with the self-cleaving intein (Shah & Muir, 2014) (see Fig. 1). The intein-fused recombinant protein should be stable during expression and purification. Therefore, it may be necessary to use an additional stabilizing protein partner, like the maltose binding protein (MBP) or glutathione transferase (GST), placed at the protein N terminus. Following purification, the stabilizing protein partner can be cleaved off using an appropriate protease. The choice of intein is also vital for preventing *in vivo* cleavage and allowing efficient cleavage once the fragment is purified. Several cloning vectors for frequently employed inteins are available from New England Biolabs (NEB), therefore, it is worthwhile to test the purity and yield of the protein thioester generated using these different inteins.

Once the fragment-intein fusion is purified, intein cleavage is initiated by adding thiol reagents. This step is typically run on the chitin affinity column. Upon thioesterification (cleavage of the intein to generate the thioester), a stable protein thioester can be obtained with minimum hydrolysis. Here the pH is critical and should be maintained at 6.5 or lower to avoid thioester hydrolysis. At neutral and basic pH (7.0 or higher), protein thioesters not only react with the selenopeptide, but also with electrophiles such as OH^- which leads to hydrolysis. This rule also applies to storage and the ligation reaction. It can be challenging to separate the hydrolyzed protein from the desired product, particularly if the two are close in molecular weight and have similar surface charges. Therefore, the purity of the protein-thioester should be assessed for each preparation using intact protein mass spectrometry, and only hydrolysis-free samples should be selected for ligation.

2.3 Selecting the ligation conditions

With a pure selenopeptide and hydrolysis-free protein-thioester at hand, the ligation can be designed. We have already discussed the need to maintain the pH at the optimal 6.5 to minimize the thioester hydrolysis. In addition, since the ligation reaction is initiated by an attack of a selenol on a protein thioester, all diselenides in the selenopeptides need to be reduced and present in the selenol-free form. Sec-containing peptides form diselenide bonds and require strong reducing reagents due to their low redox potential (Gieselman, Xie, & van Der Donk, 2001; Santi, Tidei, Scalera, Piroddi, & Galli, 2013). However, the most often used reductants, tris(2-carboxyethyl) phosphine (TCEP) and tris(hydroxypropyl)phosphine (THP), were reported to lead to selenium elimination. Thus, low concentrations should be selected to avoid the conversion of selenocysteine to dehydroalanine (Dery et al., 2015). THP is an excellent reagent because its pH does not need to be adjusted (most impurities in TCEP preparations are phosphate ions). However, TCEP at a 25mM or lower concentration is also an excellent reducing agent (if its pH is carefully adjusted).

The ideal ratio of protein-thioester to selenopeptide depends on the specific ligation. We have seen examples where five equivalents of selenopeptides to the protein thioesters sufficed for complete ligation. Therefore, it is worth comparing pilot ligation with different molar peptide ratios to protein thioesters and evaluating their yield and purity using intact protein mass spectrometry. However, if optimization is not desired, we recommend using selenopeptide concentration in the range of 1–10mM, at a 20–40 equivalent of the protein thioester concentration.

The concentration of the reducing agent should be kept between 10 and 20-fold to that of the selenopeptide.

Following the ligation, the preferred analytical method to characterize the ligated selenoprotein is mass spectrometry, because of its ability to detect thioesters, hydrolysis products, Sec oxidation, and the presence of mixed S-Se bonds due to thiol reagents. Before analysis, the ligated selenoprotein is reduced and alkylated with N-ethylmaleimide (NEM), or a similar reducing reagent, to prevent diselenide bond formation during analysis. The presence of the alkylating agent (mass adduct) is also used to confirm the presence of Sec. The choice of suitable alkylating reagents and the respective alkylation conditions is discussed at length in this review (Hansen & Winther, 2009). The presence of selenium in the ligated product should also be established by intact protein mass spectrometry, Tandem MS sequencing, or inductively coupled plasma mass spectrometry (ICP-MS).

3. Materials

3.1 Reagents for molecular cloning

1. The expression vector pTWIN1 (NEB). TWIN1 was designed to produce fusion proteins with C-terminal Mxe GyrA intein and a chitin binding domain (CBD) from *Bacillus circulans* to facilitate purifications.
2. The gene of *Homo sapiens* selenos (UniProtKB: Q9BQE4) was codon-optimized for *E. coli* expression and synthesized by GenScript. Selenos was cloned into the pMAL-C5X vector (NEB) as a fusion with an N-terminal MBP using BbvcI and NcoI restriction sites (sequences optimized for expression in *E. coli* are provided below, and additional details can be found in the references (Liu et al., 2013; Liu, Srinivasan, Pham, & Rozovsky, 2012)). A TEV protease site was inserted at the C-terminus of MBP to allow its removal. MBP-TEV-selenos 1-187 was used as a template to generate the soluble cselenos construct, MBP-TEV-selenos 52-187, by deleting selenos 1-51. MBP-TEV-selenos 1-187 and MBP-TEV-selenos 52-187 are cloned into the pTWIN1 vector (NEB) using NdeI and SapI cloning sites.
3. Restriction enzymes Ndel, SapI, and Salt-T4® DNA Ligase are from NEB.
4. Q5™ site-directed mutagenesis kit is from NEB.
5. Primers for cloning are from MilliporeSigma.

MBP Ndel forward primer: GCCGCATATGAAAATCCATCACCATCACCATCACGAAG

Selenos 187 SapI reverse primer: ATATATGCTCTTCCGCAGCCGCCTGACGACGGACCACGACG

DNA sequence for MBP-TEV-selenos 52-187:

ATGAAAATCCATCACCATCACCATCACGAAGAAGGTAAACTGGTAATCTGGATTAACGGCG
ATAAAGGCTATAACGGTCTCGCTGAAGTCGGTAAGAAATTCGAGAAAGATACCGGAATTAA
AGTCACCGTTGAGCATCCGGATAAACTGGAAGAGAAATTCCCACAGGTTGCGGCAACTGG
CGATGGCCCTGACATTATCTTCTGGGCACACGACCGCTTTGGTGGCTACGCTCAATCTGG
CCTGTTGGCTGAAATCACCCCGGACAAAGCGTTCCAGGACAAGCTGTATCCGTTTACCTG
GGATGCCGTACGTTACAACGGCAAGCTGATTGCTTACCCGATCGCTGTTGAAGCGTTATCG
CTGATTTATAACAAAGATCTGCTGCCGAACCCGCCAAAAACCTGGGAAGAGATCCCGGCG
CTGGATAAAGAACTGAAAGCGAAAGGTAAGAGCGCGCTGATGTTCAACCTGCAAGAACCG
TACTTCACCTGGCCGCTGATTGCTGCTGACGGGGGTTATGCGTTCAAGTATGAAAACGGC
AAGTACGACATTAAAGACGTGGGCGTGGATAACGCTGGCGCGAAAGCGGGTCTGACCTTC
CTGGTTGACCTGATTAAAAACAAACACATGAATGCAGACACCGATTACTCCATCGCAGAAG
CTGCCTTTAATAAAGGCGAAACAGCGATGACCATCAACGGCCCGTGGGCATGGTCCAACA
TCGACACCAGCAAAGTGAATTATGGTGTAACGGTACTGCCGACCTTCAAGGGTCAACCATC
CAAACCGTTCGTTGGCGTGCTGAGCGCAGGTATTAACGCCGCCAGTCCGAACAAAGAGCT
GGCAAAAGAGTTCCTCGAAAACTATCTGCTGACTGATGAAGGTCTGGAAGCGGTTAATAAA
GACAAACCGCTGGGTGCCGTAGCGCTGAAGTCTTACGAGGAAGAGTTGGTGAAAGATCCG
CGTATTGCCGCCACTATGGAAAACGCCCAGAAAGGTGAAATCATGCCGAACATCCCGCAG
ATGTCCGCTTTCTGGTATGCCGTGCGTACTGCGGTGATCAACGCCGCCAGCGGTCGTCAG
ACTGTCGATGAAGCCCTGAAAGACGCGCAGACTAATTCGAGCTCTGAAAACCTGTATTTTC
AGTCCTCGTCTGCGCGCGCTGCGTCAGCGTCAACTGGATCGTGCGGCGGCGGCGGTGGA
ACCGGACGTCGTGGTTAAACGTCAGGAAGCCCTGGCGGCGGCACGTCTGAAAATGCAGG
AAGAACTGAACGCACAAGTCGAAAAACACAAGGAAAAACTGAAGCAGCTGGAAGAAGAAA
AACGTCGCCAAAAGATTGAAATGTGGGATTCAATGCAGGAAGGCAAATCGTATAAGGGTAA
CGCTAAAAAGCCGCAAGAAGAAGATTCTCCGGGTCCGAGCACCAGCAGCGTGCTGAAGC
GTAAGTCCGACCGTAAACCGCTGCGTGGCGGTGGCTACAATCCGCTGAGCGGTGAAGGT
GGCGGTGCGTGTTCTTGGCGTCCGGGTCGTCGTGGTCCGTCGTCAGGCGGC

(Black: MBP; Orange: linker; Red: TEV cleavage site; Blue: selenos 52-187)

ATGAAAATCCATCACCATCACCATCACGAAGAAGGTAAACTGGTAATCTGGATTAACGGCG
ATAAAGGCTATAACGGTCTCGCTGAAGTCGGTAAGAAATTCGAGAAAGATACCGGAATTAA
AGTCACCGTTGAGCATCCGGATAAACTGGAAGAGAAATTCCCACAGGTTGCGGCAACTGG
CGATGGCCCTGACATTATCTTCTGGGCACACGACCGCTTTGGTGGCTACGCTCAATCTGG
CCTGTTGGCTGAAATCACCCCGGACAAAGCGTTCCAGGACAAGCTGTATCCGTTTACCTG
GGATGCCGTACGTTACAACGGCAAGCTGATTGCTTACCCGATCGCTGTTGAAGCGTTATCG
CTGATTTATAACAAAGATCTGCTGCCGAACCCGCCAAAAACCTGGGAAGAGATCCCGGCG
CTGGATAAAGAACTGAAAGCGAAAGGTAAGAGCGCGCTGATGTTCAACCTGCAAGAACCG
TACTTCACCTGGCCGCTGATTGCTGCTGACGGGGGTTATGCGTTCAAGTATGAAAACGGC
AAGTACGACATTAAAGACGTGGGCGTGGATAACGCTGGCGCGAAAGCGGGTCTGACCTTC
CTGGTTGACCTGATTAAAAACAAACACATGAATGCAGACACCGATTACTCCATCGCAGAAG
CTGCCTTTAATAAAGGCGAAACAGCGATGACCATCAACGGCCCGTGGGCATGGTCCAACA
TCGACACCAGCAAAGTGAATTATGGTGTAACGGTACTGCCGACCTTCAAGGGTCAACCATC
CAAACCGTTCGTTGGCGTGCTGAGCGCAGGTATTAACGCCGCCAGTCCGAACAAAGAGCT
GGCAAAAGAGTTCCTCGAAAACTATCTGCTGACTGATGAAGGTCTGGAAGCGGTTAATAAA
GACAAACCGCTGGGTGCCGTAGCGCTGAAGTCTTACGAGGAAGAGTTGGTGAAAGATCCG
CGTATTGCCGCCACTATGGAAAACGCCCAGAAAGGTGAAATCATGCCGAACATCCCGCAG
ATGTCCGCTTTCTGGTATGCCGTGCGTACTGCGGTGATCAACGCCGCCAGCGGTCGTCAG
ACTGTCGATGAAGCCCTGAAAGACGCGCAGACTAATTCGAGCTCTGAAAACCTGTATTTTC
AGTCCATGGAACGCCAAGAAGAATCGCTGTCCGCCCGCCCGGCACTGGAAACCGAAGGT
CTGCGCTTTCTGCATACCACGGTCGGCAGTCTGCTGGCCACCTATGGTTGGTACATTGTGT
TTTCATGCATCCTGCTGTATGTGGTTTTCCAGAAACTGAGCGCTCGTCTGCGCGCGCTGCG
TCAGCGTCAACTGGATCGTGCGGCGGCGGCGGTGGAACCGGACGTCGTGGTTAAACGTC
AGGAAGCCCTGGCGGCGGCACGTCTGAAAATGCAGGAAGAACTGAACGCACAAGTCGAA
AAACACAAGGAAAAACTGAAGCAGCTGGAAGAAGAAAAACGTCGCCAAAAGATTGAAATGT
GGGATTCAATGCAGGAAGGCAAATCGTATAAGGGTAACGCTAAAAAGCCGCAAGAAGAAG
ATTCTCCGGGTCCGAGCACCAGCAGCGTGCTGAAGCGTAAGTCCGACCGTAAACCGCTGC
GTGGCGGTGGCTACAATCCGCTGAGCGGTGAAGGTGGCGGTGCGTGTTCTTGGCGTCCG
GGTCGTCGTGGTCCGTCGTCAGGCGGC

(Black: MBP; Orange: linker; Red: TEV cleavage site; Blue: selenos 1-187)

3.2 Reagents for protein expression and purification

1. LB medium: 10 g/L tryptone, 5 g/L yeast extract, 5 g/L sodium chloride (NaCl), 2 g/L glucose. Autoclave the medium and add antibiotics ampicillin to 100 μg/mL after it is cooled to room temperature.
2. LB agar plate: 10 g/L tryptone, 5 g/L yeast extract, 10 g/L NaCl, 15 g/L agar. Autoclave the growth medium and add 100 μg/mL of ampicillin after it has cooled below 40 °C but before solidification. Pour LB agar into individual Petri dishes, seal the plate with Parafilm™ after complete solidification, and store at 4 °C for up to 1 week.
3. Lysis buffer: 50 mM sodium phosphate, 200 mM NaCl, 2 mM ethylenediaminetetraacetic acid (EDTA), 1 mM phenylmethylsulfonyl fluoride (PMSF), 1 mM benzamidine, pH 7.5. For purifications of full-length selenos, include 0.2% Triton X-100. Protease inhibitors and detergents should be added just before lysis.
4. Amylose binding buffer: 50 mM sodium phosphate, 200 mM NaCl, 2 mM EDTA, pH 7.5. For purifications of full-length selenos, include 0.067% n-dodecyl-β-D-maltoside (DDM). Detergents should be added just before use.
5. Amylose elution buffer: 50 mM sodium phosphate, 200 mM NaCl, 2 mM EDTA, 20 mM maltose, pH 7.5. For purifications of full-length selenos, include 0.067% DDM. Detergents should be added just before use.
6. Chitin binding buffer: 25 mM 2-(*N*-morpholino)ethanesulfonic acid (MES), 200 mM NaCl, 2 mM EDTA, pH 6.5. For purifications of full-length selenos, include 0.067% DDM. Detergents should be added just before use.
7. Chitin cleavage buffer: 25 mM MES, 200 mM NaCl, 75 mM sodium 2-mercaptoethanesulfonate (MESNA), 2 mM EDTA, pH 6.5. For purifications of full-length selenos, include 0.067% DDM. Detergents should be added just before use.
8. Isopropyl β-D-1-thiogalactopyranoside (IPTG).
9. Amylose resin and chitin resin are from NEB.
10. Tobacco etch virus (TEV) protease is from NEB or made in house following the procedure described by Fox et. al. (Blommel & Fox, 2007).

3.3 Materials for selenopeptide synthesis, purification, and quantification

1. Dichloromethane (DCM)
2. Dimethylformamide (DMF)

3. Acetonitrile (ACN)
4. *N,N*-Diisopropylethylamine (DIEA, or DIEPA)
5. Piperidine
6. 1-Hydroxy-7-azabenzotriazole (HOAT)
7. 2,4,6-Trimethylpyridine (collidine, or TMP)
8. 1-[Bis(dimethylamino)methylene]-1H-1,2,3-triazolo[4,5-b] pyridinium-3-oxide hexafluorophosphate (HATU)
9. Trifluoroacetic acid (TFA)
10. 2,2'-Dithiobis (5-nitropyridine) (DTNP)
11. Triisopropyl silane (TIPS)
12. 2-Chlorotrityl chloride resin from Chem-Impex
13. Fmoc-glycine from Chem-Impex
14. Fmoc-(S)-4-methoxybenzyl selenocysteine from MilliporeSigma
15. Deuterated hydrochloric acid (DCl)
16. Deuterated water (D_2O)
17. Sodium acetate
18. Fritted reaction vessels from Image Molding
19. Rotator mixer
20. Vacuum pump compatible with organic solvents
21. SpeedVac vacuum concentrator
22. NMR sample tubes
23. NMR spectrometer and probe with ^1H capabilities

3.4 Reagents for expressed protein ligation

1. Ligation buffer: 25 mM MES, 200 mM NaCl, 1 M MESNA, pH 6.5. For purifications of full-length selenos, include 0.067% DDM. Detergents should be added just before use.
2. Tris(2-carboxyethyl)phosphine (TCEP).
3. pH meter with a semimicro electrode.
4. 1,4-dithio-DL-threitol (DTT).
5. HiTrap™ desalting column from Cytiva.
6. Storage buffer: 50 mM sodium phosphate, 200 mM NaCl, pH 7.5. For purifications of full-length selenos, include 0.067% DDM. Detergents should be added just before use.

3.5 Materials for characterization by mass spectrometry

1. *N*-ethylmaleimide (NEM).
2. Zeba™ desalt spin columns from Pierce.

3. Trypsin-ultra at a mass spectrometry grade from NEB.
4. Trypsin digestion buffer: 50 mM Tris-hydrochloride and 20 mM calcium chloride, pH 8.0. For purifications of full-length selenos include 0.067% DDM in the digestion buffer.
5. Mass spectrometry buffer: 50 mM ammonium bicarbonate, pH 8.0. For membrane protein, include 2 CMC DDM (Laganowsky, Reading, Hopper, & Robinson, 2013).
6. Mass spectrometry autosampler vials from Waters.
7. Mass spectrometer capable of acquiring intact protein spectra and tandem MS, such as a Waters Xevo G2-S QTof and Thermo Scientific Orbitrap.

3.6 Materials for characterization by circular dichroism spectroscopy

1. CD buffer: 10 mM potassium phosphate, 50 mM sodium sulfate (Na_2SO_4), pH 7.5. For characterization of full-length selenos, include 0.067% DDM. Detergents should be added just before use.
2. Circular dichroism spectropolarimeter.

4. Protocol

4.1 Cloning and expression of MBP-TEV-cselenos (52-187)-GyrA and MBP-TEV-selenos (1-187)-GyrA

1. Clone MBP-TEV-cselenos (52-187) and MBP-TEV-selenos (1-187) into vector pTWIN1 using restriction sites NdeI and SapI to create the vector pTWIN-MBP-TEV-cselenos (1-187)-GyrA and pTWIN-MBP-TEV-cselenos (52-187)-GyrA for the expression of intein fusion protein MBP-TEV-cselenos (52-187)-GyrA and MBP-TEV-cselenos (1-187)-GyrA.
2. Transform pTWIN-MBP-TEV-cselenos (52-187)-GyrA and pTWIN-MBP-TEV-cselenos (1-187)-GyrA expression vector into *E. coli* strain BL21(DE3). Plate the transformed cells evenly onto LB agar plates supplemented with 100 μg/mL ampicillin and incubate overnight at 37 °C.
3. Pick up colonies from the plates to grow overnight in 200 mL LB medium supplemented with 100 μg/mL ampicillin at 37 °C.
4. Transfer 10 mL of overnight culture to 1 L fresh LB medium supplemented with 100 μg/mL ampicillin. Grow *E. coli* cells while shaking at 250 rpm at 37 °C.

5. After 2–3 h, measure the growth medium's optical density (OD) at 600 nm wavelength. When the OD reaches 0.5, lower the incubation temperature to 18 °C.
6. Allow the cells to recuperate at 18 °C for 30 min, add 0.5 mM IPTG to the growth medium to induce expression at 18 °C. The OD at 600 nm should not exceed 0.8 when IPTG is added. Grow cells overnight at 18 °C while shaking at 250 rpm.
7. Harvest cells after 18–20 h by centrifugation at 5,000 × *g* at 4 °C for 20 min. Decant the supernatant.
8. Resuspend the cell pellet in lysis buffer. Flash-freeze the cell pellets in liquid nitrogen and store them at –80 °C or move on to purification immediately.

4.2 Purification of cselenos (52–187) thioester and selenos (1–187) thioester

The purification of cselenos (52-187) thioester and selenos (1-187) thioester is carried out using two chromatographic steps (Fig. 3). First, the MBP-TEV-cselenos (52-187)-GyrA and MBP-TEV-selenos (1-187)-GyrA fusion proteins are purified by amylose affinity chromatography through MBP binding. Then, the fusion partner MBP is cleaved off by TEV protease. The cleavage mixture is then incubated with chitin beads, allowing cselenos (52-187)-GyrA and selenos (1-187)-GyrA immobilization through the chitin binding domain. MBP and TEV protease do not bind the chitin resin and hence can be readily removed. After the resin is washed to remove non-bound proteins, the GyrA intein is cleaved off by incubating the resin in a MESNA-containing buffer, releasing the cselenos (52-187) thioester and selenos (1-187) thioester. The proteins with thioesters are collected in the column flow-through by washing the chitin column a second time. Procedures for cselenos and selenos preparation are similar except that the buffer for selenos contains detergents throughout the process.

1. Lyse the cells using a high-pressure homogenizer (e.g., Microfluidizer® high shear fluid processor from Microfluidics), operating continuously at 9000 psi for at least 5 cycles. Keep the cell lysis samples, buffers, and homogenizer at 4 °C (e.g., on ice) to minimize heat and thus protein denaturation.
2. Immediately remove cell debris by centrifugation at 12,000 × *g* at 4 °C for 30 min. Collect the supernatant.
3. Carry out amylose purification at 4 °C. See *Note 1* for guidelines on how to estimate the amylose resin binding capacity. Equilibrate the amylose columns with 5 column volumes (CV) of amylose binding

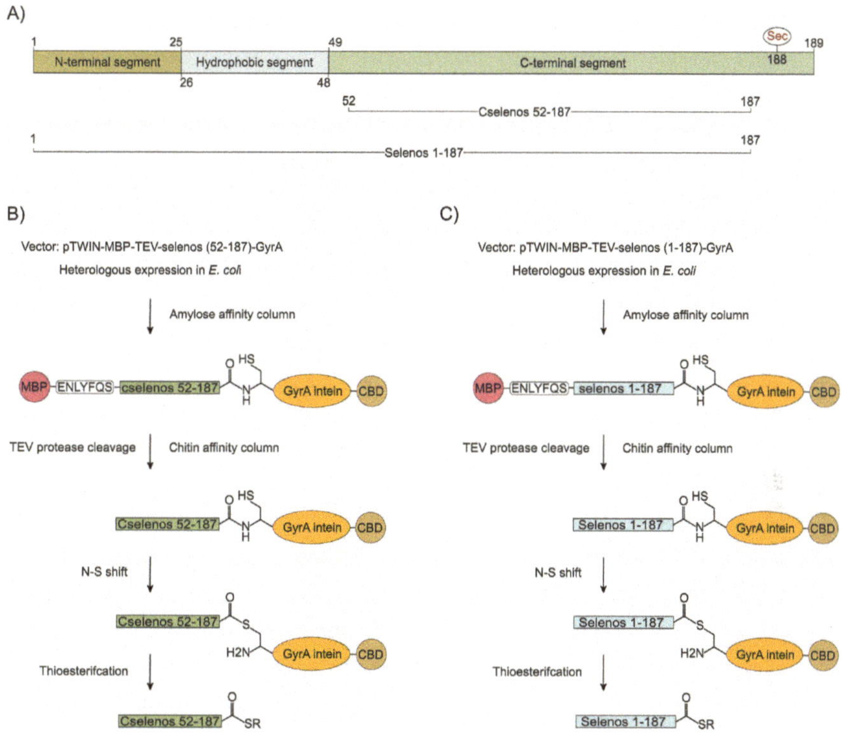

Fig. 3 Schematics illustrating selenos domain structure (in A) and the preparation of protein thioesters for cselenos (52-187) (in B) and selenos (1-187) (in C). MBP-TEV-cselenos (52-187)-GyrA and MBP-TEV-selenos (1-187)-GyrA were cloned into the pTWIN vector and over-expressed in *E. coli*. The recombinant proteins are purified using amylose affinity chromatography followed by TEV protease cleavage and chitin affinity chromatography. Protein thioesters are released from the chitin resin after intein cleavage. A detergent is required throughout the purification of selenos (1-187) thioester (in C) for solubility and stabilization.

buffer. Load the supernatant onto the column by gravity. Collect and save the flow-through until purification is finalized in case binding was not complete.

4. Wash the amylose column with 10 CV of amylose binding buffer.
5. Elute with 3 CV of amylose elution buffer and collect the eluent fusion protein MBP-TEV-cselenos (52-187)-GyrA and MBP-TEV-selenos (1-187)-GyrA.
6. Measure the concentration of eluent fusion proteins (e.g., by BCA assay).
7. Perform TEV cleavage of MBP-TEV-cselenos (52-187)-GyrA and MBP-TEV-selenos (1-187)-GyrA by adding the desired amount of TEV protease to the target protein in the elution buffer. The molar ratio of TEV protease to fusion protein is around 1:20. The TEV cleavage

reaction should proceed at 4 °C for 12–14 h while gently stirring. Confirm that cleavage is complete by analyzing the cleavage mixture by electrophoresis using a 10% Tris-Glycine gel. Once the TEV cleavage is complete, the sample is clarified by centrifuging the mixture at 20,000 × *g* for 10 min or filtration using a 0.22 μm membrane syringe filter.

8. Equilibrate chitin column with 5 CV of chitin loading buffer. Load the column with the TEV cleavage mixture and collect the flow-through containing MBP and TEV. See *Note 2* for guidelines on how to estimate the chitin resin binding capacity.
9. Wash the chitin column with 10 CV of chitin loading buffer.
10. Wash the chitin columns with 3 CV of chitin cleavage buffer. Cap the column when the level of the cleavage buffer is about 3 cm above the resin bed. Allow intein cleavage to proceed at 25 °C overnight.
11. Elute the cselenos (52-187) thioester or selenos (1-187) thioester with 2 CV of chitin binding buffer.
12. Apply 3 CV of the chitin elution buffer and collect the elution. This buffer contains MESNA and will be used to continue the cleavage of any uncleaved fusion proteins still bound to the resin.
13. Repeat steps 10 to 12 to fully cleave any cselenos (52-187)-GyrA and selenos (1-187)- GyrA still bound to the chitin column to obtain the highest yield of the thioester (see *Note 3* regarding the stability of the thioester).
14. Analyze fractions from all purification steps by SDS-PAGE. A typical purification of cselenos (52-187) thioester is shown in Fig. 4.
15. Determine the purity of the cselenos (52-187) thioester (Fig. 5A) or selenos (1-187)-GyrA (Fig. 6A) using intact protein mass spectrometry (additional instructions are provided in Section 5.1).
16. Measure the protein concentration by BCA assay (see Note 1 and Note 2 for typical yield).
17. Unless the ligation follows immediately, flash freeze cselenos (52-187) and selenos (1-187) thioesters in liquid nitrogen and keep the protein thioesters at -80 °C.

4.3 Synthesis and purification of the UG selenopeptide

1. Place 250 mg of 2-chlorotrityl chloride resin in a SPPS reaction vessel. 250 mg of trityl-based resin can support 0.25–0.5 mmol scale synthesis of the UG selenopeptide.
2. Fill the vessel with DCM until the resin is immersed. Swell the resin on a rotator mixer for 30 min at room temperature and then remove DCM by filtration under a vacuum in a chemical fume hood. The solvent is

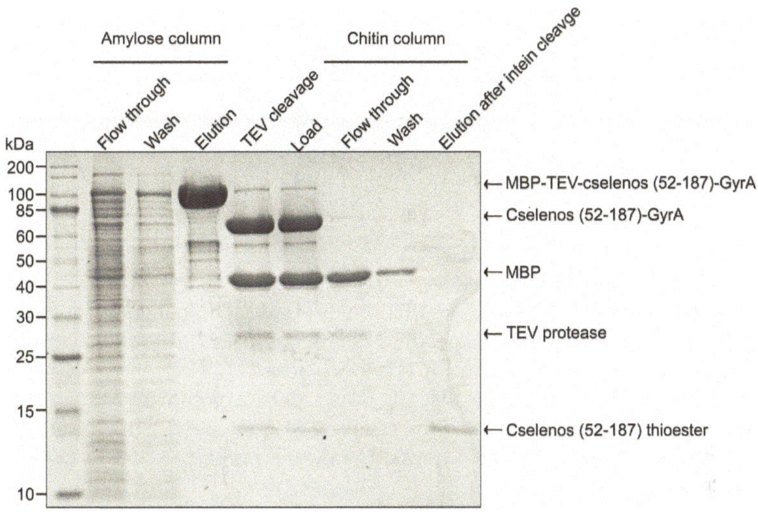

Fig. 4 Purification of cselenos (52-187) thioester characterized by gel electrophoresis. The lanes from left to right contain the following: protein molecular weight marker, amylose column flow-through, amylose column wash fraction, amylose column elution fraction, TEV protease cleavage mixture, supernatant of TEV protease cleavage mixture (loaded to chitin column), chitin column flow-through, chitin column wash fraction, and chitin column elution fraction after incubation with MESNA to cleave the intein. The elution after intein cleavage has a single band corresponding to the purified cselenos (52-187) thioester.

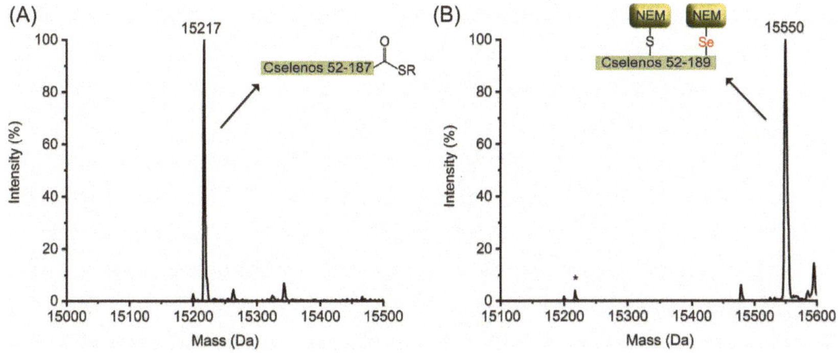

Fig. 5 Characterization of cselenos (52-187) thioester and the ligation product cselenos. (A) shows the deconvoluted ESI-MS spectrum of cselenos (52-187) thioester derived from GyrA intein. The observed mass is 15217 Da (calculated 15217 Da). (B) shows the deconvoluted ESI-MS spectrum of cselenos (52-189) ligated from UG with the cselenos (52-187) thioester in (A). The observed mass is 15550 Da (calculated 15550 Da). Asterisk (*) indicates unreacted cselenos (52-187) thioester at 15217 Da.

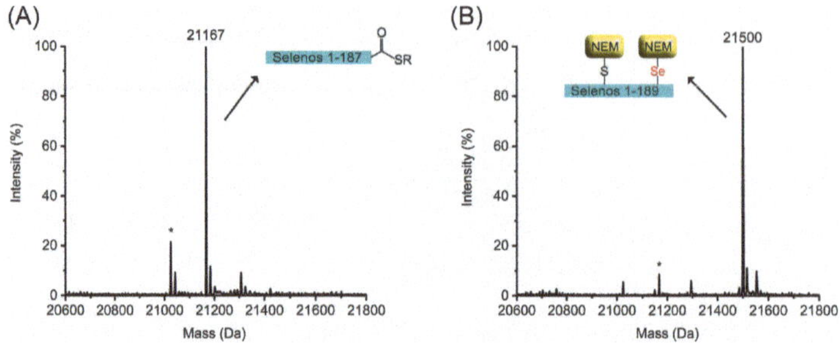

Fig. 6 Characterization of selenos (1-187) thioester and the ligation product selenos. (A) shows deconvoluted ESI-MS spectrum of selenos (1-187)-thioester derived from GyrA intein. The observed mass is 21167 Da (calculated 21167 Da). Asterisk (*) indicates hydrolyzed selenos (1-187) thioester. (B) shows deconvoluted ESI-MS spectrum of selenos (1-189) ligated from UG and the selenos (1-187)-thioester. The observed mass is 21500 Da (calculated 21500 Da). Asterisk (*) indicates unreacted selenos (1-187) thioester at 21167 Da.

drained from the reaction vessel using a side-arm flask with the side-arm connected to a vacuum pump.

3. Weigh 0.75 mmol of Fmoc-glycine and dissolve it in 3 mL of DMF with 266.5 μL DIEA (the concentration of DIEA in DMF is 8–20%) in a 15 mL disposable centrifuge tube. Activate the Fmoc-glycine by vortexing the solution for 1 min at room temperature.
4. To couple the first amino acid (here Fmoc-glycine), add the Fmoc-glycine/DMF/DIEA solution to the resin, gently mix the suspension using a rotator mixer for 1.5 h at room temperature. During SPPS, coupling the first amino acid differs from that of subsequent amino acids, see *Note 4*.
5. Wash the resin three times with 5 mL DCM. Drain the solvent by filtration using the vacuum pump.
6. To deprotect the first amino acid (here Fmoc-glycine), prepare a 20% piperidine/DMF deprotection solution. Transfer 4 mL of the deprotection solution into the reaction vessel. Gently mix using a rotator mixer for 1.5 h at room temperature. See *Note 5* about determining the duration of the reaction.
7. Drain the deprotection solution by filtration using the vacuum pump. Wash the resin three times with 5 mL DMF. Remove the DMF using the vacuum pump.
8. Weigh 0.75 mmol of Fmoc-(*S*)-4-methoxybenzyl selenocysteine and 0.75 mmol of HATU, and dissolve in 3 mL of 1:1 DMF/DCM in a 15 mL disposable centrifuge tube. Add 0.75 mmol of HOAT and

0.75 mmol TMP. Activate the amino acid by vortexing the solution for 1 min at room temperature.

9. To couple the second amino acid (here Fmoc-(*S*)-4-methoxybenzyl selenocysteine), add the Fmoc-(*S*)-4-methoxybenzyl selenocysteine/HOAT/TMP/DMF/DCM solution to the resin and allow it to mix using a rotator mixer for 2 h at room temperature.
10. Wash the resin three times with 5 mL DCM. Drain the solvent using the vacuum pump.
11. To deprotect the second amino acid, transfer 4 mL of the 20% piperidine/DMF deprotection solution into the reaction vessel. Gently mix it using a rotator mixer for 1.5 h at room temperature.
12. Drain the deprotection solution.
13. Wash the resin three times with 5 mL DMF. Drain solvent under vacuum. Wash with three times with 5 mL DCM and drain. Wash three times with 5 mL DMF, drain, and allow the resin to dry. Store the resin with bound peptide at -80 °C unless you proceed immediately to purification.
14. To cleave the selenopeptide off the resin, prepare 10 mL TFA cleavage cocktail consisting of 90% TFA, 5% H_2O, and 5% TIPS, add the TFA cleavage cocktail and 0.25–0.5 mmol DTNP (1–2 equiv.) to the dried resin. Gently agitate it on a rotator mixer for 3 h at room temperature. *Note 6* explains how much TFA should be used for cleavage and additional considerations about the DTNP concentration.
15. Filter the cleavage mixture into a 15 mL disposable centrifuge tube.
16. Remove the TFA as much as possible by purging with nitrogen or using a SpeedVac vacuum concentrator to facilitate its evaporation. For an alternative procedure, see *Note* 7.
17. Purify the UG peptide using hydrophilic interaction liquid chromatography (HILIC). Inject the crude products to HPLC-SQD2 MS (Waters) system using a BEH Amide OBD preparative column with a 4 min linear gradient from 10% H_2O (90% ACN) to 60% H_2O (40% ACN). The UG monomer elutes at about 1.2 min (m/z = 226), while the UG dimer elutes at about 1.5 min (m/z = 225). Aim for purifying the UG dimer, see *Note 8*.
18. Vacuum dry the purified UG peptide using a SpeedVac vacuum concentrator. Then, proceed to peptide solubilization and quantification for ligation. If the peptide is not immediately used for the ligation reaction, store the UG peptide powder at –20 °C.

4.4 NMR quantification of selenopeptide purity and concentration

1. To determine the concentration and purity of the UG dimer, dissolve the purified UG peptide powder in a minimum volume of deuterated hydrochloric acid (DCl).
2. Dissolve sodium acetate in deuterated water (D_2O) at a concentration of 10 mM.
3. Mix 20 μL of the solubilized UG peptide with 480 μL of 10 mM sodium acetate and load 500 μL of that mixed sample into an NMR tube.
4. Acquire ^{1}H NMR spectra of the sample on an NMR spectrometer. Use a standard Bruker water presaturation pulse program to suppress the signal of –OH protons.
5. Analyze the ^{1}H NMR spectra data using MNova or TopSpin. Integrate the area below the peak of -CH_3 protons from sodium acetate and the intensity of -CH_2 protons from the UG selenopeptide. See Fig. 7 for an example of ^{1}H NMR spectrum of the UG dimer.

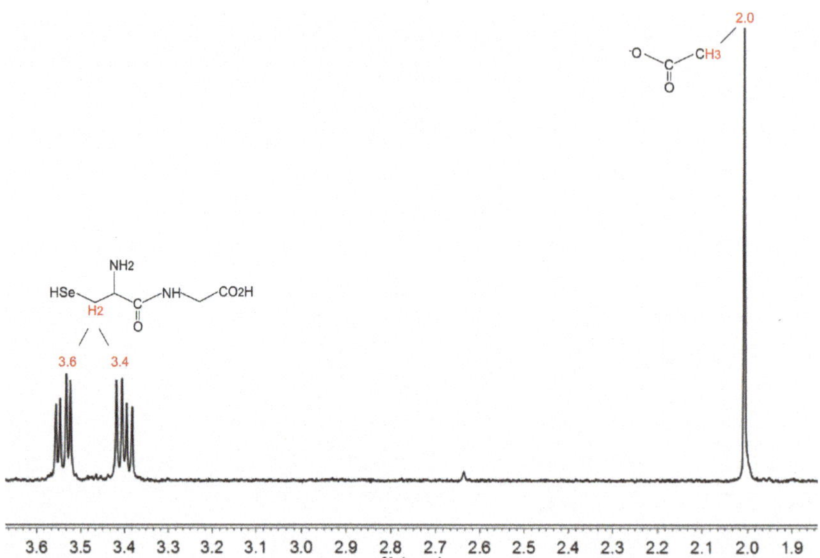

Fig. 7 1H NMR spectroscopy of solubilized selenopeptide UG after its synthesis and purification. The chemical shift split at 3.6 and 3.4 ppm corresponds to β protons in UG (colored in red), and the chemical shift at 2.0 ppm corresponds to protons in sodium acetate. The concentration of solubilized selenopeptide UG can be calculated using sodium acetate at a known concentration as a standard for quantification.

6. Calculate the ratio of the areas of -CH_3 protons from sodium acetate and the integrals of -CH_2 protons from the UG peptide. Then, based on the concentration of sodium acetate, calculate the concentration of UG peptide in the mixed sample. The typical yield for the purified UG dimer from a 0.25 mmol scale synthesis is 25 mg. Finally, calculate the concentration of solubilized UG peptide in DCl. Aim for a UG concentration between 10 and 100 mM.
7. The dissolved peptide should be used for ligation immediately. Long term storage in DCl is not recommended.

4.5 Ligating the protein thioester with the UG selenopeptide

1. Carry out the ligation reaction by adding components in a 1.5 mL microcentrifuge tube. First, dilute the UG peptide from Section 4.4 (solubilized in DCl) to 5 mM using the chitin cleavage buffer. A typical volume at this point is 150 μL. Adjust the pH of the peptide solution to 6.5 by adding 3–5 μL of 10 M NaOH 6.5, measure the pH using a semimicro pH electrode.
2. Add 200 μL of 100 μM cselenos (52-187) thioester or selenos (1-187) thioester prepared in Section 4.2, supply with 40 μL of ligation buffer.
3. Add 10 μL of 1 M TCEP (at pH 6.5) to the ligation mixture. Adjust the pH of the ligation mixture if needed to ensure it is 6.5 using a semimicro pH electrode. The ligation mixture consists of 50 μM cselenos (52-187) or selenos (1-187) thioester, 2 mM UG peptide, 135 mM MESNA, 25 mM TCEP, pH 6.5 (supplemented with 0.067% DDM for preparations of full-length selenos) at a volume of approximately 400 μL. The molar ratio of selenopeptide to protein thioester is 40:1. The ligation reaction is scalable if the concentrations of the components remain as specified in the steps above.
4. Incubate the ligation mixture in a 37 °C water bath, allowing the ligation reaction to complete. Monitor the progress using intact protein mass spectrometry. Ligation is typically over 95% complete within 12 h. Examples of the mass spectra of ligated cselenos and selenos are shown in Figs. 5 and 6, respectively.
5. When the ligation is complete, add 200 μL of 1M DTT to the reaction mixture (the final concentration of DTT is 200 mM) and incubate at room temperature for 30 min. *Note 9* explains why the

presence of the reducing agent DTT is helpful following TCEP assisted ligation.

6. Exchange the buffer of ligated products to the storage buffer using a HiTrap™ desalting column. No further purification is required. The protein can be stored long term in this buffer.
7. Determine the ligated cselenos (52-189) and selenos (1-189) protein concentration (e.g., by BCA assay).
8. To document the product purity, acquire the mass spectrum of the intact, ligated cselenos (52-189) and selenos (1-189) according to Section 5.1.
9. Verify the presence of selenium using protein sequencing by tandem MS and from selenium's unique isotopic abundance (Section 5.2).

5. Analysis of the product

5.1 Intact protein mass spectrometry

During the preparation of protein thioesters and for determining the purity of the ligation products, the molecular weights of precursors and products are detected by mass spectrometry of intact proteins using different sample preparation methods.

1. To verify thioester formation and the level of hydrolysis if present, submit 0.2 mg/mL of cselenos (52–187) and selenos (1–187) thioester in chitin cleavage buffer for intact mass detection. Do not exchange the buffer or use reducing or alkylating reagents. When calculating the expected mass, recall that the thioester will have an MESNA adduct.
2. Take 0.2 mg/mL of ligated cselenos (52-189) and incubate with 1 mM TCEP at 37 °C for 30 min, and then alkylate with 20 mM NEM for 10 min. Employ a Zeba™ desalt spin desalting column to remove excess NEM and desalt into mass spectrometry buffer.
3. Acquire a mass spectrum of the intact alkylated proteins. For example, using a Xevo G2-S QTOF on a Waters ACQUITY UPLC Protein BEH C4 reverse-phase column (300 Å, 1.7 μm, 2.1 mm x 50 mm). Use an acetonitrile gradient from 5% to 95% with 0.1% formic acid over a run time of 5 min and a constant flow rate of 0.5 mL/min at 40 °C. Deconvolute the *m/z* spectra to obtain the mass spectrum.

Representative mass spectra are shown in Figs. 5 and 6. Typical ligation efficiencies are over 95%. See *Note 10* to evaluate ligation efficiency and *Note 11* for suggestions on how to separate hydrolyzed side products if desired.

5.2 Protein sequencing by tandem mass spectroscopy

Note that when trypsin is used for digesting selenos, the Sec-containing peptide (GPSSGGUG) does not ionize well. Instead, only the RGPSSGGUG peptide that is present due to missed cleavage by trypsin, can be detected with an appreciable signal to noise.

1. Reduce 10 μg of selenos in 100 μL of trypsin digest buffer by adding 5 mM DTT. Incubate at 95 °C for 5 min to reach full reduction and denaturation. Cool the sample to room temperature, add 20 mM NEM, and incubate for 10 min to alkylate the Sec and Cys residues as well as the remaining DTT.
2. Add 0.5 μg of trypsin enzyme into 100 μL of alkylated selenos (1 mg/mL) solution and digest at 37 °C for 12 h. Use a 15 μm, 200 Å pore size ZipTip™ to desalt the trypsin digested protein sample, following the vendor's procedure.
3. Sequence the RGPSSGGUG peptide using a Thermo Scientific Orbitrap (Fig. 8). For example, a Thermo Scientific Q-Exactive Orbitrap interfaced with a high-performance liquid chromatography system. Load the sample to an ultracore super C18 reverse-phase column (75 mm x 2.1 mm) from ACE using an acetonitrile gradient of 5–95% with 0.1% formic acid at a flow rate of 0.2 mL/min over 45 min at room temperature. Under these conditions, the RGPSSGGUG peptide elutes around 9 min.
4. The Q-Exactive mass spectrometer was operated in data-dependent mode with one full MS scan at R = 70,000 (m/z = 200) followed by ten HCD MS/MS scans at R = 17,500 (m/z = 200) using a stepped normalized collision energy of 28, 30, and 35 eV. Analyze the sequencing results by Proteome Discoverer (Thermo Scientific).
5. Examine the isotopic pattern of the Sec-containing peptide in the digested sample using Thermo Xcalibur Qual Browser. Compare the detected isotopic pattern with the calculated isotopic pattern (MS isotope) to determine the presence of selenium. A representative m/z spectrum of the RGPSSGGUG peptide showing its isotopic pattern is displayed in Fig. 8.

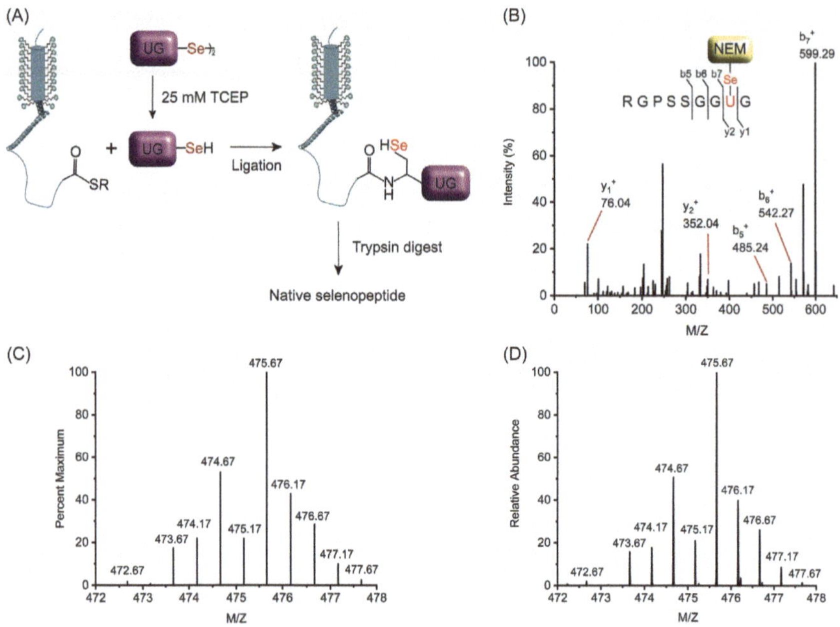

Fig. 8 Characterization of full-length selenos (1-189) prepared by EPL. (A) shows a scheme for Sec-EPL preparation of full-length selenos in detergent and the subsequent characterization of Sec insertion by detecting the native selenopeptide after trypsin digest. (B) shows tandem MS sequencing of the Sec-containing peptide from a trypsin digest of the ligated full-length selenos after NEM alkylation. Note that only the RGPSSGGUG peptide generated by an incomplete trypsin digest ionizes appreciably. (C) shows predicted isotopic pattern of the Sec-containing peptide RGPSSGGUG by MS isotopes. (D) shows detected isotopic pattern of the Sec-containing peptide RGPSSGGUG from a trypsin digest of the ligated full-length selenos after NEM alkylation.

5.3 Circular dichroism spectroscopy

1. Exchange the ligated protein and control proteins into the CD buffer using dialysis or a HiTrap™ desalting column.
2. Transfer 400 μL of protein to a 1 mm path-length cell.
3. Record the far-UV spectra for the 190–250 nm region at 20 °C on a spectropolarimeter. Take three accumulation scans for the baseline and take the average of eight accumulation scans for each sample.
4. Plot the mean residual ellipticity against wavelength for each sample. A representative CD spectrum of cselenos prepared by Sec-mediated ligation is shown in Fig. 9.

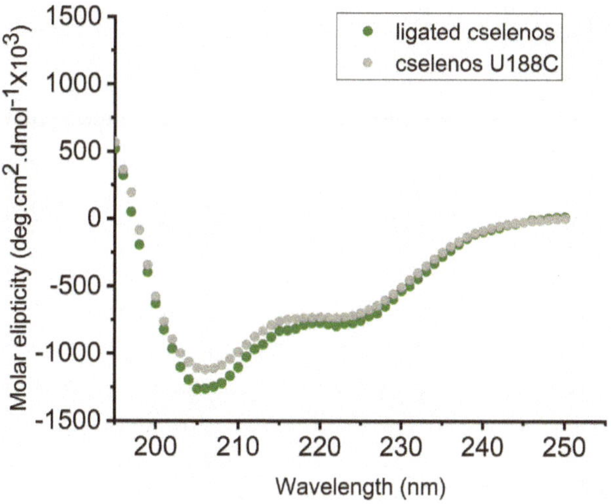

Fig. 9 CD spectrum of ligated cselenos (52-189) and heterologously expressed cselenos U188C (52-189).

6. Summary

We have described an EPL method to prepare cselenos and selenos at high yield and purity. By employing the high reactivity of Sec, it was possible to minimize side reactions such as hydrolysis to obtain pure selenos at high yield, even in detergent micelles. The presence of Sec following ligation is confirmed using an intact protein mass spectrum, peptide sequencing by tandem MS, and by comparing the calculated and observed isotopic distributions of the Sec-containing peptide.

This EPL procedure can be readily adapted for other native or artificial selenoproteins using other selenopeptides, since the coupling and decoupling steps for canonical amino acids are the same. Strategies for Sec protection, deprotection, solubilization, and reduction are as described above, and the procedure described here is already optimized to minimize side reactions to increase the ligation success. Additionally, preparation of proteins' thioesters by intein technology is by now routine, therefore, this method is broadly applicable to both soluble and membrane-bound proteins.

Notes

1. The binding capacity of the amylose resin is around 6 mg MBP-TEV-cselenos (52-187)-GyrA per mL resin. The yield of cselenos (52-187)-GyrA from 1 L is about 60 mg per L growth medium. Thus, we recommend 20 mL of amylose resin for 1 L of MBP-TEV-cselenos (52-187)-GyrA growth medium.

   The binding capacity of the amylose resin is around 3 mg MBP-TEV-selenos (1-187)-GyrA solubilized in DDM per mL of resin. The yield of MBP-TEV-selenos (1-187)-GyrA is about 25 mg per L growth medium. Therefore, we recommend using 10 mL of amylose resin for 1 L of MBP-TEV-selenos (1-187)-GyrA culture.
2. The binding capacity of chitin resin is around 3 mg per mL of resin for cselenos (52-187)-GyrA. For selenos (1-187)-GyrA solubilized in DDM, it is about 1.5 mg per mL of resin. To purify protein thioesters from 1 L growth medium, we use 40 mL of chitin resin cselenos (52-187)-GyrA, and 20 mL of chitin resin for selenos (1-187)-GyrA.
3. Cselenos (52-187) thioester and selenos (1-187) thioester are stable during intein cleavage and storage in the chitin cleavage buffer. We observed less than 5% of hydrolysis. Maintain the pH of buffers containing cselenos (52-187) thioester or selenos (1-187) thioester at or below 6.5, as basic pH accelerates thioester hydrolysis.
4. Attachment of the first amino acid to the resin does not require a coupling reagent. A base is needed to deprotonate the carboxylic group so that it can bind the resin.
5. The deprotection time can be experimentally determined by monitoring the amount of free Fmoc in the solution by measuring absorbance at 290 nm wavelength.
6. The volume of TFA cleavage cocktail depends on the amount of resin/peptide and peptide solubility. Typically, 5–10 mL is needed for a 0.25 mmol scale synthesis.

   The concentration of DTNP should be below four equivalents to avoid forming selenylsulfide bonds. For 1–2 equiv., the reaction time should be extended to 3 h.
7. For peptides that contain polar amino acids, an ester can be used to precipitate the peptides out. However, the removal of TFA is optional. If the purification is carried out immediately after peptide cleavage from resin, keep the peptide soluble in TFA in a small volume (for no longer than 1 day) to reduce the number of injections, it is not necessary to fully remove TFA or precipitate the peptide out.

8. UG peptide oxidizes and forms dimers via Se-Se bond quickly (within 30 min) after sidechain deprotection. During HPLC-MS purification, most eluted UG peptides are in the dimeric form, which has a different elution time from the monomer.
9. TCEP can reduce selenopeptides but is used at low concentrations in the ligation mixture to avoid selenium elimination. Following the ligation reaction, DTT at 200 mM can simplify analysis by removing undesired or non-specific modifications from the protein, such as adducts formed through selenylsulfide or disulfide bonds.
10. The ligation efficiency can be evaluated by quantifying the amount of Sec containing protein compared to the unligated thioester or hydrolyzed form by intact protein mass spectrometry. These protein forms will ionize at similar efficiency, and their ratio in the sample can be assessed reliably from the mass spectrum. During acquisition, avoid oversaturating the detector to ensure reliable capture of protein species.
11. If desired, it is possible to separate hydrolyzed cselenos (52-187) from fully ligated product cselenos (52-189) by size exclusion (see reference (Liu et al., 2013)). However, due to the presence of detergents in purifications of full-length selenos, it is challenging to separate hydrolyzed selenos (1-187) from fully ligated product selenos (1-189).

Acknowledgments

Funding: The research reported in this work was supported by the National Institute of General Medical Sciences (NIGMS) of the National Institutes of Health under award number GM121607 and award MCB-1616178 from the National Science Foundation. Instrumentation was supported by NIGMS awards P20GM104316 and P30GM110758.

References

Amblard, M., Fehrentz, J. A., Martinez, J., & Subra, G. (2006). Methods and protocols of modern solid phase peptide synthesis. *Molecular Biotechnology*, *33*(3), 239–254.

Aravindhan, S., Singh, H. B., Zeller, M., & Butcher, R. J. (2019). Synthesis of selenopeptides: an alternative way of incorporating selenocystine. *Amino Acids*, *51*(4), 661–667.

Blommel, P. G., & Fox, B. G. (2007). A combined approach to improving large-scale production of tobacco etch virus protease. *Protein Expression and Purification*, *55*(1), 53–68.

Capelle, C. M., Zeng, N., Daneleviciute, E., Rodrigues, S. F., Ollert, M., Balling, R., et al. (2021). Identification of VIMP as a gene inhibiting cytokine production in human CD4 + effector T cells. *iScience*, *24*(4), 102289.

Cheng, R., Liu, J., Wang, L., Forstner, M. B., & Rozovsky, S. (2021). Reengineering the site-specific incorporation of selenocysteine into proteins. In J. Jez (Ed.), *Vol. 3. Encyclopedia of Biological Chemistry III* (3rd ed., pp. 757–765). Elsevier.

Conibear, A. C., Watson, E. E., Payne, R. J., & Becker, C. F. W. (2018). Native chemical ligation in protein synthesis and semi-synthesis. *Chemical Society Reviews*, *47*(24), 9046–9068.

Dawson, P. E., & Kent, S. B. (2000). Synthesis of native proteins by chemical ligation. *Annual Review of Biochemistry*, *69*, 923–960.

Dawson, P. E., Muir, T. W., Clark-Lewis, I., & Kent, S. B. (1994). Synthesis of proteins by native chemical ligation. *Science*, *266*(5186), 776–779.

Dery, S., Reddy, P. S., Dery, L., Mousa, R., Dardashti, R. N., & Metanis, N. (2015). Insights into the deselenization of selenocysteine into alanine and serine. *Chemical Science*, *6*(11), 6207–6212.

Dery, L., Reddy, P. S., Dery, S., Mousa, R., Ktorza, O., Talhami, A., et al. (2017). Accessing human selenoproteins through chemical protein synthesis. *Chemical Science*. https://doi.org/10.1039/c6sc04123j.

Eckenroth, B., Harris, K., Turanov, A. A., Gladyshev, V. N., Raines, R. T., & Hondal, R. J. (2006). Semisynthesis and characterization of mammalian thioredoxin reductase. *Biochemistry*, *45*(16), 5158–5170.

Eckenroth, B. E., Lacey, B. M., Lothrop, A. P., Harris, K. M., & Hondal, R. J. (2007). Investigation of the C-terminal redox center of high-Mr thioredoxin reductase by protein engineering and semisynthesis. *Biochemistry*, *46*(33), 9472–9483.

Flemer, S., Jr. (2015). Fmoc-Sec(Xan)-OH: Synthesis and utility of Fmoc selenocysteine SPPS derivatives with acid-labile sidechain protection. *Journal of Peptide Science*, *21*(1), 53–59.

Gieselman, M. D., Xie, L., & van Der Donk, W. A. (2001). Synthesis of a selenocysteine-containing peptide by native chemical ligation. *Organic Letters*, *3*(9), 1331–1334.

Gladyshev, V. N., Arner, E. S., Berry, M. J., Brigelius-Flohe, R., Bruford, E. A., Burk, R. F., et al. (2016). Selenoprotein gene nomenclature. *Journal of Biological Chemistry*, *291*(46), 24036–24040.

Hansen, R. E., & Winther, J. R. (2009). An introduction to methods for analyzing thiols and disulfides: Reactions, reagents, and practical considerations. *Analytical Biochemistry*, *394*(2), 147–158.

Laganowsky, A., Reading, E., Hopper, J. T., & Robinson, C. V. (2013). Mass spectrometry of intact membrane protein complexes. *Nature Protocols*, *8*(4), 639–651.

Lilley, B. N., & Ploegh, H. L. (2004). A membrane protein required for dislocation of misfolded proteins from the ER. *Nature*, *429*(6994), 834–840.

Liu, J., Chen, Q., & Rozovsky, S. (2017). Utilizing selenocysteine for expressed protein ligation and bioconjugations. *Journal of the American Chemical Society*, *139*(9), 3430–3437.

Liu, J., Cheng, R., & Rozovsky, S. (2018). Synthesis and semisynthesis of selenopeptides and selenoproteins. *Current Opinion in Chemical Biology*, *46*, 41–47.

Liu, J., Li, F., & Rozovsky, S. (2013). The intrinsically disordered membrane protein selenoprotein S is a reductase in vitro. *Biochemistry*, *52*(18), 3051–3061.

Liu, J., & Rozovsky, S. (2013). Contribution of selenocysteine to the peroxidase activity of selenoprotein S. *Biochemistry*, *52*(33), 5514–5516.

Liu, J., & Rozovsky, S. (2015). Membrane-bound selenoproteins. *Antioxidants & Redox Signaling*, *23*(10), 795–813.

Liu, J., Srinivasan, P., Pham, D. N., & Rozovsky, S. (2012). Expression and purification of the membrane enzyme selenoprotein K. *Protein Expression and Purification*, *86*(1), 27–34.

Mitchell, N. J., Sayers, J., Kulkarni, S. S., Clayton, D., Goldys, A. M., Ripoll-Rozada, J., et al. (2017). Accelerated protein synthesis via one-pot ligation-deselenization chemistry. *Chem*, *2*(5), 703–715.

Santi, C., Tidei, C., Scalera, C., Piroddi, M., & Galli, F. (2013). Selenium containing compounds from poison to drug candidates: A review on the GPx-like activity. *Current Chemical Biology*, 7(1), 25–36.

Schroll, A. L., Hondal, R. J., & Flemer, S., Jr. (2012). The use of 2,2'-dithiobis(5-nitropyridine) (DTNP) for deprotection and diselenide formation in protected selenocysteine-containing peptides. *Journal of Peptide Science*, *18*(3), 155–162.

Shah, N. H., & Muir, T. W. (2014). Inteins: Nature's gift to protein chemists. *Chemical Science*, *5*(1), 446–461.
Thompson, R. E., & Muir, T. W. (2020). Chemoenzymatic semisynthesis of proteins. *Chemical Reviews*, *120*(6), 3051–3126.
Turanov, A. A., Shchedrina, V. A., Everley, R. A., Lobanov, A. V., Yim, S. H., Marino, S. M., et al. (2014). Selenoprotein S is involved in maintenance and transport of multiprotein complexes. *Biochemical Journal*, *462*(3), 555–565.
Ye, Y. H., Shibata, Y., Yun, C., Ron, D., & Rapoport, T. A. (2004). A membrane protein complex mediates retro-translocation from the ER lumen into the cytosol. *Nature*, *429*(6994), 841–847.

CHAPTER EIGHT

Chemoproteomic interrogation of selenocysteine by low-pH isoTOP-ABPP

Daniel W. Bak* and Eranthie Weerapana*
Department of Chemistry, Boston College, Chestnut Hill, MA, United States
*Corresponding authors: e-mail address: bakd@bc.edu; eranthie@bc.edu

Contents

Abstract

Selenoproteins comprise a small group of selenocysteine (Sec) containing proteins, often involved in redox homeostasis. While Sec is functionally similar to cysteine (Cys), with both acting as protein-centered nucleophiles, chemoproteomic strategies employing electrophilic probes have often failed to rigorously identify Sec residues, due to their relatively low abundance with respect to Cys across a proteome. To improve the enrichment and detection of selenoproteins, herein we describe a chemoproteomic strategy that relies on the unique properties of Sec as compared to Cys, such as reduced pK_a and the unique isotopic distribution of selenium. Low pH electrophilic probe labeling of mouse proteomes reduces Cys reactivity, resulting in increased identification of most soluble selenoproteins. This quantitative chemoproteomic platform provides a method to reliably measure changes in selenoprotein abundance across growth conditions as well as quantify inhibition by selenoprotein specific inhibitors, such as Auranofin.

Methods in Enzymology, Volume 662
ISSN 0076-6879
https://doi.org/10.1016/bs.mie.2021.10.003

1. Introduction

A small subset of mammalian proteins requires the micronutrient selenium for the production and incorporation of the rare amino acid selenocysteine (Sec; U) (Kryukov et al., 2003; Labunskyy, Hatfield, & Gladyshev, 2014; Lu & Holmgren, 2009). These selenoproteins employ Sec functionally in a manner analogous to the standard amino acid cysteine (Cys; C), with selenium replacing sulfur within the nucleophilic thiol. Incorporation of Sec into a growing polypeptide chain occurs co-translationally through ribosomal insertion of Sec at an internal UGA (stop) codon. This complex amino acid insertion mechanism is achieved with the aid of a host of unique protein factors as well as 3′ signaling sequences and stem-loop structures on the target selenoprotein mRNA. The details of this process have been thoroughly described elsewhere, and recent reviews can be found which summarize the cogent details (Donovan & Copeland, 2010; Labunskyy et al., 2014; Xu et al., 2006). Due to the presence of an internal UGA codon as well as the requisite 3′ signaling sequence, the complete selenoproteome of many organisms, including mammals, has been bioinformatically determined (Mariotti, Lobanov, Guigó, & Gladyshev, 2013). In mice for example, there are 24 selenoproteins (25 in humans), with all but two containing only a single Sec residue (Fig. 1). Around half of these selenoproteins have characterized functions, mostly involved in pathways related to cellular redox homeostasis, including thiol reduction (***Txnrds***), oxygen detoxification (***Gpxs***), and reversal of methionine oxidation (***Msrb1***) (Kryukov et al., 2003; Labunskyy et al., 2014; Lobanov, Hatfield, & Gladyshev, 2009). Many of the other selenoproteins have poorly defined functions, but likely also participate in redox processes as they contain predicted redox-active CxxU (or similar) motifs. Intriguingly, some organisms can employ Cys-containing orthologs or completely forego the use of Sec, bringing into question the functional necessity of Sec utilization (Lobanov et al., 2007; Lobanov, Hatfield, & Gladyshev, 2008). A current theory is that Sec is more resistant than Cys to irreversible oxidation (Ingold et al., 2018; Marques et al., 2017; Snider, Ruggles, Khan, & Hondal, 2013) and may be an evolutionary adaptation to high oxygen environments (Hondal & Ruggles, 2011; Zhang, Romero, Salinas, & Gladyshev, 2006). Despite the many questions surrounding the utilization and function of Sec, it is clear that selenoproteins play important roles in health and disease, as many have been implicated in heart disease

Symbol	Name	Protein Diagram	Selenocysteine (U) Postion
Txnrd1	Thioredoxin Reductase 1	CU	612
Txnrd2	Thioredoxin Reductase 2	CU	523
Txnrd3	Thioredoxin Reductase 3	CU	651
Gpx1	Glutathione Peroxidase 1	U	47
Gpx2	Glutathione Peroxidase 2	U	40
Gpx3	Glutathione Peroxidase 3	U	73
Gpx4	Glutathione Peroxidase 4	U	73
Dio1	Iodothyronine Deiodinase 1	CxU	126
Dio2	Iodothyronine Deiodinase 2	U U	130, 263
Dio3	Iodothyronine Deiodinase 3	CxU	170
MsrB1	Methione-R-Sulfoxide Reductase	U	95
Sephs2	Selenophosphate Synthase 2	UxC	63
SelenoF	Selenoprotein F	CxU	96
SelenoH	Selenoprotein H	CxxU	38
SelenoI	Selenoprotein I	U	388
SelenoK	Selenoprotein K	U	92
SelenoM	Selenoprotein M	CxxU	48
SelenoN	Selenoprotein N	CU	428
SelenoO	Selenoprotein O	CxxU	665
SelenoP	Selenoprotein P	UxxC U U UxC CxU UxU U UxU	59, 259, 277, 318, 330, 352, 366, 368, 375, 377
SelenoS	Selenoprotein S	U	189
SelenoT	Selenoprotein T	CxxU	49
SelenoV	Selenoprotein V	CxxU	255
SelenoW	Selenoprotein W	CxxU	13

Fig. 1 The mouse selenoproteome. There are 24 selenoproteins in mice, each harboring at least one Sec (U) residue. Mouse selenoprotein P and iodothyronine deiodinase 2 have 10 and 2 Sec residues, respectively. Many Sec residues participate in redox active CxxU (or similar) motifs with a nearby Cys (C) residue. Selenoproteins that are secreted to the extracellular space are underlined in black, those that are integral membrane proteins are underlined in red.

(Handy & Loscalzo, 2016), inflammation (Kaushal, Gandhi, Nelson, Narayan, & Prabhu, 2011), and cancer (Davis, Tsuji, & Milner, 2012; Hatfield, Tsuji, Carlson, & Gladyshev, 2014). Specifically, several selenoproteins display dysregulated expression during cancer progression, for example, reduced levels of selenoprotein P have been observed in cancerous cell lines (Al-Taie et al., 2009; Calvo et al., 2002). Conversely, Txnrd1 shows increased expression (Cadenas et al., 2010; Fernandes et al., 2009), which may in part protect malignant cells from oxidative damage (Yoo et al., 2010).

As selenoproteins are poised as future therapeutic targets for cancer and other diseases, proteomic techniques that can broadly interrogate the selenoproteome are required to provide crucial information on Sec redox state, inhibitor susceptibility, and function. The global exploration of the reactivity, redox sensitivity, and inhibitor susceptibility of Cys residues in a proteomic context has developed dramatically over the past 10 years with the introduction of isoTOP-ABPP platforms (Backus et al., 2016; Bak, Pizzagalli, & Weerapana, 2017; Martell et al., 2016; Weerapana et al., 2010; Whitby, Obach, Simon, Hayward, & Cravatt, 2017; Zhou et al., 2016). Typical isoTOP-ABPP platforms rely on Cys reactive electrophilic probes, such as iodoacetamide (IA)-alkyne, to modify and enrich reactive Cys residues. As Sec is a structural isoster of Cys (Fig. 2A) and a nucleophile, it is amenable to modification by the same electrophilic probes used to enrich Cys residues (Fig. 2B) (Huber & Criddle, 1967), yet in isoTOP-ABPP studies Sec residues are rarely identified (Whitby et al., 2017). This poor Sec enrichment is the result of the low proteome abundance of Sec compared to Cys in a given organism (e.g., in mice there are only 34 Sec residues but over 100,000 Cys residues (Giron, Dayon, & Sanchez, 2010)) (Fig. 2C). Traditional strategies employed to overcome this low abundance limitation require metabolic incorporation of radioactive isotopes of selenium (^{75}Se) and visualization by gel autoradiography (Kühbacher et al., 2009). These strategies unfortunately provide no information on selenoprotein inhibition susceptibility, redox state, or reactivity, and also are complicated by the fact that many selenoproteins are of a similar molecular weight and cannot be easily separated by gel. An alternative strategy would be the employment of Sec specific electrophilic probes. Unfortunately, to date, no probes exist which are capable of the targeted enrichment of selenocysteine residues for MS analysis, though recent reports of dinitrophenyl ethers and sulfonamides have shown promising selectivity (Li, Feng, Zhai, & Feng, 2017; Zhang et al., 2015). In lieu of Sec-targeted

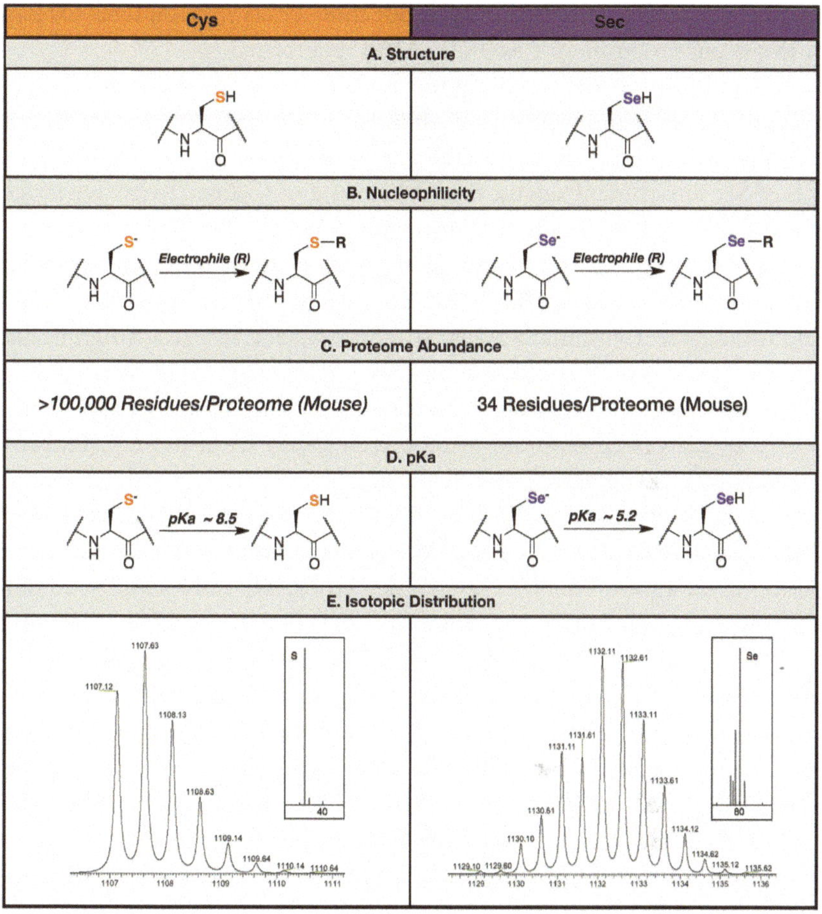

Fig. 2 A comparison of the biophysical properties of Sec and Cys residues. (A) The chemical structure of the isosteric amino acids Cys and Sec, which contain a terminal sulfur or selenium atom, respectively. (B) Both Cys and Sec are nucleophiles and can react in a similar manner with a host of electrophilic species to form chemical adducts. (C) In contrast to Cys, which can be found in greater than 100,000 positions across the proteome, Sec is much less common and is only found at 34 distinct locations (on 24 proteins) throughout the mouse proteome. (D) Free Cys has a pK_a of ~8.5, whereas free Sec has a significantly reduced pK_a of ~5.2. (E) The isotopic distribution of sulfur is dominated by ^{32}S (>95%), which results in the isotopic distribution of Cys-containing peptides dominated by the presence of ^{13}C. In contrast, selenium exists in a number of stable isotopes, ^{74}Se (0.86%), ^{76}Se (9.23%), ^{77}Se (7.60%), ^{78}Se (23.69%), ^{80}Se (49.80%), and ^{82}Se (8.82%). This gives Sec-containing peptides a characteristic isotopic envelope that is strongly influence by the presence of these selenium isotopes.

probes, alternative enrichment strategies are necessary to improve coverage of the selenoproteome for isoTOP-ABPP platforms.

To address the challenges enumerated above, we recently developed a strategy to enrich and identify low abundance Sec-containing proteins from a complex proteome (Bak, Gao, Wang, & Weerapana, 2018). To achieve robust Sec enrichment against the background of Cys reactivity, we looked toward the unique biophysical properties of Sec. Sec and Cys residues display only minimal differences in electronegativity, reduction potential, and ionic radius (Arnér, 2010), which are not useful for chemically distinguishing the two amino acids. In contrast, Sec displays a significantly reduced pK_a (~5.2) compared to Cys (~8.5) (Fig. 2D) (Arnér, 2010; Huber & Criddle, 1967). Sec is also more nucleophilic than Cys (Arnér, 2010), and has been reported to be reactive with IA at pH 2.0 (Huber & Criddle, 1967), suggesting that Sec is capable of reacting with IA even in its protonated Selenol (-SeH) state. This is in contrast to Cys, whose thiol (-SH) reactivity with IA is muted. We also took note of the isotopic distribution of selenium, which exists as a range of stable isotopes (^{74}Se: 0.86%, ^{76}Se: 9.23%, ^{77}Se: 7.60%, ^{78}Se: 23.69%, ^{80}Se: 49.80%, and ^{82}Se: 8.82%), as compared to sulfur (^{32}S: >95%), resulting in a unique isotopic distribution pattern for Sec-containing peptides (Fig. 2E). With these differences between Sec and Cys in mind, we determined IA-labeling conditions (pH ~ 5.75) where the bulk of Cys residues in the proteome would be protonated (Fig. 3A). At this reduced pH, only Sec and unprotonated Cys residues would be labeled by our IA-alkyne probes (Fig. 3B). Importantly, a number of functional Cys residues are known to have dramatically depressed pK_a values (e.g., the catalytic Cys in human gluataredoxin; pK_a ~ 3.5) (Dyson et al., 1997; Jao et al., 2006; Kanzok et al., 2001). This low-pH (5.75) IA-alkyne labeling strategy resulted in a significant global decrease in Cys reactivity, while Sec reactivity was largely unperturbed (Bak et al., 2018). Another report also recently demonstrated that low pH (4.0) IA labeling resulted in decreased Cys and increased Sec labeling as compared to IA labeling at high pH (8.0) (Guo et al., 2018). To further increase detection, a targeted mass list was employed during MS analysis to preferentially isolate Sec-containing peptides for fragmentation and identification. Finally, identifications were validated by manual examination of the unique isotopic envelope associated with Sec-containing peptides, conferred by the presence of natural abundance selenium.

Here we present three variations of our standard Sec-specific low-pH isoTOP-ABPP workflow. First, for the non-quantitative identification of

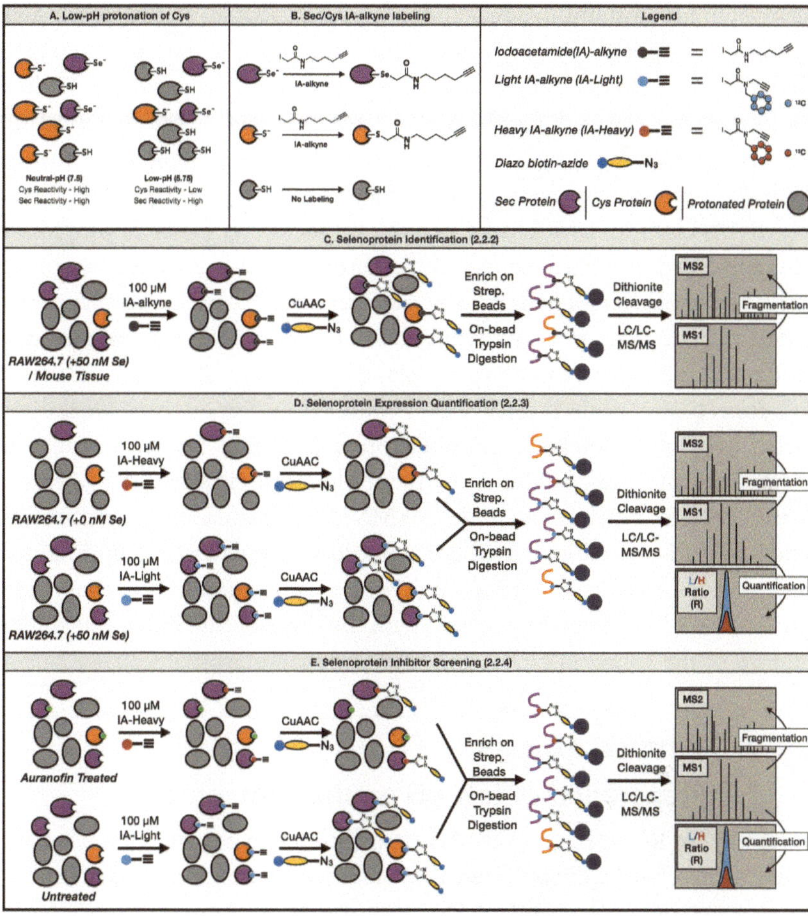

Fig. 3 Low-pH isoTOP-ABPP workflows for Sec identification and quantification. (A) As compared to neutral pH (7.50), low pH (5.75) conditions result in the protonation of Cys residues (orange proteins). In contrast, due to their lower pK_a, Sec residues are not expected to show an appreciable increase in protonation at reduced pH values. (B) IA-alkyne labeling at low pH will enrich Sec residues, as well as a reduced number of Cys residues that remain deprotonated due to the local environment of the folded protein. Protonated Cys residues will not be readily labeled by IA-alkyne at reduced pH. (C) Low-pH ABPP workflow for the identification of Sec residues from a single proteome. (D) Low-pH isoTOP-ABPP workflow for the quantification of changes in selenoprotein abundance between two proteomes (RAW264.7 cells supplemented with either 0 or 50 nM $NaSeO_3$). (E) Low-pH isoTOP-ABPP workflow for the screening of selenoprotein inhibitors (RAW264.7 cell treated with either vehicle or 100 μM Auranofin).

Sec-containing proteins in a proteome (Fig. 3C), low-pH Sec labeling with a non-isotopic IA-alkyne probe is performed followed by copper assisted azide-alkyne cycloadditon (CuAAC) with a chemically cleavable diazo biotin-azide tag. Second, for the quantification of selenoprotein abundance changes between two growth conditions (Fig. 3D), proteomes with high and low selenoprotein expression are labeled at low-pH with isotopically distinct IA-Light and IA-Heavy probes, respectively. Labeled samples are then appended to diazo biotin-azide through CuACC and combined. Third, for quantifying changes in Sec reactivity (e.g., upon inhibitor treatment) (Fig. 3E), low-pH IA-Light and IA-Heavy labeling is performed on vehicle and inhibitor treated proteomes, respectively, followed by CuACC with diazo biotin-azide and pairwise combination. For all three workflows, labeled protein samples are enriched on streptavidin beads and digested on-bead with trypsin, before subsequent chemical cleavage and release of IA-labeled Sec-containing peptides for MS identification and quantification. Here, detailed procedures for these workflows are described, highlighting the key steps, expected results, and future opportunities for the chemoproteomic study of selenoproteins, with the hope of furthering our understanding of the function of Sec residues within the proteome.

2. Technical aspects

2.1 Preparation of low-pH mouse proteomes

Proteomes for low-pH isoTOP-ABPP analysis were prepared from isolated mouse tissues (liver, kidney, intestines, brain, and testes) as well as the mouse macrophage cell line, RAW264.7. Tissue samples from male C57BL/6J mice were obtained frozen and stored at −80 °C until ready for harvesting. These tissues samples must then be mechanically homogenized to break down tougher tissue before continuing with the low pH lysis procedure. RAW264.7 cells were cultured under standard conditions (37 °C, 5% CO_2) in DMEM/High glucose media supplemented with 10% fetal bovine serum and 1% penicillin/streptomycin. Excess selenite was included in the growth media to induce robust selenoprotein expression as most commercial medias are deficient in selenium (Karlenius, Shah, Yu, & Hawkes, 2011). In order to enrich selenocysteine containing proteins through isoTOP-ABPP analysis it is essential that labeling be performed at reduced pH and that protein precipitation and denaturation be minimized. Therefore, cell lysis and clarification, both for tissue homogenates and RAW264.7 cells, is performed in low pH (5.75) selenocysteine enrichment (SE) buffer at 4 °C. Soluble proteomes can then be stored in SE buffer at −80 °C for extended periods of time and then thawed on ice immediately prior to use.

2.1.1 Materials

Male C57BL/6J mouse-tissue samples: liver, kidney, large intestines, brain, and testes (The Jackson Laboratories)

RAW264.7 cells (ATCC, catalog number TIB-71)

DMEM/High Glucose media (GE Lifesciences, catalog number SH30243.01)

Hyclone Penicillin-streptomycin (Fisher Scientific, catalog number SV30010)

Fetal Bovine Serum (Atlanta Biologicals, catalog number S11150)

SE buffer: 50 mM MES, 50 mM NaH_2PO_4, 100 mM NaCl, pH 5.75 (made in house)

Sodium selenite (Sigma-Aldrich, catalog number S5261)

Auranofin (10 mM stock in DMSO) (Sigma-Aldrich, catalog number A6733)

Weigh boats

Razor blade

1.5 mL Eppendorf tubes

15 mL conical tubes

Ultracentrifuge tubes

Cell scraper

Glass dounce homogenizer (Cole-Parmer, catalog number UX-44468-10)

Ultrasonicator (Cole-Parmer Ultrasonic Processor)

Refrigerated bench-top centrifuge (Thermo Scientific Sorvall ST16R Centrifuge)

Microcentrifuge (Eppendorf Centrifuge 5418)

Ultracentrifuge (Thermo Scientific Sorvall MX120 Micro-Ultracentrifuge)

BioRad DC protein assay kit (BioRad, catalog number 5000112)

2.1.2 Preparation of mouse tissue homogenates (timing—3 h)

1. Thaw tissue samples in a plastic weigh boat and rinse sample four times with ~5 mL of ice-cold selenoprotein-enrichment (SE) Buffer to remove any remaining blood.
2. Mince tissue samples with a razor blade to improve subsequent homogenization (particularly important for tougher tissues).
3. Transfer tissue sample into a pre-chilled glass dounce homogenizer, washing weigh boat with ~5 mL volumes of ice-cold SE buffer. The

ratio of tissue to buffer should be around 1:5 (w:v). **Note:** This ratio may vary for different tissues and should be optimized for both total protein yield and proteome concentration (2 mg/mL is required for subsequent experiments).

4. Homogenize tissue with a Teflon pestle (~10 strokes). Homogenate should be free of large tissue particles. **Note:** Additional homogenization may be necessary for tougher tissues, but small pieces of fat or connective tissue that won't break down should be manually removed.
5. Divide homogenate into 1.5 mL eppendorf tubes and sonicate on ice (3×10 pulses at 75% power).
6. Transfer homogenate to one or more ultracentrifuge tubes and spin at $45{,}000 \times g$ for 45 min at 4 °C. If homogenate was divided into multiple 1.5 mL Eppendorf tubes, these should be combined into a single "soluble" proteome fraction. **Note:** If a "membrane" proteome fraction is desired, the pellets should be combined and washed twice by resuspension in ice-cold SE buffer and ultracentrifugation ($45{,}000 \times g$ for 45 min at 4 °C). After discarding the supernatant, the pellet should be resuspended a final time in ~1 mL ice-cold SE buffer.
7. Measure protein concentration by BioRad DC protein assay according to the manufacturer's instructions.
8. Protein should be diluted to 2 mg/mL with the appropriate volume of ice-cold SE buffer.
9. If not proceeding directly to IA-labeling (Section 2.2), then aliquot proteome into 500 μL fractions and store at −80 °C in order to avoid unnecessary freeze-thaw cycles.

2.1.3 Preparation of RAW264.7 cell lysates (timing—2–3 h)

1. Passage RAW264.7 cells to 10% confluency using DMEM/High glucose media (with 10% fetal bovine serum and 1% penicillin/streptomycin) supplemented with $NaSeO_3$ (50 nM, unless another concentration specified).
2. Replace media every second day with fresh $NaSeO_3$ supplemented media until the cells are fully confluent.

If cells are treated with inhibitor (e.g., auranofin) continue with steps 3–5, otherwise skip to step 6:

3. Add 100 μL of 10 mM auranofin (100 μM final concentration) or vehicle (DMSO) to 10 mL of $NaSeO_3$ supplemented media.
4. Remove spent media and replace with auranofin or vehicle containing media.
5. Incubate cells for 2 h.

6. Remove spent media and wash cells once with 10 mL of ice-cold SE buffer.
7. Add 10 mL ice-cold SE buffer and harvest cells by scraping with a cell scraper.
8. Transfer cell suspension to conical tube and centrifuge at 1800 × *g* for 5 min at 4 °C.
9. Discard supernatant and wash cell pellet twice with 10 mL ice-cold SE buffer
10. Resuspend cell pellet in 1 mL SE buffer and transfer into a 1.5 mL eppendorf tube and sonicate on ice (3 × 10 pulses at 75% power).
11. Centrifuge lysates at 10,000 × *g* for 5 min at 4 °C and transfer supernatant to clean 1.5 mL Eppendorf tube. **Note:** This represents an "unfractionated" sample (both soluble and membrane proteins). To obtain isolated "soluble" or "membrane" fractions proceed with ultracentrifugation (Section 2.1.2: step 6).
12. Measure protein concentration by BioRad DC protein assay according to the manufacturer's instructions
13. Protein should be diluted to 2 mg/mL with the appropriate volume of ice-cold SE buffer
14. If not proceeding directly to IA-labeling (Section 2.2), then aliquot proteome into 500 μL fractions in 1.5 mL Eppendorf tubes and store at −80 °C in order to avoid unnecessary freeze-thaw cycles

2.2 Low-pH isoTOP-ABPP workflows

The following low-pH isoTOP-ABPP workflow represents a variation on standard platforms for identifying and quantifying changes in cysteine reactivity (Abo, Li, & Weerapana, 2018; Qian et al., 2013; Weerapana et al., 2010; Weerapana, Speers, & Cravatt, 2007). Uniquely, the proteome is labeled with a Cys(Sec)-reactive electrophilic IA-probe, at reduced pH (5.75), in order to improve the ratio of Sec to Cys labeling and subsequent MS identification. Three unique workflows for selenocysteine enrichment are presented: (1) Identification of Sec residues in a proteomic sample using IA-alkyne (Section 2.2.2; Fig. 3C), (2) Quantification of changes in selenoprotein abundance between two proteomic samples using IA-Light and IA-Heavy probes (Section 2.2.3; Fig. 3D), and (3) Quantification of changes in Sec reactivity across a proteome (e.g., upon inhibitor treatment) using IA-Light and IA-Heavy probes (Section 2.2.4; Fig. 3E). This third workflow could also be employed to examine non-inhibitor-based changes in Sec reactivity, such as oxidation or post-translation modification. If necessary, selenoprotein levels can be monitored by western blot (not described here), to verify that measured changes in Sec reactivity are occurring independently of any difference in protein expression induced by the inhibitor treatment conditions. Alternatively, inhibitor/vehicle treatment can be performed in separate aliquots from the same batch of low-pH lysates. For all three workflows it is essential to use freshly made reagents to ensure efficient peptide recovery by the end of the workflow. All IA probes (IA-alkyne, IA-Light, IA-Heavy) and diazo biotin-azide should be diluted to the appropriate concentration, from a solid stock, in DMSO. These reagents should then be divided into aliquots (~50 μL) and frozen at −20 °C to avoid multiple freeze thaw cycles that will result in unnecessary degradation. At the conclusion of the workflow, the acidified peptide sample can be stored at −20 °C for a number of months until ready for MS analysis.

2.2.1 Materials

IA-alkyne probe (10 mM solution in DMSO) (Made in house) (Weerapana, Simon, & Cravatt, 2008)

IA-Light (IAL) probe (10 mM solution in DMSO) (Made in house) (Abo et al., 2018)

IA-Heavy (IAH) probe (10 mM solution in DMSO) (Made in house) (Abo et al., 2018)

Diazo biotin-azide (5 mM solution in DMSO) (Click Chemistry Tools, catalog number 1041–25)

Tris-2-carboxyethyl phosphine (TCEP) (50 mM solution in water) (Sigma-Aldrich, catalog number C4706)

Tris[(1-benzyl-1H-1,2,3-triazol-4-yl)methyl]amine ("ligand") (1.7 mM solution in DMSO, t-butanol 1:4) (Sigma-Aldrich, catalog number 678937)

Copper (II) sulfate (100 mM solution in water) (Sigma-Aldrich, catalog number C1297-10G)

Methanol, Optima LC/MS grade (Fisher Scientific, catalog number A456-4)

DPBS (Fisher Scientific, catalog number MT-21-031-CM)

Sodium dodecyl sulfate (SDS) (0.2% and 1.2% solutions in DPBS) (BioRad, catalog number 1610302)

Streptavidin agarose beads (Thermo Scientific, catalog number 20353)

Urea (2 M and 6 M solutions in DPBS) (Sigma-Aldrich, catalog number AC424580025)

Ultrapure Dithiothreitol (DTT) (200 mM solution in water) (Invitrogen, catalog number 15508-013)

Iodoacetamide (400 mM solution in water) (Acros Organics, catalog number 122270050)

Calcium chloride (100 mM solution in water) (Sigma-Aldrich, catalog number C1016-100G)

Trypsin solution: 20 μg of sequencing grade trypsin reconstituted in 40 μL of trypsin buffer (provided in package) (Promega, catalog number 161527)

Sodium hydrosulfite (50 mM solution in DPBS) (Sigma-Aldrich, catalog number 161527)

Formic acid, Optima LC/MS Grade (Fisher Scientific, catalog number A117-50)

Refrigerated bench-top centrifuge (Thermo Scientific Sorvall ST16R Centrifuge)

Microcentrifuge (Eppendorf Centrifuge 5418)

Ultrasonicator (Cole-Parmer Ultrasonic Processor)

Screw-cap centrifuge tubes (Thermo Scientific, catalog number 3471S)

1.5 mL Eppendorf tubes

15 mL conical tubes

Benchtop Rotator

Benchtop Vortexer

Heat block

2.2.2 IA-alkyne labeling and click chemistry for selenoprotein identification (timing—3 h)

1. Label two 500 μL aliquots of RAW264.7 cell or mouse tissue lysates (2 mg/mL in SE buffer) by the addition of 5 μL of 10 mM IA-alkyne probe (100 μM final concentration). Vortex tubes briefly and incubate for 1 h at 25 °C.
2. To each aliquot add, in order, 10 μL of 5 mM diazo biotin-azide (100 μM final concentration), 10 μL of freshly prepared 50 mM TCEP (1 mM final concentration), 30 μL of 1.7 mM TBTA "ligand" (100 μM final concentration), and 10 μL of 100 mM copper sulfate (2 mM final concentration). Vortex and incubate click chemistry reaction for 1 h at 25 °C, vortexing reactions every 15 min. **Note:** the addition of copper will result in precipitation of proteins, which necessitates resuspension by vortexing to ensure efficient click chemistry occurs.

3. After click chemistry reaction, combine tubes pairwise and centrifuge at 6500 × *g* for 5 min at 25 °C to pellet protein. Discard supernatant.
4. Resuspend pellet in 500 μL of ice-cold methanol by gently sonication (5–10 pulses at 50% power) and centrifuge at 6500 × *g* for 5 min at 4 °C. Discard supernatant. **Note:** Discarded methanol will appear quite yellow in color. This is as a result of removing unreacted diazo biotin-azide tag, which if not removed would reduce yields during subsequent streptavidin enrichment.
5. Repeat methanol wash of pellet, discarding supernatant after centrifugation.
6. Carry protein pellet through remaining streptavidin enrichment, reduction and alkylation, trypsinization and cleavage as described in Section 2.2.5.

2.2.3 IA-Light and IA-Heavy labeling and click chemistry for quantification of Selenoprotein expression (timing—3 h)

1. Label two 500 μL aliquots of RAW264.7 lysates (2 mg/mL in SE buffer) from cells cultured in media supplemented with 50 nM selenium (high selenoprotein expression condition) by the addition of 5 μL of 10 mM IA-Light probe (100 μM final concentration). In parallel, label two 500 μL aliquots of RAW264.7 lysates (2 mg/mL in SE buffer) from cells cultured in media supplemented with 0 nM selenium (low selenoprotein expression condition) by the addition of 5 μL of 10 mM IA-Heavy probe (100 μM final concentration). Vortex tubes briefly and incubate for 1 h at 25 °C. **Note:** Always label the sample that is expected to have a greater abundance of selenoproteins (50 nM Se supplementation) with IA-Light. This ensures the best chance of identifying Sec-containing peptides when using an IA-Light targeted mass list (see Section 2.3.3). As a control, the same labeling experiment can be performed where both IA-Light and IA-Heavy probes are used to labeled RAW264.7 cells cultured in media supplemented with 50 nM selenium.
2. To each aliquot add, in order, 10 μL of 5 mM diazo biotin-azide (100 μM final concentration), 10 μL of freshly prepared 50 mM TCEP (1 mM final concentration), 30 μL of 1.7 mM TBTA "ligand" (100 μM final concentration), and 10 μL of 100 mM copper sulfate (2 mM final concentration). Vortex and incubate click chemistry reaction for 1 h at 25 °C, vortexing reactions every 15 min. **Note:** addition of copper will result in precipitation of proteins, which necessitates resuspension by vortexing to ensure efficient click chemistry occurs.

3. After click chemistry reaction, combine tubes pairwise (1 × IA-Light labeled tube with 1 × IA-Heavy labeled tube) and centrifuge at 6500 × *g* for 5 min at 25 °C to pellet protein. Discard supernatant. You should now have two 1.5 mL Eppendorf tubes, each with an identical protein pellet.
4. Resuspend both pellets in 500 μL of ice-cold methanol by gently sonication (5–10 pulses at 50% power), combine tubes pairwise again and centrifuge at 6500 × *g* for 5 min at 4 °C. Discard supernatant. You should now have a single 1.5 mL Eppendorf tube with a pellet. **Note:** Discarded methanol will appear quite yellow in color. This is as a result of removing unreacted diazo biotin-azide, which if not removed would reduce yields during subsequent streptavidin enrichment.
5. Repeat methanol wash of pellet, discarding supernatant after centrifugation.
6. Carry protein pellet through remaining streptavidin enrichment, reduction and alkylation, trypsinization and cleavage as described in Section 2.2.5.

2.2.4 IA-Light and IA-Heavy labeling and click chemistry for selenocysteine inhibitor screening (timing—3 h)

1. Label two 500 μL aliquots of RAW264.7 lysates (2 mg/mL in SE buffer) from cells treated with DMSO for 2 h (control condition) by the addition of 5 μL of 10 mM IA-Light probe (100 μM final concentration). In parallel, label two 500 μL aliquots of RAW264.7 lysates (2 mg/mL in SE buffer) from cells treated with 100 μM Auranofin for 2 h (inhibitor condition) by the addition of 5 μL of 10 mM IA-Heavy probe (100 μM final concentration). Vortex tubes briefly and incubate for 1 h at 25 °C. **Note:** Always label the sample that is expected to have a greater Sec reactivity (vehicle treated) with IA-Light. This ensures the best chance of identifying Sec-containing peptides when using an IA-Light targeted mass list (see Section 2.3.3).
2. To each aliquot add, in order, 10 μL of 5 mM diazo biotin-azide (100 μM final concentration), 10 μL of freshly prepared 50 mM TCEP (1 mM final concentration), 30 μL of 1.7 mM TBTA "ligand" (100 μM final concentration), and 10 μL of 100 mM copper sulfate (2 mM final concentration). Vortex and incubate click chemistry reaction for 1 h at 25 °C, vortexing reactions every 15 min. **Note:** addition of copper will result in precipitation of proteins, which necessitates resuspension by vortexing to ensure efficient click chemistry occurs.

3. After click chemistry reaction, combine tubes pairwise (1 × IA-Light labeled tube with 1 × IA-Heavy labeled tube) and centrifuge at 6500 × *g* for 5 min at 25 °C to pellet protein. Discard supernatant. You should now have two 1.5 mL Eppendorf tubes, each with an identical protein pellet.
4. Resuspend pellet in 500 μL of ice-cold methanol by gently sonication (5–10 pulses at 50% power), combine tubes pairwise again and centrifuge at 6500 × *g* for 5 min at 4 °C. Discard supernatant. You should now have a single 1.5 mL Eppendorf tube with a pellet. **Note:** Discarded methanol will appear quite yellow in color. This is as a result of removing unreacted diazo biotin-azide, which if not removed would reduce yields during subsequent streptavidin enrichment.
5. Repeat methanol wash of pellet, discarding supernatant after centrifugation.
6. Carry protein pellet through remaining streptavidin enrichment, reduction and alkylation, trypsinization and cleavage as described in Section 2.2.5.

2.2.5 Streptavidin enrichment, denaturation, reduction, alkylation, on-bead tryptic digestion, and reductive cleavage (timing—2 days)

1. Resuspend protein pellet (from Sections 2.2.2–2.2.4) in 1 mL of 1.2% SDS in DPBS by gentle sonication (5–10 pulses at 50% power) and subsequent heating at ~90 °C for 5 min. **Note:** After sonication and heating, the solution should be relatively clear, indicating that the protein pellet has been successfully resolubilized.
2. Centrifuge at 6500 × *g* for 5 min at 25 °C in order to pellet excess copper.
3. Transfer 1 mL of supernatant to a 15 mL conical tube with 5 mL of DPBS (0.2% SDS final concentration).
4. Aliquot 200 μL of streptavidin agarose slurry (~100 μL of beads) into a 1.5 mL Eppendorf and add 500 μL of DPBS. Pellet beads by centrifugation at 1400 × *g* for 1 min at 25 °C and remove supernatant. Wash beads 2 more times with 500 μL DPBS.
5. Add washed streptavidin beads to 15 mL conical tube with sample. **Note:** Use ~100 μL of sample to resuspend beads for transfer. Repeat as needed until all beads are moved from 1.5 mL Eppendorf to 15 mL conical tube.
6. Incubate sample, with rotation, overnight at 4 °C.

7. Transfer samples to 25 °C and incubate with rotation for 2h to resolubilize SDS.
8. Centrifuge samples at 1400 × *g* for 5 min at 4 °C and remove supernatant from beads.
9. Wash beads one time with 5 mL of 0.2% SDS in DPBS, three times with 5 mL of DPBS, and three times with 5 mL of water. Between washes pellet beads by centrifugation at 1400 × *g* for 5 min at 4 °C and discard supernatant.
10. Transfer beads to screw-cap Eppendorf tubes using three sequential 166 μL aliquots of 6 M urea in DPBS (total of 500 μL). **Note:** Screw-cap eppendorf tubes are necessary as the high urea concentration causes the traditional snap cap eppendorf tubes to leak during the subsequent rotation steps, resulting in loss of sample.
11. Add 25 μL of freshly prepared 200 mM dithiothreitol (10 mM final concentration) and heat sample for 20 min at 65 °C, mixing by inversion after 10 min. **Note:** This step results in reduction of any remaining cysteine disulfide bonds.
12. Remove from heat and sample let cool for a couple minutes. Add 25 μL of freshly prepared 400 mM iodoacetamide (20 mM final concentration) and incubate samples with rotation at 37 °C for 30 min. **Note:** This step is necessary to cap any reduced cysteine residues, preventing reoxidation and disulfide bond formation that would hinder MS peptide identification.
13. Add 950 μL of DPBS (2 M urea final concentration) and centrifuge samples at 1400 × *g* for 5 min at 25 °C. Remove supernatant completely.
14. Resuspend beads in 200 μL of 2 M Urea in DPBS.
15. Add 2 μL of 100 mM $CaCl_2$ (1 mM final concentration) and 4 μL (2 μg) of sequence grade modified trypsin solution.
16. Mix by inversion and incubate with rotation overnight at 37 °C.
17. Centrifuge samples at 1400 × *g* for 1 min at 25 °C and remove supernatant. **Note:** This supernatant will contain unlabeled tryptic peptides, but any IA-alkyne, IA-Light, or IA-Heavy labeled Sec-containing peptides will remain bound to the streptavidin agarose beads. Removal of these unlabeled tryptic peptides results in increased enrichment and MS detection of labeled Sec-containing peptides.
18. Wash beads three times with 500 μL of DPBS and three times with water. Between washes pellet beads by centrifugation at 1400 × *g* for 5 min at 4 °C and remove supernatant.

19. Elute labeled peptides from beads by addition of 50 μL of freshly prepared 50 mM sodium hydrosulfite in DPBS and incubation with rotation for 1 h at 25 °C. **Note:** Sodium hydrosulfite will reductively cleave the diazobenzene linker, releasing the IA-alkyne, IA-Light, or IA-Heavy labeled Sec-containing peptides from the streptavidin agarose beads.
20. Centrifuge samples at 1400 × *g* for 1 min at 25 °C and transfer supernatant to fresh 1.5 mL eppendorf tube.
21. Incubate beads twice more for 1 h at 25 °C with 75 μL of freshly prepared 50 mM sodium hydrosulfite in DPBS for 1 h at 25 °C. Each time, after centrifugation at 1400 × *g* for 1 min at 25 °C, transfer supernatant to same Eppendorf tube with previous fractions.
22. Wash beads twice more with 75 μL water, and after centrifugation at 1400 × *g* for 1 min at 25 °C, combine supernatant with the previous fractions.
23. To the total sample volume (350 μL), add 17.5 μL of formic acid and mix by inversion.
24. Store samples at −20 °C until ready for mass spectrometry analysis.

2.3 Sec-targeted MudPIT LC/LC-MS/MS analysis

Low-pH isoTOP-ABPP peptide samples were analyzed on an Orbitrap LTQ Discovery with a multi-dimensional proteomic identification technology (MuDPIT) approach, which has been employed successfully in the past for the quantification of reactivity Cys residues (Weerapana et al., 2007; Weerapana et al., 2010). The use of alternative separation strategies (i.e., high-pH RP-LC off-line fractionation) and mass spectrometers (i.e., Qxactive, Exploris, etc.) is not expected to impact the identification of Sec-containing peptides, and newer more sensitive instrumentation will likely improve Sec identification. Importantly, the generation and use of a targeted mass list is necessary to enhance identification of Sec-containing peptides, resulting in improved quantification downstream. As with all MS experiments, all reagents and buffers should be mass spectrometry grade.

2.3.1 Materials

Fused-silica capillary tubing (100 and 250 μm internal diameter) (Agilent, catalog number 50-928-099 and 50–928-081)

Inline microfilter assembly (IDEX health and science, catalog number M-520-1)

Aqua C18 reverse-phase resin (Phenomenex, catalog number 04A-4299)

Partisphere strong cation exchange (SCX) resin (Phenomenex, catalog number 00G-4398-E0)

PC77 Pressure Injection Cell (Next Advance)

Model P-2000 Tip Puller (Sutter Instruments)

Water, Optima LC/MS Grade (Fisher Scientific, catalog number W6-4)

Acetonitrile, Optima LC/MS Grade (Fisher Scientific, catalog number A955-4)

Formic acid, Optima LC/MS Grade (Fisher Scientific, catalog number A117-50)

Ammonium acetate, Optima LC/MS Grade (Fisher Scientific, catalog number A11450)

Buffer A: 95% water, 5% acetonitrile, 0.1% formic acid (Made in house)

Buffer B: 20% water, 80% acetonitrile, 0.1% formic acid (Made in house)

Buffer C: 95% water, 5% acetonitrile, 0.1% formic acid, 500 mM ammonium acetate (Made in house)

Orbitrap LTQ Discovery (Thermo Scientific)

Agilent 1200 series HPLC (Agilent)

Xcalibur software package (Thermo Scientific)

2.3.2 Packing a MudPit column (timing—3–4 h)

1. Make a 5 μm tip by cutting ~45 cm of 100 μm fused-silica capillary tubing. Burn off ~3 cm of polyimide coating in the center of the capillary tubing and wipe off coating with a methanol soaked chemwipe. Use a laser tip puller to pull capillary tubing at cleaned center point to make two 5 μm tips. **Note:** when burning off polyimide coating, be careful not to melt the glass column underneath.
2. Use a pressure injection cell to load a 5 μm tip with 10 cm of Aqua C18 reverse-phase resin. Next, use pressure injection cell to load 4 cm of strong cation exchange (SCX) resin. **Note:** It may be necessary to gently scratch the surface of the tip with a piece of ceramic to start flow of material into the capillary tubing. Be careful not to fully break the tip.
3. Equilibrate 5 μm tip by pressure loading first buffer B for 10 min and then buffer A for 20 min. Store tip side down, being carefully not to scratch or damage tip.
4. Attach a 12 cm length of 250 μm fused-silica capillary tubing to an inline micro-filter assembly unit to create a desalting column.

5. Use a pressure injection cell to load desalting column with 4 cm of Aqua C18 reserve-phase resin.
6. Equilibrate desalting column by pressure loading first buffer B for 10 min and then buffer A for 20 min.
7. Load entire ~367.5 μL peptide sample (from Section 2.2.5) onto equilibrated desalting column using pressure inject cell.
8. Store tip and desalting column at 4 °C if not using the same day.

2.3.3 Generation of IA-Alkyne/IA-Light labeled sec-containing peptide targeted mass list (timing—1–2 h)

1. For each soluble mouse selenoprotein, determine the predicted tryptic Sec-containing peptides (Table 1). Trypsin cleaves after lysine (K) or arginine (R) residues, so the Sec-containing peptide is the sequence of amino acids between the two trypsin cleavage sites (or the start/end of the protein) to either side of the Sec residue. Trypsin will not cleave if the K/R residue is followed by proline (P). Additionally, trypsin may not cleave if there are multiple sequential K/R residues or a labeled Sec or Cys residue. In these cases, include both successful and missed cleavage versions of the peptide. **Note:** it may be useful to include additional missed cleavage versions of the peptide if none of the above missed cleavage rules apply, especially if the tryptic peptide is particularly short (<6 aa). Additionally, separate or combined targeted mass lists for Sec-containing peptides from membrane or secreted selenoproteins can also be generated but in general it is good to keep a mass list relatively short in length (~50 masses).
2. For each predicted tryptic peptide, use the Peptide Mass Calculator at: http://rna.rega.kuleuven.be/masspec/pepcalc.htm. Paste in the peptide sequence and set the program to calculate "ESI series positive mode" using "monoisotopic masses."
3. Set a user defined residue, **1**, to the mass of an alkylated Cys (**+160.0306**). Replace Cys (C) residues in the sequence (if any) with **1**. **Note:** User defined mass includes both the standard mass of Cys (**+103.0091**) plus the additional mass of the modification (**+57.0215**).
4. Set a user defined residue, **2**, to the mass of an IA-alkyne labeled Sec (**+466.1231**) for selenoprotein ID experiments (Section 2.2.2) or IA-Light labeled Sec (**+514.1231**) for selenoprotein expression and inhibitor screening experiments (Sections 2.2.3–2.2.4). Replace the Sec residue in the sequence with **2**. **Note:** Only an IA-Light and not an IA-Heavy targeted mass list will be applied, as it is not necessary to

Table 1 IA-alkyne and IA-Light labeled selenopeptide targeted mass lists.

Protein	Peptide	Charge	IA-alkyne mass list	IA-Light mass list
Gpx1	VLLIENVASLUGTTIR	+2	1033.0395	1057.0395
		+3	689.0289	705.0289
		+4	517.0236	529.0236
		+5	413.8204	423.4204
Gpx2	AVLIENVASLUGTTTR	+2	1005.9978	1029.9978
		+3	671.0011	687.0011
		+4	503.5028	515.5028
		+5	403.0038	412.6038
Gpx4	GFVCIVTNVASQUGK	+2	973.4469	997.4469
		+3	649.3005	665.3005
		+4	487.2273	499.2273
Txnrd1	SGGDILQSGCUG	+2	758.7917	782.7917
		+3	506.1971	522.1971
	RSGGDILQSGCUG	+2	836.8423	860.8423
		+3	558.2308	574.2308
		+4	418.9250	430.9250

Continued

Table 1 IA-alkyne and IA-Light labeled selenopeptide targeted mass lists.—cont'd

Protein	Peptide	Charge	IA-alkyne mass list	IA-Light mass list
Txnrd2	SGLEPTVTGCUG	+2	772.3097	796.3097
		+3	515.2091	531.2091
		+2	850.3603	874.3603
		+3	567.2428	583.2428
		+4	425.6840	437.6840
Txnrd3	SSGLDITQKGCUG	+2	844.8523	868.8523
		+3	563.5708	579.5708
		+4	422.9300	434.9300
Msrb1	FUIFSSSLK	+2	697.8226	721.8226
		+3	465.5510	481.5510
Sephs2	GUGCK	+2	444.1589	468.1589
	GUGCKVPQETLLK	+2	898.4254	922.4254
		+3	599.2862	615.2862
		+4	449.7166	461.7166
	LTSFSGMKGUGCK	+2	869.8694	893.8694
		+3	580.2489	596.2489
		+4	435.4386	447.4386

SelenoF	LYAGAILEVCGUK	+2	880.4092	904.4092
		+3	587.2754	603.2754
		+4	440.7085	452.7085
	KLYAGAILEVCGUK	+2	944.4567	968.4567
		+3	629.9737	645.9737
		+4	472.7322	484.7322
SelenoH	LAEGATVVIEHCTSUR	+2	1054.9765	1078.9765
		+3	703.6536	719.6536
		+4	527.9922	539.9922
		+5	422.5953	432.1953
SelenoM	VETCGGUQLNR	+2	800.3341	824.3341
		+3	533.8920	549.8920
		+4	400.6709	412.6709
SelenoO	STEEQSSYSNRPPLWAAELCVTUSS	+2	1583.1783	1607.1783
		+3	1055.7882	1071.7882
		+4	792.0931	804.0931
		+5	633.8760	643.4760

Continued

Table 1 IA-alkyne and IA-Light labeled selenopeptide targeted mass lists.—cont'd

Protein	Peptide	Charge	IA-alkyne mass list	IA-Light mass list
Seleno V	VMYCGLUSYGLR	+2	892.8798	916.8798
		+3	595.5891	611.5891
		+4	446.9438	458.9438
Seleno W	VVYCGAUGYKPK	+2	854.3830	878.3830
		+3	569.9246	585.9246
		+4	427.6954	439.6954

identify both the light and heavy peptides for subsequent cimage calculation of a L/H ratio. The inclusion of the IA-Heavy masses would only dilute the targeting and the experiment should hopefully have been designed in such a way that the IA-Light labeled peptide signal will be more intense than the IA-Heavy labeled peptide signal (see Sections 2.2.3–2.2.4).

5. Click "GO" and locate the "electrospray series (positive mode)" section of the resultant pop-up window. Record all *m/z* values between 400 and 1800 that are associated with charge states between +2 and +5. **Note:** 400–1800 m/z is the detection window for the Orbitrap LTQ. This may vary on other mass spectrometers. It is unlikely that charge states higher than +5 will be identified, as these m/z values are often too small unless the peptide is very large. The same is true for +1 charge states, most peptides will harbor more than 1 charge and will have too high of an m/z value for identification.
6. Repeat for every predicted tryptic Sec-containing peptide, as determined in step 1. This should generate a mass list of ~50 m/z values (Table 1).

2.3.4 Setting and running a MudPIT analysis on an Orbitrap LTQ (timing—12 h)

1. Assemble tip and desalting column together and position on a nanospray stage attached to Orbitrap LTQ Discovery mass spectrometry equipped with an Agilent 1200 series HPLC.
2. Set up a 5-step MudPIT experiment using the elution profiles indicated in Table 2, with a flow rate of 0.25 μL/min and a spray voltage of 2.75 kV. A full MS scan (FTMS; 400–1800) is followed by 8 data-dependent scans (ITMS).
3. For each data-dependent scan the previously generated Sec-containing peptide targeted mass lists should be uploaded (IA-alkyne Sec-containing peptide masses for selenoprotein ID experiments (Section 2.2.2) and IA-Light labeled Sec-containing peptide masses for selenoprotein expression and inhibitor screening experiments (Sections 2.2.3–2.2.4).
4. The 8 data-dependent scans should be set to represent the nth most intense ions from the supplied mass list or if no more list masses are present, then the nth most intense ions otherwise. Dynamic exclusion should be disabled to improve coverage and identification of Sec-containing peptides. **Note:** Applying the above options will significantly reduce the number of labeled Cys-containing peptides identified.

Table 2 Solvent system for MudPIT LC/LC-MS/MS analysis.

Time (min)	% Buffer A	% Buffer B	% Buffer C
Step 1: 0% Buffer C			
0.00	100	0	0
5.00	100	0	0
60.00	55	45	0
70.00	0	100	0
100.00	0	100	0
Step 2: 50% Buffer C			
0.00	100	0	0
6.00	100	0	0
6.10	45	5	50
8.00	45	5	50
8.10	95	5	0
15.00	85	15	0
35.00	75	25	0
75.00	45	55	0
80.00	45	55	0
Step 3: 80% Buffer C			
0.00	100	0	0
5.00	100	0	0
5.10	15	5	80
8.00	15	5	80
8.10	95	5	0
18.00	85	15	0
63.00	75	25	0
115.00	45	55	0
120.00	45	55	0
Step 4: 100% Buffer C			
0.00	100	0	0
4.00	100	0	0

Table 2 Solvent system for MudPIT LC/LC-MS/MS analysis.—cont'd

4.10	0	0	100
20.00	0	0	100
20.10	93	7	0
25.00	85	15	0
100.00	70	30	0
184.00	0	100	0
194.00	0	100	0
195.00	100	0	0
200.00	100	0	0
Step 5: 100% Buffer C			
0.00	100	0	0
4.00	100	0	0
4.10	0	0	100
14.00	0	0	100
14.10	93	7	0
30.00	70	30	0
50.00	0	100	0
55.00	0	100	0
56.00	100	0	0
60.00	100	0	0

If coverage of both the Cys and Sec resides is desired, then an identical sample should be run with the 8 data-dependent scans representing the nth most intense ions from the MS1 scan. For these runs dynamic exclusion should be enabled (30 s, repeat of 1).

5. Run MudPit LC/LC-MS/MS analysis. **Note:** peptides are eluted from the desalting column to the SCX resin in the tip during the first step of the 5-step MudPIT procedure and therefore virtually no peptides will be identified in this run. In the next four MudPIT steps, increasing salt pushes (buffer C) at the beginning of the gradient push subpopulations of peptides off the SCX resin and onto the 10 cm C18 resin for separation with a gradient of increasing buffer B (Table 2).

2.4 Selenoproteomics data analysis

The data analysis procedure detailed below describes the use of the Sequest algorithm (Eng, McCormack, & Yates, 1994), employing an IPI mouse protein database, and the cimage program (Weerapana et al., 2010) to search for and quantify, respectively, IA-labeled Sec-containing peptides. The use of alternative proteomic identification and quantification software (e.g., Proteome Discoverer, MaxQaunt, etc.) and protein databases (e.g., Swiss-Prot) will necessitate different user generated analysis workflows. Validation of Sec-containing peptide isotopic envelopes can be performed using the Xcalibur software package, where both "experimental" and "theoretical" isotopic envelopes can be generated for comparison.

2.4.1 Materials

Xcalibur software package (Thermo Scientific)

RAW Xtract (Version 1.9.9.2; 2004 release—http://fields.scripps.edu/downloads.php)

Sequest (Yates Lab—http://fields.scripps.edu/downloads.php)

DTAselect 2.0 (Yates Lab—http://fields.scripps.edu/downloads.php)

CIMAGE (Weerapana et al., 2010)

2.4.2 Selenopeptide searching by Sequest (timing—2–3 h)

1. Convert the 5 MudPIT .raw files into 5 .ms2 files using the RAW Xtract program.
2. Search the tandem MS data (.ms2 files). with the SEQUEST algorithm (Eng et al., 1994) for either IA-alkyne, IA-Light, or IA-Heavy labeled Sec-containing peptides as appropriate against a mouse reverse-concatenated non-redundant FASTA database combining International Protein Index (IPI) and Ensembl identifiers. **Note:** Sequest does not automatically recognize the one letter code for Sec (U), so it is necessary to use a database, such as IPI where "U" has been replaced with "C" for all selenoproteins.
3. Set the precursor-ion mass tolerance at 50 ppm and the fragment-ion mass tolerance to the default setting of 0.
4. Set a static modification on Cys residues (+**57.0215** *m/z*, iodoacetamide alkylation) with no assumed enzyme specificity.
5. For Sec identification (Section 2.2.2), specify a differential mass on Cys for IA-alkyne labeling (**+306.0925** m/z) and perform a single SEQUEST search. For selenoprotein expression quantification or inhibitor screening

(Sections 2.2.3–2.2.4), independently perform two SEQUEST searches for the specified differential mass on Cys for either IA-Light (**+354.0925** m/z) or IA-Heavy (**+360.1126** m/z). **Note:** Since "U" has been replaced with "C" across the protein database, the specified differential masses on Cys also include the mass difference between selenium and sulfur (**+47.9445**) As such, these differential masses represent IA-labeled Sec residues, and not IA-labeled Cys residues. As a result, true IA-labeled Cys-containing peptides will not be identified. If necessary, separate SEQUEST searches can be performed to identify modified Cys residues using differential masses for IA-alkyne (**+258.1481** m/z), IA-Light (**+306.1481** m/z), or IA-Heavy (**+312.1682** m/z) that don't contain the added mass of selenium.

6. Assemble MS2 spectral matches into protein identifications and filter using DTASelect2.0 (Tabb, McDonald, & Yates, 2002) to generate a list of hits with a peptide false-discovery rate of 5%, with the –trypstat and –modstat options applied. Peptides were restricted to fully tryptic (−y 2) with a found modification (−m 0) and a delta-CN score greater than 0.06 (−d 0.06). Single peptides per locus were also allowed (−p 1) as were redundant peptide identifications from multiple proteins, but the database contained only a single consensus splice variant for each protein

2.4.3 *Manual validation of selenopeptide isotopic envelopes (timing—2–3 h)*

1. For each identified Sec-containing peptide in the filtered dtaselect results files, calculate the chemical formula for the IA-alkyne, IA-Light, or IA-Heavy labeled tryptic peptide. **Note:** Make sure to calculate the chemical formula for the Sec and not the Cys version of the peptide.
2. In the Xcalibur software package calculate the theoretical isotopic envelope for the appropriate charge state of the peptide using the previously determined chemical formula. **Note:** It is necessary to return to the filtered dtaselect results file and determine the charge state of the identified peptide and add the appropriate number of hydrogen atoms to the chemical formula and indicate the charge state to be determined (e.g., for a + 3 peptide, add 3 hydrogen atoms to the chemical formula and search for the +3 charge state).
3. Save this calculate m/z ion distribution as a "theoretical" isotopic envelope.
4. For each identified Sec-containing peptide in the filtered dtaselect results files, search the full MS TIC for the identified peptide m/z value to

generate an extracted ion chromatogram (calculate this from the value in the ***CalcM*+*H*+** column using the formula (***CalcM*$^+$*H*$^+$** + (**charge** − 1)/**charge**, where **charge** is the charge state of the identified peptide). Set the mass precision to four decimal place and the mass tolerance to 50 ppm.

5. Select an appropriate full MS scan from the most intense portion of the extracted ion peak and center the view window on the appropriate mass range (±10–20 m/z around the extracted ion value).
6. Save this view as the "experimental" isotopic envelope.
7. Compare, the "theoretical" and "experimental" isotopic envelopes for both the location and relative intensity of the isotopic peaks. These should be superimposable.

2.4.4 Selenopeptide quantification by Cimage (timing—2–3 h)

1. Convert the 5 MudPIT .raw files into 5 .mzxml files.
2. In the filtered dtaselect results files, manually reconvert Cys (C) residues back to Sec (U) for each identified peptide sequence. **Note:** This is only necessary for the 34 annotated mouse Sec residues, and not for other identified Cys residues, which are likely to be misidentifications.
3. Calculate peptide light to heavy (L:H) ratios with the cimage quantification package (Weerapana et al., 2010) using the modified DTASelect result files and the previously generated .mzxml as inputs. **Note:** In order to quantify L:H ratios from Sec-containing peptides an additional column and row for selenium and selenocysteine, respectively, was included into the cimage.params table.

3. Results and discussion

3.1 Identification of selenoproteins in mouse tissues and cell lines

As a demonstration of the utility of our low-pH isoTOP-ABPP platform, soluble mouse liver lysates were prepared at both neutral (7.5) and low pH (5.75). These were then labeled with IA-alkyne, appended to diazo biotin-azide by CuACC and enriched for MS analysis as described previously. Of the 24 mouse selenoproteins, only Gpx1 and Txnrd2 were identified from the resultant fragmentation data by neutral-pH IA-alkyne labeling (Fig. 4A). In contrast, IA-alkyne labeling at low pH resulted in the identification of 6 soluble mouse selenoproteins, Gpx1, Gpx4, Txnrd1, Txnrd2, Msrb1, and Sephs2 (Fig. 4A). Despite this improved selenoprotein coverage, only a single MS2 fragmentation spectrum (spectral count) was

A. Selenoprotein Spectral Counts

Selenopeptide		pH 7.50		pH 5.75	
		- ML	+ ML	- ML	+ ML
Gpx1	*K.VLLIENVASLUGTTIR.D*	22	123	44	2317
Gpx4	*R.GFVCITNVASQUGK.T*	-	-	1	16
Txnrd1	*R.SGGDILQSGCUG.-*	-	3	3	7
Txnrd2	*R.SGLEPTVTGCUG.-*	1	-	1	5
MsrB1	*R.FUIFSSSLK.F*	-	-	1	29
Sephs2	*K.GUGCKVPQETLLK.L*	-	-	4	23
SelenoW	*R.VVYCGAUGYKPK.Y*	-	-	-	10

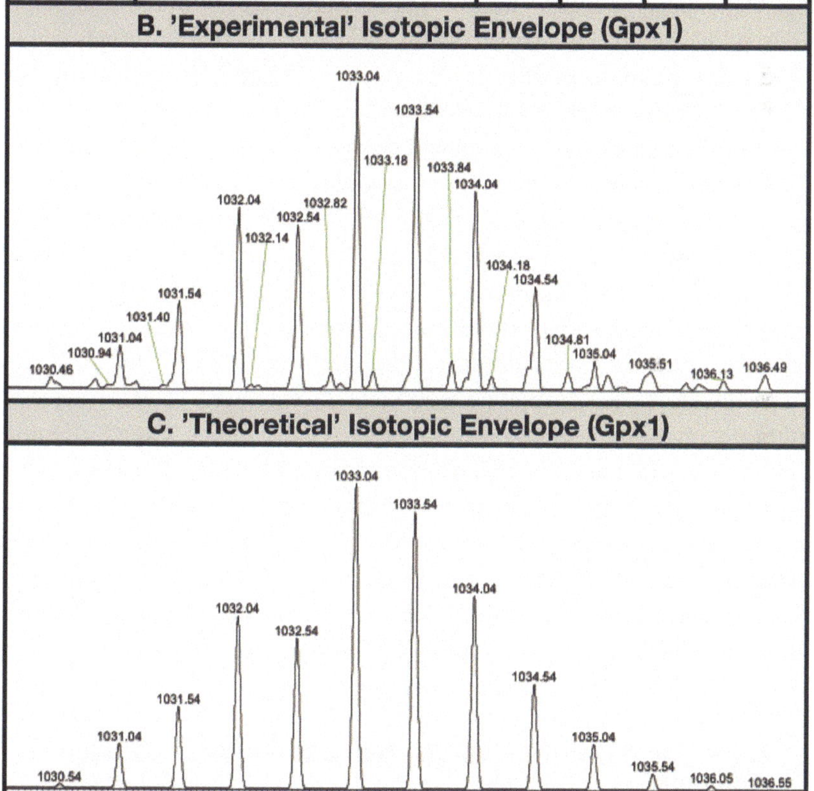

Fig. 4 Identification of Sec-containing peptides in mouse liver. (A) Sec-peptides were identified by LC/LC-MS/MS analysis in mouse liver cells by IA-alkyne labeling under both neutral (pH 7.5) and reduced (pH 5.75) pH conditions, with or without the inclusion of targeted mass list (ML). (B) "Experimental" and (C) "theoretical" isotopic envelopes for the Gpx1 Sec-containing peptide, K.VLLIENVASLUGTTIR.D.

observed for half of the low-pH identified selenoproteins. To improve the reliability of selenoprotein identification a Sec-containing peptide targeted mass list was applied. This significantly improved the number of fragmentation spectra generated for each of the observed selenoproteins (>5 spectra/peptide), as well as resulted in the identification of a seventh selenoprotein, SelenoW (Fig. 4A). The application of a targeted mass list to the neutral pH labeled sample did result in an increase in the number of fragmentation spectra generated for Gpx1, but did not improve overall selenoprotein coverage (Fig. 4A). To validate Sec-containing peptide identifications, the unique isotopic envelope of Sec residues was experimentally verified (Fig. 4B) and manually compared to a theoretically generated isotopic envelope for the corresponding peptide (Fig. 4C). The application of a low-pH isoTOP-ABPP workflow in combination with a Sec-specific targeted mass list results in significant improvement in the coverage and reliability of Sec-containing peptide identification compared to a standard isoTOP-ABPP workflow.

To further expand our coverage of the mouse selenoproteome, a panel of 5 mouse tissues as well as the mouse macrophage cell line, RAW264.7, were examined by low-pH isoTOP-ABPP (Fig. 5). A strong correlation was observed between selenoprotein gene-expression data (Su et al., 2002; Wu et al., 2016) and identification by low-pH isoTOP-ABPP. Specifically, a number of selenoprotein identifications demonstrated tissue specificity, including Gpx2, a gastrointestinal Gpx, Txnrd3, a testes specific Txnrd, and SelenoM, a selenoprotein predominantly expressed in brain tissue. In contrast many housekeeping selenoproteins (Gpx1, Gpx4, Txnrd1, Txnrd2, Msrb1) were identified across all or most tissues and the RAW264.7 cell line. This result suggests that selenoprotein identification is mostly limited by low expression, which can be overcome by low-pH isoTOP-ABPP in those tissues that achieve a critical abundance threshold. In total, 12 of the 14 soluble mouse selenoproteins could be identified (all except SelenoV and SelenoO), demonstrating the utility of the low-pH isoTOP-ABPP platform to globally profile selenoproteins within a complex proteome.

3.2 Quantification of selenoprotein expression and Sec reactivity

We next applied our platform to quantitatively examine changes in selenoprotein expression as a function of selenium availability. Certain selenoproteins are well known to display selenium-dependent expression changes (Gpx1/4, Msrb1, and SelenoM), while selenoproteins like Txnrd1/2 appear to be insensitive to the intracellular levels of selenium

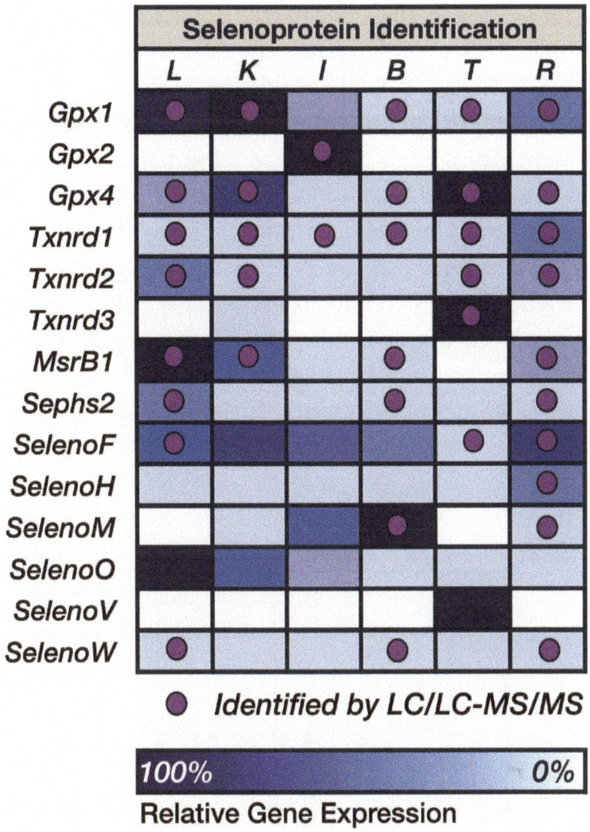

Fig. 5 Identification of selenoproteins in mouse tissue and RAW264.7 cells. IA-alkyne labeled Sec-containing peptides were identified from the soluble proteomes of 5 mouse tissues, liver (L), kidney (K), intestines (I), brain (B), and testes (T), as well as the mouse macrophage cell line, RAW264.7 (R). Purple dots indicate identification of a selenoprotein (row) in specified tissue (column). Blue shading is indicative of relative tissue specific gene expression levels (obtained from BioGPS; Su et al., 2002; Wu, Jin, Tsueng, Afrasiabi, & Su, 2016).

(Legrain, Touat-Hamici, & Chavatte, 2014; Sunde & Raines, 2011). To globally quantify changes in selenoprotein expression, Sec-containing peptides from RAW264.7 cell lysates cultured in 50 nM (light) or 0 nM (heavy) $NaSeO_3$ supplemented media were enriched by IA-Light or IA-Heavy labeling (as a relative measure of selenoprotein abundance). To quantify the extent of IA-Light and IA-Heavy labeled Sec-containing peptides, the cimage quantification software (Weerapana et al., 2010) was adapted for the presence of selenium. Light: heavy (L/H) ratios, were calculated from the relative intensities of the extracted ion chromatograms of co-eluting IA-light (50 nM $NaSeO_3$) and IA-heavy (0 nM $NaSeO_3$) labeled Sec-containing peptide signals. A L/H ratio (R) value ~1 indicates no significant change in IA labeling (suggesting no protein abundance increase upon selenium supplementation), while an R value >1 indicates increased IA labeling (protein abundance) upon selenium supplementation. Sec-containing peptides from 5 selenoproteins were quantified (Fig. 6). Txnrd1 displayed an R value of 1.61, consistent with its position as one of the least selenium-dependent selenoproteins. Conversely, Gpx4, Msrb1, SelenoM, and SelenoW all displayed *R* values >2, which corresponds with their known selenium-dependent expression profiles (Legrain et al., 2014; Sunde & Raines, 2011). In contrast no significant selenium-dependent increase was observed for Cys-containing peptides, and all 5 Sec-containing peptides

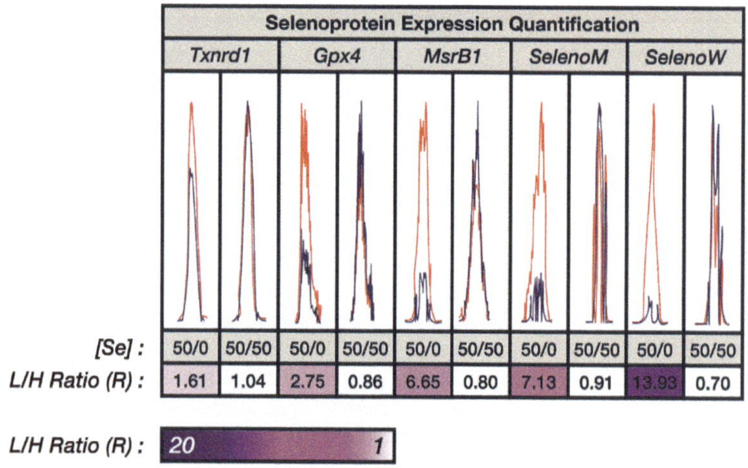

Fig. 6 Quantification of selenoprotein abundance changes in response to selenium availability in RAW264.7 cells. The extracted light (red) and heavy (blue) ion chromatograms are shown for 5 Sec-containing peptides from RAW264.7 cells grown with or without 50 nM $NaSeO_3$ supplementation. The indicated L/H ratio (R) for these Sec-containing peptides is a presumed measure of the change in the corresponding selenoprotein abundance as a function of supplemental selenium. The left panel for each protein represents the change in reactivity for IA-Light (50 nM $NaSeO_3$) versus IA-Heavy (0 nM $NaSeO_3$) labeled Sec-containing peptides. The right panel for each protein represents a control for the change in reactivity with IA-Light (50 nM $NaSeO_3$) versus IA-Heavy (50 nM $NaSeO_3$) labeled Sec-containing peptides. Purple shading is indicative of the relative sensitivity for induction of selenoprotein expression upon $NaSeO_3$ supplementation.

had R values ~1 in control experiments where 50 nM $NaSeO_3$ supplemented RAW264.7 cells were used for labeling with both IA-Light and IA-Heavy.

Sec reactivity can also be monitored in the absence of predicted protein abundance changes, for example as a function of inhibitor binding. In our previous study (Bak et al., 2018) we examined the effect of auranofin treatment on RAW264.7 cells. The cellular targets of auranofin, a gold-base therapeutic and known inhibitor of Txnrd1 (Marzano et al., 2007; Rigobello, Folda, Baldoin, Scutari, & Bindoli, 2009), have yet to be fully characterized at the proteome level. In order to quantify the specificity of auranofin for Txnrd1 against other Sec and Cys residues in the proteome, low-pH isoTOP-ABPP was applied to lysates from vehicle (IA-Light) or auranofin (IA-Heavy) treated RAW264.7 cells. A L/H ratio (R) value >1 indicates a significant loss of IA labeling and sensitivity to auranofin treatment. This analysis identified Txnrd1 (R: 2.52) as a target of auranofin, but also identified a number of other Sec and Cys residues with R values >2; Gpx4 (R: 4.41), Msrb1 (R: 3.34), Txndc5 (R: 23.88) and Clic1 (R: 6.18). Identified Sec and Cys residues with high R values were enriched in predicted functional disulfides, suggesting a common binding mode. This result demonstrates the utility of our low-pH isoTOP-ABPP to screen Sec-specific inhibitor against a complex proteome.

3.3 Future applications and directions

This chapter has focused on the application and technical aspects of low-pH isoTOP-ABPP for the identification and quantification of Sec residues, but it is important to highlight limitations in the workflow as well as alternative strategies to globally study the selenoproteome. The main limitation is the requirement for IA labeling at reduced pH values, as many proteins may lose their structural integrity and activity outside of a narrow neutral pH range. This issue is not particularly relevant for the identification and quantification of selenoprotein abundance, as active protein is not required for Sec labeling. It is however relevant when considering measurements of Sec reactivity, such as during inhibitor treatment and oxidation. As a control, it was demonstrated that auranofin inhibition of Txnrd1 still occurred at reduced pH (Bak et al., 2018), suggesting that inhibitor screening is still a viable application of low-pH isoTOP-ABPP. In the future, screening of potential Sec selective electrophiles will hopefully result in the identification of a probe capable of enriching Sec residues for proteomic interrogation in the neutral pH range. Such a Sec selective probe for isoTOP-ABPP would be a valuable tool for identifying therapeutic inhibitors of disease relevant selenoproteins, such as Txnrd1, as well as an important probe to functionally interrogate poorly characterized selenoproteins. Alternative proteomic strategies for the identification of Sec residues have relied on computational methods to assist in the detection of selenium-containing peptide signals based on the unique isotopic proprieties of selenium (Gao et al., 2018). Importantly, computational approaches are complementary to our low-pH isoTOP-ABPP workflow, and when combined result in improved identification of selenoproteins. Finally, while we have described the application of our platform to the characterization of known selenoproteins, recent work has demonstrated that Sec can be incorporated at non-canonical sites in place of Cys residues under high selenium growth conditions (Hoffman et al., 2019; Jedrychowski et al., 2020). Our understanding of what drives this misincorporation is poor, and proteomic strategies such as low-pH isoTOP-ABPP will be useful in providing important insight into the dynamics of Sec incorporation at both canonical and non-canonical sites within the proteome.

References

Abo, M., Li, C., & Weerapana, E. (2018). Isotopically-labeled iodoacetamide-alkyne probes for quantitative cysteine-reactivity profiling. *Molecular Pharmaceutics, 15*, 743–749.

Al-Taie, O. H., Uceyler, N., Eubner, U., Jakob, F., Mork, H., Scheurlen, M., et al. (2009). Expression profiling and genetic alterations of the selenoproteins GI-GPx and SePP in colorectal carcinogenesis. *Nutrition and Cancer, 48*, 6–14.

Arnér, E. S. J. (2010). Selenoproteins—What unique properties can arise with selenocysteine in place of cysteine? *Experimental Cell Research, 316*, 1296–1303.

Backus, K. M., Correia, B. E., Lum, K. M., Forli, S., Horning, B. D., González-Páez, G. E., et al. (2016). Proteome-wide covalent ligand discovery in native biological systems. *Nature, 534*, 570–574.

Bak, D. W., Gao, J., Wang, C., & Weerapana, E. (2018). A quantitative chemoproteomic platform to monitor selenocysteine reactivity within a complex proteome. *Cell Chemical Biology, 25*(9), 1157–1167.

Bak, D. W., Pizzagalli, M. D., & Weerapana, E. (2017). Identifying functional cysteine residues in the mitochondria. *ACS Chemical Biology, 12*, 947–957.

Cadenas, C., Franckenstein, D., Schmidt, M., Gehrmann, M., Hermes, M., Geppert, B., et al. (2010). Role of thioredoxin reductase 1 and thioredoxin interacting protein in prognosis of breast cancer. *Breast Cancer Research, 12*, R44.

Calvo, A., Xiao, N., Kang, J., Best, C. J. M., Leiva, I., Emmert-Buck, M. R., et al. (2002). Alterations in gene expression profiles during prostate cancer progression: Functional correlations to tumorigenicity and down-regulation of selenoprotein-P in mouse and human tumors. *Cancer Research, 62*, 5325–5335.

Davis, C. D., Tsuji, P. A., & Milner, J. A. (2012). Selenoproteins and cancer prevention. *Annual Review of Nutrition, 32*, 73–95.

Donovan, J., & Copeland, P. R. (2010). Threading the needle: Getting selenocysteine into proteins. *Antioxidants & Redox Signaling, 12*, 881–892.

Dyson, H. J., Jeng, M.-F., Tennant, L. L., Slaby, I., Lindell, M., Cui, D.-S., et al. (1997). Effects of buried charged groups on cysteine thiol ionization and reactivity in Escherichia coli thioredoxin: Structural and functional characterization of mutants of asp 26 and Lys 57. *Biochemistry, 36*, 2622–2636.

Eng, J. K., McCormack, A. L., & Yates, J. R. (1994). An approach to correlate tandem mass spectral data of peptides with amino acid sequences in a protein database. *Journal of the American Society for Mass Spectrometry, 5*(11), 976–989.

Fernandes, A. P., Capitanio, A., Selenius, M., Brodin, O., Rundlöf, A. K., & Björnstedt, M. (2009). Expression profiles of thioredoxin family proteins in human lung cancer tissue: Correlation with proliferation and differentiation. *Histopathology, 55*, 313–320.

Gao, J., Yang, F., Che, J., Han, Y., Wang, Y., Chen, N., et al. (2018). Selenium-encoded isotopic signature targeted profiling. *ACS Central Science, 4*(8), 960–970.

Giron, P., Dayon, L., & Sanchez, J.-C. (2010). Cysteine tagging for MS-based proteomics. *Mass Spectrometry Reviews, 30*, 366–395.

Guo, L., Yang, W., Huang, Q., Qiang, J., Hart, J. R., Wang, W., et al. (2018). Selenocysteine-specific mass spectrometry reveals tissue-distinct selenoproteomes and candidate selenoproteins. *Cell Chemical Biology, 25*(11), 1380–1388.

Handy, D. E., & Loscalzo, J. (2016). Selenoproteins in cardiovascular redox pathology. In D. Hatfield, U. Schweizer, P. Tsuji, & V. Gladyshev (Eds.), *Selenium. Its molecular biology and role in human health* (pp. 463–474). Springer.

Hatfield, D. L., Tsuji, P. A., Carlson, B. A., & Gladyshev, V. N. (2014). Selenium and selenocysteine: Roles in cancer, health, and development. *Trends in Biochemical Sciences, 39*, 112–120.

Hoffman, K. S., Vargas-Rodriguez, O., Bak, D. W., Mukai, T., Woodward, L. K., Weerapana, E., et al. (2019). A cysteinyl-tRNA synthetase variant confers resistance against selenite toxicity and decreases selenocysteine misincorporation. *The Journal of Biological Chemistry, 294*(34), 12855–12865.

Hondal, R. J., & Ruggles, E. L. (2011). Differing views of the role of selenium in thioredoxin reductase. *Amino Acids, 41*, 73–89.

Huber, R. E., & Criddle, R. S. (1967). Comparison of the chemical properties of selenocysteine and selenocystine with their sulfur analogs. *Archives of Biochemistry and Biophysics, 122*, 164–173.

Ingold, I., Berndt, C., Schmitt, S., Doll, S., Poschmann, G., Buday, K., et al. (2018). Selenium utilization by GPX4 is required to prevent hydroperoxide-induced ferroptosis. *Cell, 172*, 409–412. e21.

Jao, S. C., English Ospina, S. M., Berdis, A. J., Starke, D. W., Post, C. B., & Mieyal, J. J. (2006). Computational and mutational analysis of human glutaredoxin (thioltransferase): Probing the molecular basis of the low pKa of cysteine 22 and its role in catalysis. *Biochemistry, 45*, 4785–4796.

Jedrychowski, M. P., Lu, G. Z., Szpyt, J., Mariotti, M., Garrity, R., Paulo, J. A., et al. (2020). Facultative protein selenation regulates redox sensitivity, adipose tissue thermogenesis, and obesity. *Proceedings of the National Academy of Sciences of the United States of America, 117*(20), 10789–10796.

Kanzok, S. M., Fechner, A., Bauer, H., Ulschmid, J. K., Muller, H.-M., Botella-Munoz, J., et al. (2001). Substitution of the thioredoxin system for glutathione reductase in Drosophila melanogaster. *Science, 291*, 643–646.

Karlenius, T. C., Shah, F., Yu, W. C., & Hawkes, H. J. (2011). The selenium content of cell culture serum influences redox-regulated gene expression. *Biotechniques, 50*, 295–301.

Kaushal, N., Gandhi, U. H., Nelson, S. M., Narayan, V., & Prabhu, K. S. (2011). Selenium and inflammation. In D. L. Hatfield, M. J. Berry, & V. L. Gladyshev (Eds.), *Selenium. Its molecular biology and role in human health* (pp. 443–456). Springer.

Kryukov, G. V., Castellano, S., Novoselov, S. V., Lobanov, A. V., Zehtab, O., Guigó, R., et al. (2003). Characterization of mammalian selenoproteomes. *Science, 300*, 1439–1443.

Kühbacher, M., Bartel, J., Hoppe, B., Alber, D., Bukalis, G., Bräuer, A. U., et al. (2009). The brain selenoproteome: Priorities in the hierarchy and different levels of selenium homeostasis in the brain of selenium-deficient rats. *Journal of Neurochemistry*, *110*, 133–142.

Labunskyy, V. M., Hatfield, D. L., & Gladyshev, V. N. (2014). Selenoproteins: Molecular pathways and physiological roles. *Physiological Reviews*, *94*, 739–777.

Legrain, Y., Touat-Hamici, Z., & Chavatte, L. (2014). Interplay between selenium levels, selenoprotein expression, and replicative senescence in WI-38 human fibroblasts. *The Journal of Biological Chemistry*, *289*, 6299–6310.

Li, M., Feng, W., Zhai, Q., & Feng, G. (2017). Selenocysteine detection and bioimaging in living cells by a colorimetric and near-infrared fluorescent turn- on probe with a large stokes shift. *Biosensors & Bioelectronics*, *87*, 894–900.

Lobanov, A. V., Fomenko, D. E., Zhang, Y., Sengupta, A., Hatfield, D. L., & Gladyshev, V. N. (2007). Evolutionary dynamics of eukaryotic selenoproteomes: Large selenoproteomes may associate with aquatic life and small with terrestrial life. *Genome Biology*, *8*, R198.

Lobanov, A. V., Hatfield, D. L., & Gladyshev, V. N. (2008). Selenoproteinless animals: Selenophosphate synthetase SPS1 functions in a pathway unrelated to selenocysteine biosynthesis. *Protein Science*, *17*, 176–182.

Lobanov, A. V., Hatfield, D. L., & Gladyshev, V. N. (2009). Eukaryotic selenoproteins and selenoproteomes. *Biochimica et Biophysica Acta*, *1790*, 1424–1428.

Lu, J., & Holmgren, A. (2009). Selenoproteins. *The Journal of Biological Chemistry*, *284*, 723–727.

Mariotti, M., Lobanov, A. V., Guigó, R., & Gladyshev, V. N. (2013). SECISearch3 and seblastian: New tools for prediction of SECIS elements and selenoproteins. *Nucleic Acids Research*, *41*, e149.

Marques, M. C., Tapia, C., Gutiérrez-Sanz, O., Ramos, A. R., Keller, K. L., Wall, J. D., et al. (2017). The direct role of selenocysteine in [NiFeSe] hydrogenase maturation and catalysis. *Nature Chemical Biology*, *13*, 544–550.

Martell, J., Seo, Y., Bak, D. W., Kingsley, S. F., Tissenbaum, H. A., & Weerapana, E. (2016). Global cysteine-reactivity profiling during impaired insulin/IGF-1 signaling in C. elegans identifies uncharacterized mediators of longevity. *Cell Chemical Biology*, *23*, 955–966.

Marzano, C., Gandin, V., Folda, A., Scutari, G., Bindoli, A., & Rigobello, M. P. (2007). Inhibition of thioredoxin reductase by auranofin induces apoptosis in cisplatin-resistant human ovarian cancer cells. *Free Radical Biology & Medicine*, *42*, 872–881.

Qian, Y., Martell, J., Pace, N. J., Ballard, T. E., Johnson, D. S., & Weerapana, E. (2013). An isotopically tagged azobenzene-based cleavable linker for quantitative proteomics. *Chembiochem*, *14*, 1410–1414.

Rigobello, M. P., Folda, A., Baldoin, M. C., Scutari, G., & Bindoli, A. (2009). Effect of Auranofin on the mitochondrial generation of hydrogen peroxide. Role of thioredoxin reductase. *Free Radical Research*, *39*, 687–695.

Snider, G. W., Ruggles, E., Khan, N., & Hondal, R. J. (2013). Selenocysteine confers resistance to inactivation by oxidation in thioredoxin reductase: Comparison of selenium and sulfur enzymes. *Biochemistry*, *52*, 5472–5481.

Su, A. I., Cooke, M. P., Ching, K. A., Hakak, Y., Walker, J. R., Wiltshire, T., et al. (2002). Large-scale analysis of the human and mouse transcriptomes. *Proceedings of the National Academy of Sciences of the United States of America*, *99*, 4465–4470.

Sunde, R. A., & Raines, A. M. (2011). Selenium regulation of the selenoprotein and non-selenoprotein transcriptomes in rodents. *Advances in Nutrition*, *2*, 138–150.

Tabb, D. L., McDonald, W. H., & Yates, J. R. (2002). DTASelect and contrast: Tools for assembling and comparing protein identifications from shotgun proteomics. *Journal of Proteome Research*, *1*(1), 21–26.

Weerapana, E., Simon, G. M., & Cravatt, B. F. (2008). Disparate proteome reactivity profiles of carbon electrophiles. *Nature Chemical Biology*, *4*(7), 405–407.

Weerapana, E., Speers, A. E., & Cravatt, B. F. (2007). Tandem orthogonal proteolysis-activity-based protein profiling (TOP-ABPP)—A general method for mapping sites of probe modification in proteomes. *Nature Protocols*, *2*(6), 1414–1425.

Weerapana, E., Wang, C., Simon, G. M., Richter, F., Khare, S., Dillon, M. B. D., et al. (2010). Quantitative reactivity profiling predicts functional cysteines in proteomes. *Nature*, *468*, 790–795.

Whitby, L. R., Obach, R. S., Simon, G. M., Hayward, M. M., & Cravatt, B. F. (2017). Quantitative chemical proteomic profiling of the in vivo targets of reactive drug metabolites. *ACS Chemical Biology*, *12*, 2040–2050.

Wu, C., Jin, X., Tsueng, G., Afrasiabi, C., & Su, A. I. (2016). BioGPS: Building your own mash-up of gene annotations and expression profiles. *Nucleic Acids Research*, *44*, D313–D316.

Xu, X.-M., Carlson, B. A., Mix, H., Zhang, Y., Saira, K., Glass, R. S., et al. (2006). Biosynthesis of selenocysteine on its tRNA in eukaryotes. *PLoS Biology*, *5*, e4.

Yoo, M.-H., Carlson, B. A., Tsuji, P., Irons, R., Gladyshev, V. N., & Hatfield, D. L. (2010). Alteration of thioredoxin reductase 1 levels in elucidating cancer etiology. In E. Gladyshev, & L. Packer (Eds.), *Thiol redox transitions in cell signaling, part B: Cellular localization and signaling* (pp. 255–275). Elsevier.

Zhang, B., Ge, C., Yao, J., Liu, Y., Xie, H., & Fang, J. (2015). Selective selenol fluorescent probes: Design, synthesis, structural determinants, and biological applications. *Journal of the American Chemical Society*, *137*, 757–769.

Zhang, Y., Romero, H., Salinas, G., & Gladyshev, V. N. (2006). Dynamic evolution of selenocysteine utilization in bacteria: A balance between selenoprotein loss and evolution of selenocysteine from redox active cysteine residues. *Genome Biology*, *8*, R94.

Zhou, Y., Wynia-Smith, S. L., Couvertier, S. M., Kalous, K. S., Marletta, M. A., Smith, B. C., et al. (2016). Chemoproteomic strategy to quantitatively monitor trans-nitrosation uncovers functionally relevant S-nitrosation sites on cathepsin D and HADH2. *Cell Chemical Biology*, *23*, 727–737.

CHAPTER NINE

SecMS analysis of selenoproteins with selenocysteine insertion sequence and beyond

Chenfang Si, Ye Cao, and Yaoyang Zhang*
Interdisciplinary Research Center on Biology and Chemistry, Shanghai Institute of Organic Chemistry, Chinese Academy of Sciences, Shanghai, China
*Corresponding author: e-mail address: zyy@sioc.ac.cn

Contents

Abstract

Selenocysteine (Sec, U) is the 21st amino acid, and proteins with selenocysteine are defined as selenoproteins. The currently known selenoproteins are all featured by the presence of selenocysteine insertion sequence (SECIS) on their mRNA, and SECIS plays an essential role in the selenocysteine insertion mechanism. However, due to the extremely low occurrences of selenoproteins in a whole proteome (e.g., only 25 selenoproteins in the human proteome) and the low sequence conservation of SECIS, analysis of selenoproteins and discovery of new selenoproteins exclusively on SECIS are intrinsically challenging. To this end, the selenocysteine-specific mass

Methods in Enzymology, Volume 662
ISSN 0076-6879
https://doi.org/10.1016/bs.mie.2021.10.022

spectrometry (SecMS) and SECIS-independent selenoprotein (SIS) database are developed, showing abilities to profile whole selenoproteomes sensitively and to discover potential new selenoproteins. Here, we detail the SecMS strategy and propose it will advance the exploration for new selenoproteins and functional studies of selenoproteins.

1. Introduction

Selenium is an essential trace element (Green, 2018) and mainly functions in the form of selenocysteine (Sec, U) and selenoprotein in living organisms. Selenoproteins have been found in all three domains of life, highlighting their essential roles in the evolutionary process.

Selenocysteine is an analog of cysteine (Cys, C), in which sulfur is replaced by selenium (Fig. 1). In chemistry, cysteine and selenocysteine side chains are both nucleophilic, but selenocysteine is more acidic (pK_a is approximately 8.3 for Cys and 5.2 for Sec; Huber & Criddle, 1967), thus selenocysteine is more prone to be deprotonated at a physiological pH with an elevated reactivity. This feature makes selenoproteins very suitable in redox regulation, including enzymes of glutathione peroxidases (Gpx) 1–4 and 6, thioredoxin reductases

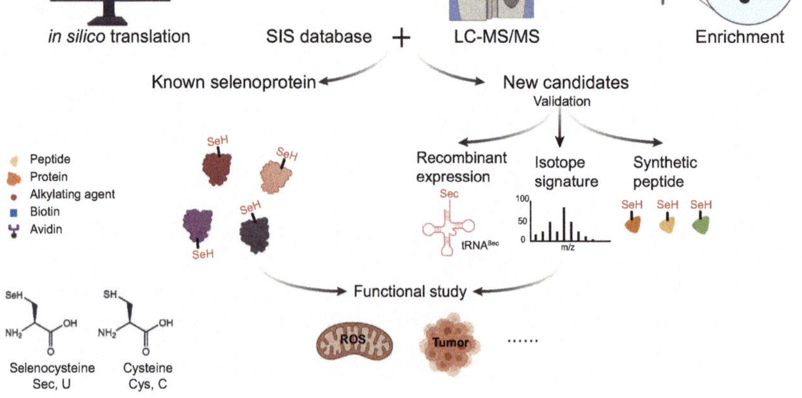

Fig. 1 Overview of SecMS and SIS database.

(Txnrd) 1–3, and methionine-R-sulfoxide reductase B1 (MsrB1) (Lee et al., 2013; Peeler & Weerapana, 2019; Tarrago, Kaya, Weerapana, Marino, & Gladyshev, 2012).

The biosynthesis of selenoproteins undertakes a unique mechanism. A conventional mRNA stop codon UGA is hijacked by the selenocysteine insertion machinery to incorporate selenocysteine instead of terminating the translation (Low & Berry, 1996). This specific machinery consists of a stem-loop structured selenocysteine insertion sequence (SECIS) on mRNA, SECIS-binding protein 2 (SBP2) in eukaryotes, selenocysteine-specific $tRNA^{Sec}$, and other elements (Mariotti et al., 2016). The currently known selenoproteins are all featured by a SECIS on their mRNA, and screening for a SECIS has been demonstrated to be a useful approach to discover new selenoproteins (Kryukov et al., 2003). Collectively, 25 human and 24 mouse selenoproteins have been identified with SECIS. SECIS-dependent selenoprotein database, SelenoDB, has been established (Castellano, Gladyshev, Guigó, & Berry, 2008; Romagné et al., 2014). This computational approach predicts the existence of several new selenoprotein candidates. However, it is unclear if there are selenoproteins without SECIS or without a canonical SECIS structure, as large-scale screen has indicated thousands of selenium-containing signals in human proteomes (Gao et al., 2018).

The lack of a robust analytical method for selenoproteins has limited our ability to systematically characterize selenoproteins and discover new selenoproteins. To this end, we develop a new selenocysteine-specific mass spectrometry-based technique, named SecMS, that allows the profiling of the selenoproteome and the discovery of new selenoproteins (Guo et al., 2018). SecMS is built upon an integrated workflow that combines a selenocysteine-specific peptide enrichment method based on selective alkylation of selenocysteine at a low pH with a SECIS-independent selenoprotein (SIS) database (Fig. 1). SIS is designed to translate the UGA codon to selenocysteine *in silico*. With SecMS, we could profile almost all selenoproteins in mice. The combination of SecMS and SIS database enabled to discover at least five new selenoprotein candidates with high confidence (Guo et al., 2018). Our discovery marks the first expansion of the mouse selenoproteome in near two decades after characterizing it using SECIS.

In this chapter, we will discuss the development and applications of SecMS and SIS for selenoprotein analysis. These methods will shed lights on the discovery of new selenoproteins and as yet undetermined SECIS structures, and open up new directions for selenium biology.

2. SecMS and SIS database

In-depth mass spectrometry-based analysis of a sub-proteome (e.g., phosphoproteome and acetylome) requires an enrichment step to isolate the target peptides or proteins. The enrichment step can be accomplished using chemical derivatization, chemical materials, or antibody-based affinity purification (Guo, Gao, Gao, Zhu, & Zhang, 2020; Zhang, Fonslow, Shan, Baek, & Yates 3rd., 2013). However, developing such an enrichment strategy for selenopeptides is difficult as the proteinogenic selenocysteine shares a very similar chemical property with cysteine but has a much lower abundance. Hence many nucleophilic reactions developed targeting the thiol group on cysteine hardly differentiate between cysteine and selenocysteine under normal conditions.

To make chemical derivatization that occurs selectively on selenocysteine, we developed the SecMS method. The basis of SecMS is that under an acidic condition, cysteine is protonated thus has limited reactivity, while selenocysteine, which has a lower p$K_{a,}$ is deprotonated. Thereafter, an alkylating agent conjugated with biotin, specifically EZ-Link iodoacetyl-PEG2-biotin (IodoAPB) in this chapter, will react preferentially with selenocysteine at a low pH. Similar experiments can also be performed by using electrophilic chemical probes followed by click-chemistry (Bak, Gao, Wang, & Weerapana, 2018). With synthetic peptides containing cysteine or selenocysteine, we showed that the alkylation could selectively occur on selenopeptides but not their cysteine counterparts under a pH of 4 (Guo et al., 2018).

The SecMS was further applied to different cell and animal systems. First, the expression levels of selenoproteins were generally up-regulated with selenium supplementation in cultured mouse microglial BV2 cells, and such enhanced selenoproteome could significantly decrease diamide-induced cellular reactive oxygen species (ROS). The SecMS analyses of multiple mouse tissues revealed a tissue-specific distribution of selenoproteome, highlighting its regulations and roles in the aging process. Altogether, SecMS allowed to identify 22 out of the 24 known mouse selenoproteins. The undetected selenoprotein DIO2 has thyroid expression specificity, which was not sampled in this study. The second undetectable GPX3 has an extremely long selenopeptide after trypsin digest. These data support that SecMS is a powerful method in characterizing whole selenoproteomes from complex organisms (Guo et al., 2018).

The most striking feature of SecMS is its ability to explore as yet unknown selenoproteins when it couples to the SIS database. The presence of SECIS has been a dogma in the field and many selenoproteins were firstly discovered based upon the prediction of SECIS (Mariotti, Lobanov, Guigo, & Gladyshev, 2013; Santesmasses, Mariotti, & Gladyshev, 2020). As the sequence conservation of SECIS of different selenoproteins is low, we question if there are new selenoproteins lacking a canonical SECIS on their mRNA. We hypothesize the UGA codon is still required for selenocysteine insertion and translate all UGA codons in a genome to selenocysteine instead of terminating translation *in silico*, regardless of whether a SECIS is predicted. The translation is only allowed to terminate at a non-UGA stop codon (UAA or UAG). By combining SIS and SecMS, at least five new selenoprotein candidates have been identified from mice (Guo et al., 2018). These proteins were further validated using selenium isotope signature and synthetic peptides. These results demonstrate that SecMS and SIS database can characterize selenoproteins irrespective of SECIS, uncovering new selenoprotein candidates.

3. Materials

3.1 Biological samples

SecMS can process different types of samples, including standard peptides, proteins, and complex cell or tissue lysates.

1. Cell culture. We take female mouse-derived BV2 microglial cell as an example in this section. BV2 cells were cultured in high-glucose Dulbecco's Modified Eagle's Medium (DMEM) containing 10% fetal bovine serum (FBS) and 1% penicillin-streptomycin (PS) at 37 °C under a 5% CO_2 atmosphere.
2. Animal tissue. All animal experimental procedures must comply with the authority's regulations and guidelines. Experiments using human tissues must conform to the highest human moral and ethical standards. The experiment described in this chapter uses C57BL/6 mice with different ages. Frozen tissues are used in this section.

3.2 Materials and chemicals

1. Dithiothreitol (DTT) (1 M stock solution)
2. EZ-Link iodoacetyl-PEG2-biotin (IodoAPB) (20 mM stock solution)
3. Iodoacetamide (IAA) (500 mM stock solution)
4. Calcium chloride dihydrate ($CaCl_2$) (100 mM stock solution)

5. Ethylenediaminetetraacetic acid (EDTA) (20 mM stock solution)
6. Monomeric avidin agarose
7. Protease inhibitor cocktail
8. BCA protein quantification kit
9. Sequencing grade trypsin
10. 30-kDa/10-kDa ultrafiltration unit (Merck Millipore)

3.3 Solutions

1. 100 mM tris HCl buffer (pH 8.5)
2. 8 M urea buffer: 8 M urea, 100 mM tris HCl buffer (pH 8.5) (make freshly)
3. Citric acid buffer: 8 M urea, 200 mM citric acid (pH 2.0) (make freshly)
4. PBSN buffer: phosphate buffered saline (PBS), 1% NP-40
5. 10% acetonitrile (ACN) (LC-MS grade)
6. Elution buffer: 50% acetonitrile (LC-MS grade), 2% trifluoroacetic acid (TFA) (HPLC grade)
7. 0.1% formic acid (FA) (LC-MS grade)
8. HPLC solvent A: 0.1% FA in 2% ACN
9. HPLC solvent B: 0.1% FA in 98% ACN

3.4 Equipment

Representative equipment is listed. Other alternative instrument models with equivalent functions are compatible with the SecMS protocol.

1. Ultrasonic homogenizer (FB120 Sonic Dismembrator, Fisher Scientific)
2. Bead-beating tissue homogenizer (Precellys)
3. Spectrophotometer (Multiskan GO, Thermo Scientific)
4. Mixer (ThermoMixer, Eppendorf)
5. Centrifuge (5810 R and 5430R, Eppendorf)
6. Freeze dryer (Labconco)
7. High-performance liquid chromatography (EASY-nLC 1000, Thermo Scientific)
8. Mass spectrometer (Q-Exactive HF, Thermo Scientific)

4. Protocols

4.1 Protein extraction

1. Cell sample

Cells grown in 10 cm Petri dishes are harvested and washed three times with cold PBS. Add 1 mL of 8 M urea buffer containing protease inhibitor cocktail to lyse cells on ice for 10 min. The resulting cell lysates are then transferred to 1.5 mL centrifuge tubes and sonicated for 30 s. The supernatants containing protein extracts are collected after centrifuging at 15,000 rpm at 4 °C for 10 min.

2. Tissue sample

Add 8 M urea solution containing 1 mM EDTA and protease inhibitor cocktail to pre-washed tissue at a ratio of ~10:1 (v/w). Tissues are lysed using a bead-beating tissue homogenizer. Proteins are collected after centrifuging at 15,000 rpm at 4 °C for 10 min.

4.2 Total protein assay

1. The protein concentrations of samples are measured using a BCA protein assay. Bovine serum albumin (BSA) standard protein is used to generate standard curves. The colorimetric absorbance at 562 nm is measured with a spectrophotometer.

4.3 Protein digestion

1. 2 mg of total protein in ~400 μL 8 M urea solution is a practical amount for SecMS analysis. However, a reduced or an increased protein quantity is also applicable. 8 μL 1 M DTT is added to 2 mg protein sample (final concentration 20 mM) in urea solution and incubated in a mixer at 37 °C for 30 min.
2. Citric acid buffer (pH 2.0) is added to dilute the sample (~1:2 v/v) to a final pH of ~4.0. Collect the supernatants after centrifuging at high speed (i.e., 13,000 × *g*) at 4 °C for 5 min.
3. IodoAPB is added (final concentration 0.1 mM) and incubated with gentle shaking at 25 °C for 60 min on a mixer in the dark.
4. Transfer the samples to 30-kDa (Option: 10-kDa filter is also applicable, but will require a longer centrifuge time) ultrafiltration units to remove extra IodoAPB. Add 300 μL 8 M urea buffer to exchange buffer four times, adjusting the pH back to ~8.5. Alkylate free cysteine by adding 3 μL 500 mM IAA (final concentration 5 mM) for 20 min at 4 °C in the dark, and remove remaining buffer by a centrifugation at 13,000 × *g* for 20 min at 4 °C.
5. Add 200 μL 100 mM tris HCl buffer to proteins retained on the 30-kDa membrane. Add 2 μL 100 mM $CaCl_2$ (final concentration 1 mM) and trypsin (1:100 w/w), and incubated with shaking at 37 °C overnight. The resulting peptides are collected by centrifuging the ultrafiltration units at 13,000 × *g* for 30 min.

4.4 Selenopeptide enrichment

1. Prepare C_8 tips for enriching biotin-tagged selenopeptides. C_8 tips are in-house prepared using standard 200 μL tips, as mentioned previously (Rappsilber, Mann, & Ishihama, 2007).
2. Place C_8 tips into 5 mL Eppendorf tubes, and wash the tips with 200 μL of ACN three times followed by 200 μL of PBSN buffer three times by centrifuging at 500 × *g*. 10 μL of monomeric avidin agarose is packed into a C_8 tip and then rinsed three times with 200 μL PBSN by centrifugation at 500 × *g*.
3. Load 200 μL of tryptic peptides to the prepared C_8 tips and then centrifuge the tips at an ultra-low speed (i.e., 50–100 g) for 60 min at 4 °C. Collect the flow-through fraction and re-load it to the tip for another low-speed centrifugation. Repeat the reloading twice, and discard the last-round flow-through fraction. When a large volume of peptides needs to be handled, the sample can be divided, and processed either sequentially or in parallel.
4. To remove non-specifically bound peptides, the agarose beads are washed with 200 μL PBSN three times, 200 μL PBS three times, and 200 μL 10% ACN three times, sequentially. Each wash step is completed by centrifuging at a low speed (i.e., 500 × *g*) for 10–15 min.
5. The bound biotin-tagged peptides are eluted by adding 100 μL elution buffer containing 50% ACN and 2% TFA by low-speed centrifugation (i.e., 100 × *g*) for 30 min. Collect the eluted fractions.
6. Concentrate the peptides. The collected peptides are freeze-dried in a vacuum dryer at 4 °C, and then re-suspended in 0.1% FA for further LC-MS/MS analysis.

4.5 LC-MS/MS analysis

1. The biotin-tagged peptides are analyzed on an LC-MS/MS system consisting of an online EASY-nLC 1000 HPLC and Orbitrap Q-Exactive HF mass spectrometer. Other instrumental configurations suitable for shotgun proteomic analysis are acceptable.
2. 15 cm capillary column (100 μm I.D.) with an ESI emitter tip is laser-pulled and packed in house with C_{18} resin (C_{18}-AQ, 1.9 μm) (Dr. Maisch, Germany). HPLC solvent A consists of 0.1% FA, 2% ACN and 98% H_2O, and HPLC solvent B consists of 0.1% FA, 2% H_2O and 98% ACN. The peptides are separated by a 120-min gradient (solvent B: 4% at 0 min, 8% at 3 min, 22% at 89 min, 35% at 113 min, 90% at 116 min, 90% at 120 min) at a flow rate of 300 nL/min. The MS data are acquired in positive mode using data-dependent acquisition method

(top-20). The MS1 parameters can be set as below: scan range (m/z) = 350–1500; resolution = 60,000; automatic gain control (AGC) target = 3e6; maximum injection time = 20 ms; dynamic exclusion = 30 s. The MS2 parameters can be set as below: resolution = 30,000; fixed first mass (m/z) = 120; AGC target = 1e5; maximum injection time = 100 ms; collision energy = 27.

4.6 SECIS-independent Selenoprotein (SIS) database

1. To establish an SIS database, all mRNA sequences are translated *in silico*. mRNA sequences are acquired from the NCBI Reference Sequences (RefSeq) database. An in-house written script is used to translate mRNA codons. All codons are translated by the standard coding rule except for the UGA stop codon. UGA codons are translated to U, and the translation only stops at a non-UGA stop codon. The generated SIS database is saved in FASTA format and used for database search. The most widely used search algorithms, such as Mascot and MaxQuant, can recognize U directly in the database search.

4.7 Identification and quantification

1. Identification

   Peptide identification can be performed by using any database search engine. In this chapter, we take Mascot (Perkins, Pappin, Creasy, & Cottrell, 1999) as a representative. MS raw files are converted to .mgf files using ProteoWizard msconvert application (Chambers et al., 2012) and then submitted to the Mascot search engine (Perkins et al., 1999). Databases used for searching include the canonical mouse SwissProt/UniProt protein database, SelenoDB 2.0 (Romagné et al., 2014), and the SIS database. For identification, the precursor mass tolerance is set to 10 ppm, and the fragment mass tolerance is 0.02 Da. Enzyme is trypsin, and the allowed maximum missed cleavage site is 3. ^{13}C was set to 1 to correct the mis-selection of the monoisotopic peak. For modifications, IodoAPB (+414.19369 Da, C/U), carbamidomethyl (+57.021464 Da, C/U), dehydroalanine (DHA) (−81.932170 Da, U) and oxidation (+15.994915 Da, M) are set as variable modifications. These modifications can be searched separately.
2. Quantification

   The Mascot search results are imported to Skyline (MacLean et al., 2010) for label-free quantification and manual inspection. The extracted ion chromatographs (XICs) of monoisotopic peak m_0, and m_1, m_2 peaks are used for quantification. More isotopic peaks can be further taken into account if necessary. The summed peak intensities are used to represent the abundance of selenopeptides, and the summed intensities of selenopeptides are used to calculate the abundance of a selenoprotein. Other platforms, such as MaxQuant (Cox & Mann, 2008), can also generate label-free quantitation on peptides.

4.8 Validation

1. Validation using selenium isotope signature

   Selenium has six major isotopes that exist in specific quantities (^{74}Se: ^{76}Se: ^{77}Se: ^{78}Se: ^{80}Se: ^{82}Se 0.89%/9.37%/7.63%/23.77%/49.61%/8.73%). As a result, peptides containing selenocysteine display a signature isotopic distribution in mass spectrometry (Fig. 2). This isotopic feature

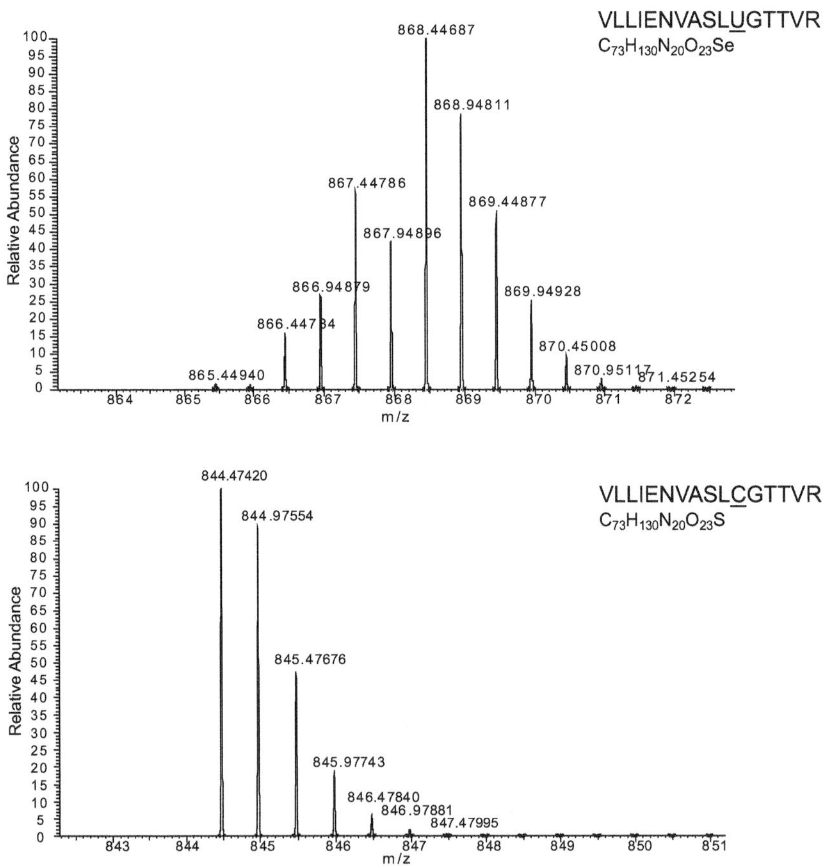

Fig. 2 Isotopic distribution of a selenopeptide in mass spectrometry. Shown is a pair of doubly charged peptides with a residue substitution between Cys and Sec. The selenopeptide is featured by a selenium isotopic signature.

allows us to validate the results of selenopeptide identifications, which is especially critical for validating newly-discovered selenopeptides to increase the identification confidence. The theoretical isotope distributions can be predicted using MS tools, such as Xcalibur Qual Browser (Thermo Scientific). The measured precursor ions in MS1, or even fragment ions containing selenocysteine in MS2, should bear the selenium isotope signature, and their isotope distributions should be the same as the theoretical ones. The comparison can be made manually, as the number of selenopeptide identifications is generally not very high.

2. Validation of newly discovered selenoproteins

A golden standard for reporting or validating a new selenoprotein is its successful overexpression, with evidence that selenocysteine is genuinely incorporated into the protein sequence. However, recombinant expression of a selenoprotein is always challenging. The selenocysteine insertion at the UGA codon involves a very complex and unique mechanism, which needs to compete with its dominating function as a stop codon. The selenocysteine insertion efficiency varies between different selenoproteins (Mehta, Rebsch, Kinzy, Fletcher, & Copeland, 2004) and can be very low under certain circumstances. It's not surprising that the selenocysteine insertion efficiency for a new selenoprotein candidate will be even worse. Therefore, expressing a new selenoprotein candidate is extremely difficult.

When overexpressing a new selenoprotein candidate in a significant quantity is unsuccessfully, a synthetic selenopeptide, either synthesized in-house or by peptide synthesis service, can be used to validate the protein at the peptide-level. The paired synthetic and endogenous selenopeptides should behave on HPLC and MS alike, such as retention time and ion fragmentation in MS2.

5. Summary

In summary, the SecMS discussed in this chapter is able to profile selenoproteins specifically and systematically. Unlike polymerase chain reaction (PCR) test targeting the mRNA or western blot analysis targeting the whole protein body, the selenocysteine-specific feature of SecMS provides a precise measurement to assess selenocysteine insertion and selenoprotein activity. More importantly, the combination of SecMS and SIS enables the discovery of new selenoprotein candidates that don't have a canonical SECIS. The biosynthesis of these new selenoproteins may involve new mechanisms. For example, there may be an unusual SECIS that is only effective under certain conditions. We will look forward to seeing more newly-claimed selenoproteins in the future. After finding a selenoprotein candidate, the later in-depth investigation on its functions, or functions rather than redox regulation for known selenoproteins, will need further extensive efforts. The emerging methods on recombinant expression of selenoproteins will be very helpful in the functional and mechanistic studies (Fu, Söll, & Sevostyanova, 2018; Novoselov et al., 2007; Peeler et al., 2020; Peeler & Weerapana, 2019). Altogether, we propose that the integrated approach of SecMS and SIS provides valuable tools to the community of selenium biology and opens up many new possibilities for selenoprotein studies.

Acknowledgment

This chapter writing was supported by the National Program on Key Basic Research Project of China to Y.Z. (2016YFA0501900 and 2016YFA0501904), the National Natural Science Foundation of China to Y.Z. (32071245 and 31671428), the Natural Science Foundation of Shanghai to Y.Z. (20ZR1474400), and the Shanghai Municipal Science and Technology Major Project (Grant No. 2019SHZDZX02) to Y.Z.

References

Bak, D. W., Gao, J., Wang, C., & Weerapana, E. (2018). A quantitative chemoproteomic platform to monitor selenocysteine reactivity within a complex proteome. *Cell Chemical Biology*, *25*(9), 1157–1167.e1154. https://doi.org/10.1016/j.chembiol.2018.05.017.

Castellano, S., Gladyshev, V. N., Guigó, R., & Berry, M. J. (2008). SelenoDB 1.0: A database of selenoprotein genes, proteins and SECIS elements. *Nucleic Acids Research*, *36*(Database issue), D332–D338. https://doi.org/10.1093/nar/gkm731.

Chambers, M. C., Maclean, B., Burke, R., Amodei, D., Ruderman, D. L., Neumann, S., et al. (2012). A cross-platform toolkit for mass spectrometry and proteomics. *Nature Biotechnology*, *30*(10), 918–920. https://doi.org/10.1038/nbt.2377.

Cox, J., & Mann, M. (2008). MaxQuant enables high peptide identification rates, individualized p.p.b.-range mass accuracies and proteome-wide protein quantification. *Nature Biotechnology*, *26*(12), 1367–1372. https://doi.org/10.1038/nbt.1511.

Fu, X., Söll, D., & Sevostyanova, A. (2018). Challenges of site-specific selenocysteine incorporation into proteins by Escherichia coli. *RNA Biology*, *15*(4–5), 461–470. https://doi.org/10.1080/15476286.2018.1440876.

Gao, J., Yang, F., Che, J., Han, Y., Wang, Y., Chen, N., et al. (2018). Selenium-encoded isotopic signature targeted profiling. *ACS Central Science*, *4*(8), 960–970. https://doi.org/10.1021/acscentsci.8b00112.

Green, D. R. (2018). An element of life. *Cell*, *172*(3), 389–390. https://doi.org/10.1016/j.cell.2018.01.003.

Guo, L., Gao, J., Gao, Y., Zhu, Z., & Zhang, Y. (2020). Aspirin reshapes acetylomes in inflammatory and cancer cells via CoA-dependent and CoA-independent pathways. *Journal of Proteome Research*, *19*(2), 962–972. https://doi.org/10.1021/acs.jproteome.9b00853.

Guo, L., Yang, W., Huang, Q., Qiang, J., Hart, J. R., Wang, W., et al. (2018). Selenocysteine-specific mass spectrometry reveals tissue-distinct selenoproteomes and candidate selenoproteins. *Cell Chemical Biology*, *25*(11), 1380–1388.e1384. https://doi.org/10.1016/j.chembiol.2018.08.006.

Huber, R. E., & Criddle, R. S. (1967). Comparison of the chemical properties of selenocysteine and selenocystine with their sulfur analogs. *Archives of Biochemistry and Biophysics*, *122*(1), 164–173. https://doi.org/10.1016/0003-9861(67)90136-1.

Kryukov, G. V., Castellano, S., Novoselov, S. V., Lobanov, A. V., Zehtab, O., Guigó, R., et al. (2003). Characterization of mammalian selenoproteomes. *Science*, *300*(5624), 1439–1443. https://doi.org/10.1126/science.1083516.

Lee, B. C., Péterfi, Z., Hoffmann, F. W., Moore, R. E., Kaya, A., Avanesov, A., et al. (2013). MsrB1 and MICALs regulate actin assembly and macrophage function via reversible stereoselective methionine oxidation. *Molecular Cell*, *51*(3), 397–404. https://doi.org/10.1016/j.molcel.2013.06.019.

Low, S. C., & Berry, M. J. (1996). Knowing when not to stop: Selenocysteine incorporation in eukaryotes. *Trends in Biochemical Sciences*, *21*(6), 203–208.

MacLean, B., Tomazela, D. M., Shulman, N., Chambers, M., Finney, G. L., Frewen, B., et al. (2010). Skyline: An open source document editor for creating and analyzing targeted proteomics experiments. *Bioinformatics*, *26*(7), 966–968. https://doi.org/10.1093/bioinformatics/btq054.

Mariotti, M., Lobanov, A. V., Guigo, R., & Gladyshev, V. N. (2013). SECISearch3 and seblastian: New tools for prediction of SECIS elements and selenoproteins. *Nucleic Acids Research*, *41*(15). https://doi.org/10.1093/nar/gkt550, e149.

Mariotti, M., Lobanov, A. V., Manta, B., Santesmasses, D., Bofill, A., Guigó, R., et al. (2016). Lokiarchaeota marks the transition between the archaeal and eukaryotic selenocysteine encoding systems. *Molecular Biology and Evolution*, *33*(9), 2441–2453. https://doi.org/10.1093/molbev/msw122.

Mehta, A., Rebsch, C. M., Kinzy, S. A., Fletcher, J. E., & Copeland, P. R. (2004). Efficiency of mammalian selenocysteine incorporation. *The Journal of Biological Chemistry*, *279*(36), 37852–37859. https://doi.org/10.1074/jbc.M404639200.

Novoselov, S. V., Lobanov, A. V., Hua, D., Kasaikina, M. V., Hatfield, D. L., & Gladyshev, V. N. (2007). A highly efficient form of the selenocysteine insertion sequence element in protozoan parasites and its use in mammalian cells. *Proceedings of the National Academy of Sciences of the United States of America*, *104*(19), 7857–7862. https://doi.org/10.1073/pnas.0610683104.

Peeler, J. C., Falco, J. A., Kelemen, R. E., Abo, M., Chartier, B. V., Edinger, L. C., et al. (2020). Generation of recombinant mammalian selenoproteins through genetic code expansion with photocaged selenocysteine. *ACS Chemical Biology*, *15*(6), 1535–1540. https://doi.org/10.1021/acschembio.0c00147.

Peeler, J. C., & Weerapana, E. (2019). Chemical biology approaches to interrogate the selenoproteome. *Accounts of Chemical Research*, *52*(10), 2832–2840. https://doi.org/10.1021/acs.accounts.9b00379.

Perkins, D. N., Pappin, D. J., Creasy, D. M., & Cottrell, J. S. (1999). Probability-based protein identification by searching sequence databases using mass spectrometry data. *Electrophoresis*, *20*(18), 3551–3567. https://doi.org/10.1002/(sici)1522-2683(19991201)20:18<3551::aid-elps3551>3.0.co;2-2.

Rappsilber, J., Mann, M., & Ishihama, Y. (2007). Protocol for micro-purification, enrichment, pre-fractionation and storage of peptides for proteomics using StageTips. *Nature Protocols*, *2*(8), 1896–1906. https://doi.org/10.1038/nprot.2007.261.

Romagné, F., Santesmasses, D., White, L., Sarangi, G. K., Mariotti, M., Hübler, R., et al. (2014). SelenoDB 2.0: Annotation of selenoprotein genes in animals and their genetic diversity in humans. *Nucleic Acids Research*, *42*(Database issue), D437–D443. https://doi.org/10.1093/nar/gkt1045.

Santesmasses, D., Mariotti, M., & Gladyshev, V. N. (2020). Bioinformatics of selenoproteins. *Antioxidants & Redox Signaling*, *33*(7), 525–536. https://doi.org/10.1089/ars.2020.8044.

Tarrago, L., Kaya, A., Weerapana, E., Marino, S. M., & Gladyshev, V. N. (2012). Methionine sulfoxide reductases preferentially reduce unfolded oxidized proteins and protect cells from oxidative protein unfolding. *The Journal of Biological Chemistry*, *287*(29), 24448–24459. https://doi.org/10.1074/jbc.M112.374520.

Zhang, Y., Fonslow, B. R., Shan, B., Baek, M. C., & Yates, J. R., 3rd. (2013). Protein analysis by shotgun/bottom-up proteomics. *Chemical Reviews*, *113*(4), 2343–2394. https://doi.org/10.1021/cr3003533.

CHAPTER TEN

An accelerated and optimized algorithm of selenium-encoded isotopic signature targeted profiling for global selenoproteome analysis

Guogeng Jia[a], Jinjun Gao[b], Fan Yang[a], Tianyu Feng[a], and Chu Wang[a,b,*]

[a]Synthetic and Functional Biomolecules Center, Beijing National Laboratory for Molecular Sciences, Key Laboratory of Bioorganic Chemistry and Molecular Engineering of Ministry of Education, College of Chemistry and Molecular Engineering, Peking University, Beijing, China
[b]Peking-Tsinghua Center for Life Sciences, Academy for Advanced Interdisciplinary Sciences, Peking University, Beijing, China
*Corresponding author: e-mail address: chuwang@pku.edu.cn

Contents

Abstract

Selenoproteins play crucial roles including protection and recovery from oxidative stress in organisms. Direct profiling of selenoproteins in proteomes is challenging due to their extremely low abundance. We have developed a computational algorithm termed selenium-encoded isotopic signature targeted profiling (SESTAR) to increase the sensitivity of detecting selenoproteins in complex proteomic samples. In this chapter, we briefly described the basic algorithm of SESTAR. We then introduced SESTAR++, an updated version of SESTAR, with accelerated computation speed and lowered false positive rate. We also provided a detailed workflow to apply SESTAR++ to proteomic profiling of selenoproteins, including the instruction of running the software and implementing it in a targeted profiling mode.

Methods in Enzymology, Volume 662
ISSN 0076-6879
https://doi.org/10.1016/bs.mie.2021.10.013

1. Introduction

Selenium was first discovered by Berzelius in 1817 and plays an essential role in life (Reich & Hondal, 2016). Like its name's meaning—moon, selenium has two faces. The excess of selenium can cause "alkali disease" and "blind staggers" (Franke, 1934), while selenium deficiency is related to diverse diseases such as Keshan's disease (Chen, 2012) and Kashin-Beck disease (Sokoloff, 1988). The major form of selenium in organisms is selenocysteine (Sec), which is called "the 21st amino acid" (Atkins & Gesteland, 2000; Böck et al., 1991). Sec is structurally identical to cysteines (Cys) only except the substitution of selenium on the site of sulfur. The selenol group of Sec is more reactive compared with the thiol group of Cys, which renders it with higher enzymatic reactivity and redox activity (Snider, Ruggles, Khan, & Hondal, 2013). Therefore, some selenoproteins play important roles in redox reactions utilizing the nucleophilic and reductive properties of Sec. For example, the first discovered selenoprotein GPX1 (Flohé, Eisele, & Wendel, 1971) was found to be responsible for the recovery from oxidative stress (Sutherland, Shankaranarayanan, Schewe, & Nigam, 2001). TRXR2 keeps redox homeostasis in cells by involving the scavenging of hydrogen peroxide (Stanley et al., 2011). The reactive selenocysteine of GPX4 is crucial for preventing hydroperoxide-induced ferroptosis (Ingold et al., 2018).

The insertion mechanism of Sec is much more complicated than the 20 canonical proteinogenic amino acids. Instead of having corresponding codons, Sec is genetically encoded by a stop codon "UGA" and inserted into selenoproteins with the guide of cis-acting Sec insertion sequence (SECIS) elements (Labunskyy, Hatfield, & Gladyshev, 2014). A total of 25 selenoproteins have been predicted in human by bioinformatics analysis of SECIS elements and UGA codons (Kryukov et al., 2003). However, direct discovery of selenoproteins from complex proteomes remains challenging because they are often with extremely low abundance. In this regard, methods have been developed to chemically enrich selenopeptides for proteomic analysis (Bak, Gao, Wang, & Weerapana, 2018; Guo et al., 2018).

In this chapter, we first introduce a targeted proteomic method that has been recently developed by our group, termed "selenium-encoded isotopic signature targeted profiling" (SESTAR) (Gao, Yang, et al., 2018), for global analysis of selenoproteins. We then describe the latest update of SESTAR algorithm with much higher computation speed and less false positive rates. We also provide a detailed workflow on how to apply SESTAR to proteomic profiling of selenoproteins, including the instruction of using the software and setting up the targeted profiling method.

2. Targeted proteomic profiling and SESTAR

Shot-gun proteomics based on mass spectrometry is a powerful tool for protein profiling and quantification in complex proteome samples. Data dependent acquisition (DDA) is a common method in shot-gun proteomics. However, DDA only selects top-N most intensive ions in a full MS scan to fragmentation (Zhang, Fonslow, Shan, Baek, & Yates 3rd, 2013), causing a bias against low-abundance peptides including selenopeptides. Targeted proteomic profiling is developed to overcome such a bias and improve the sensitivity of detecting low-abundance peptides, which can select specific peptides in a user-provided list for MS/MS fragmentation regardless of their relative abundance (Stergachis, MacLean, Lee, Stamatoyannopoulos, & MacCoss, 2011). In particular, specific isotope-encoded patterns can be used to direct the target proteomic analysis (Fig. 1A). In this methodology, elements with unique isotopic distribution are incorporated into peptides to create distinct isotopic patterns, making them distinguishable from proteogenic peptides in full mass scans. A pioneering work by the Aebersold group employed a dichloride tag to identify low-abundance cysteine-containing peptides (Goodlett et al., 2000). Recently, a computational method called "isoStamp" (Palaniappan et al., 2011) was developed to identify the isotopic pattern of dibrominated tags and applied in targeted glycoproteomics (Woo, Iavarone, Spiciarich, Palaniappan, & Bertozzi, 2015). In addition to natural elements, artificial isotopic patterns with $^{12}C/^{13}C$ or H/D mixture were also applied in targeted proteomic profiling (Gao, Mfuh, Amako, & Woo, 2018; Woo et al., 2017, 2018).

Selenium has six natural stable isotopes with detectable abundance (^{74}Se, 0.89%; ^{76}Se, 9.37%; ^{77}Se, 7.63%; ^{78}Se, 23.77%; ^{80}Se, 49.61%; ^{82}Se, 8.73%), making a unique isotopic pattern when it is incorporated into target peptides (Fig. 1B). Such properties in principle can be utilized in the targeted proteomic profiling to overcome DDA's bias against low-abundance selenopeptides. To test this hypothesis, we developed a targeted method for selenopeptide profiling termed "selenium-encoded isotopic signature targeted profiling" (SESTAR) (Gao, Yang, et al., 2018). In this algorithm,

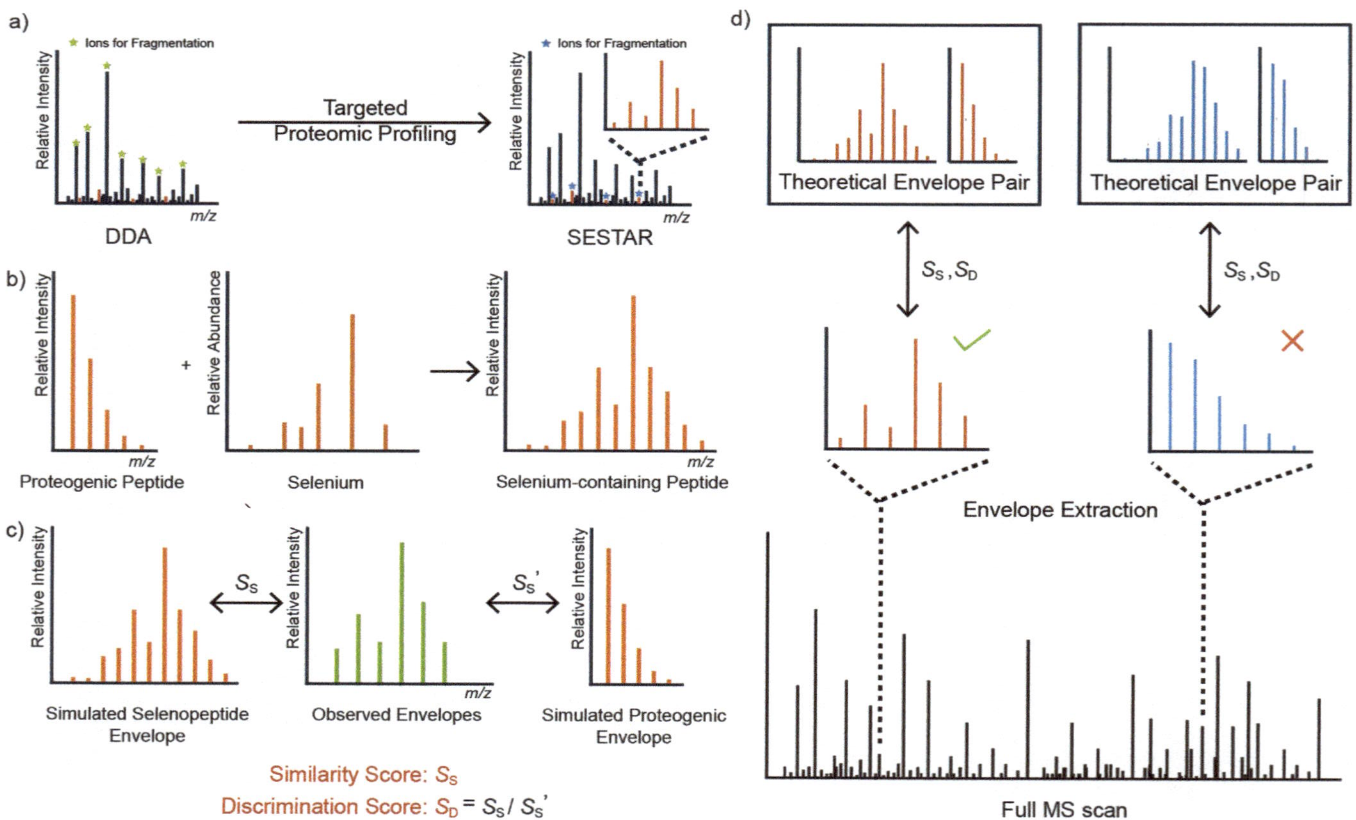

Fig. 1 See figure legend on opposite page.

all isotopic envelopes in a full MS scan are extracted, and for each envelope, the "averagine" assumption (Senko, Beu, & McLaffertycor, 1995) is employed to estimate the element composition (Fig. 1D) of the corresponding peptide, which only relies on the peptide's mass but not the fragmentation data. Two theoretical envelopes are then simulated for the peptide, one for its selenium-containing form and the other for its non-selenium-containing counterpart. The experimentally observed envelope is then compared to these two simulated envelopes to calculate two scores in order to judge whether the peptide contains selenium. One is the similarity score (S_S), reflecting the similarity between the observed envelope and the theoretically simulated selenium-containing envelope; the other is the discrimination score (S_D), reflecting the uniqueness of the similarity to the selenopeptide envelope but not the non-selenium-containing envelope (Fig. 1C). These two scores ensure that positive hits are similar to the selenium-containing envelope but distinct from the proteogenic envelope. The formulas for S_S and S_D are provided as below:

$$S_S = \cos^{-1}\left(\frac{E_O \bullet E_{Se}}{|E_O||E_{Se}|}\right)$$

$$S_D = \cos^{-1}\left(\frac{E_O \bullet E_{Se}}{|E_O||E_{Se}|}\right) / \cos^{-1}\left(\frac{E_O \bullet E_N}{|E_O||E_N|}\right)$$

where E_O, E_{Se} and E_N refer to the vectorized intensities of the experimentally observed envelope, theoretically simulated selenium-containing envelope and non-selenium-containing envelope, respectively.

Fig. 1 Concept and algorithm of SESTAR. (A) Isotope-directed targeted proteomic profiling can overcome the abundance limitation of the traditional data-dependent acquisition (DDA). DDA only selects the top-N most intensive ions (shown as green stars) in a full MS scan to MS/MS fragmentation, ignoring the low-abundance ones (shown as red peaks). SESTAR selects ions with selenium-specific isotopic patterns (shown as blue stars) for targeted fragmentation and achieves better detection sensitivity for selenium-containing peptides. (B) Selenium can introduce significant change to the isotopic distribution of a target peptide. (C) For each observed envelope, the selenium-containing envelope and the proteogenic envelope are theoretically simulated based on the peptide mass. Two scores, similarity score "S_S" and discrimination score "S_D" are calculated. (D) The overall workflow of SESTAR algorithm. After isotopic envelopes are extracted from full MS scans, S_S and S_D are calculated in order to determine whether the observed envelope corresponds to a selenium-containing peptide or not. The positive hits from SESTAR might be used to guide a following round of targeted proteomic analysis.

SESTAR was written in R and operated in Linux. The software can be readily applied to scan and discover selenium-containing patterns from proteomic data in the open-source MS1 format (McDonald et al., 2004). To demonstrate its application scenarios, we run SESTAR across proteome datasets of tissues and cell types downloaded from the Human Proteome Map database and discovered additional selenoproteins that were missed by traditional database search approaches. We also combined SESTAR with targeted analysis modes by using an inclusion list and/or parallel reaction monitoring (PRM), and found that the sensitivity of detecting selenopeptides in both whole proteomes and selenopeptide-enriched proteomes could be dramatically enhanced as compared to the regular DDA analysis. In addition, SESTAR is also applicable to improve the detection of chemically tagged selenium-containing peptides, suggesting that selenium-tag might be an alternative for chemically-directed proteomics. Detailed data description and analysis of SESTAR could be found in the reference (Gao, Yang, et al., 2018).

3. SESTAR++: Accelerated SESTAR algorithm

We recently updated SESTAR to SESTAR++, a C++ version that runs >100 times faster (Fig. 2A). SESTAR++ can finish one sample analysis in several minutes which usually takes hours to complete by SESTAR. It's therefore possible to schedule the targeted profiling immediately after the regular DDA analysis by using the information provided by SESTAR++, which ensures that the retention time of each peptide remains as unchanged as possible in two consecutive LC runs. We also created a user-friendly graphic interface for SESTAR++ on the Windows operating system to make it more accessible.

We estimated the false positive rate (FPR) of SESTAR and SESTAR++ using the *Arabidopsis thaliana* proteome which is predicted to contain no selenoproteins (Fig. 2B). We did the estimation in different ranges of peptide masses and the results showed that FPR of SESTAR increases with the increasing peptide masses. In practice, the envelope of a selenopeptide usually can't be detected completely in full MS scans because certain isotopes of selenium such as ^{74}Se are of relatively low abundance, and SESTAR therefore accepts partially detected envelopes to make a judgment. To reduce the possibility of identifying false positive hits, SESTAR requires that a "critical zone" containing six peaks that represent the most unique part of selenium-containing envelope must be detected. Although the full

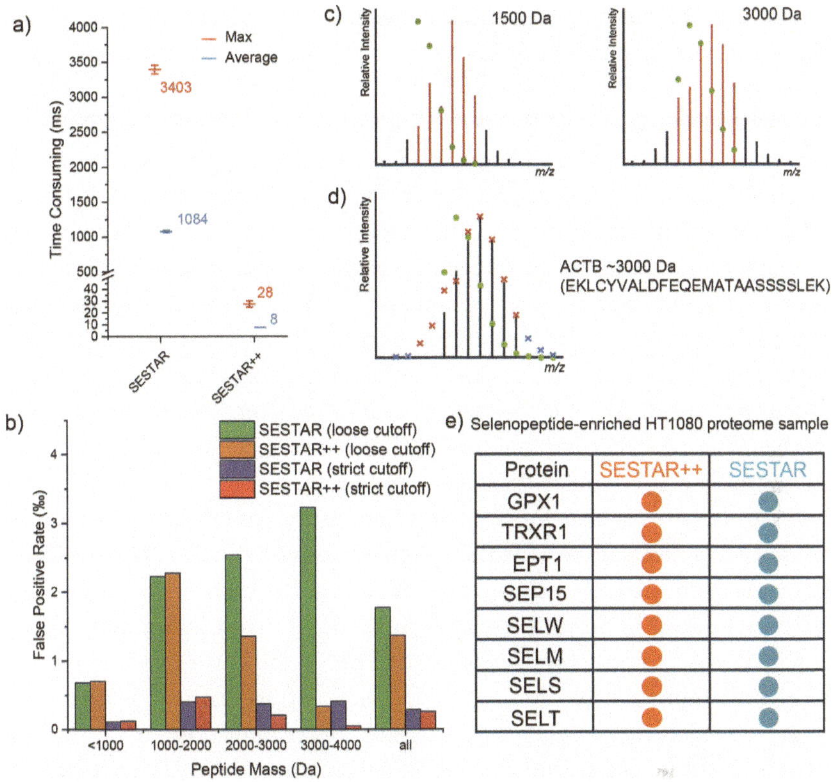

Fig. 2 The performance of SESTAR++. (A) Time consumption for analyzing one spectrum by SESTAR vs SESTAR++. SESTAR and SESTAR++ calculations were performed 3 times on the data collected from a selenopeptide-enriched HT1080 proteome sample, and the CPU run time for analyzing each full-MS spectrum was recorded. The red mark represents the maximal run time spent on a spectrum, and the blue mark represents the averaged run time spent on a spectrum. The error bars stand for the standard deviations. (B) Comparison of false positive rates of SESTAR and SESTAR++ in different peptide mass ranges. Loose cutoff refers to $S_S < 15$, $S_D < 1/4$ and 6–14 peaks in one envelope while strict cutoff refers to $S_S < 10$, $S_D < 1/6$ and 6–14 peaks in one envelope. (C) Illustration of the "critical zone" (shown as red peaks) of theoretically simulated envelopes with different peptide masses as originally implemented in SESTAR. The corresponding simulated proteogenic envelope is shown as green dots. The envelope of "critical zone" becomes more similar to the proteogenic envelope when the peptide's mass increases. (D) A false positive hit of SESTAR is filtered by the optimized "critical zone" in SESTAR++. The observed envelope is represented by black peaks. The simulated selenium-containing envelope is shown by crosses with those referring to the "critical zone" in SESTAR++ colored in red. The simulated proteogenic envelope is shown by green dots. The observed envelope lacks two peaks on the left and is therefore filtered by SESTAR++. (E) Similar performance of detecting selenoproteins in the HT1080 cellular proteome by SESTAR and SESTAR++.

selenium-containing envelopes are unique, the envelopes of such "critical zone" become more and more similar to the proteogenic envelopes when the target peptide mass increases (Fig. 2C), which will result in false identification and the rise of FPR.

We reasoned that the FPR of SESTAR could be further lowered if the number of isotopic peaks in the "critical zone" can vary depending on the actual mass of the target peptide. Inspired by a previous study of identifying selenopeptide's envelope computationally (Kim et al., 2012), we optimized the algorithm of SESTAR++ to enlarge the "critical zone" by including more isotopic peaks when a peptide with higher mass is analyzed. More specifically, a list of all possible combinations of truncated selenium-containing envelopes (i.e., peaks are truncated from the envelope based on the order of intensity from low to high) for a specific mass are theoretically simulated. Then the similarity scores between each of the truncated envelopes in the list and the non-selenium-containing envelope are calculated. The "critical zone" for this specific mass is defined as the truncated envelope with the least number of peaks and a similarity score higher than the preset cutoff. The FPR estimation results showed that while FPR of SESTAR++ is slightly higher than that of SESTAR in the low-mass range, its FPR is significantly reduced by the new algorithm in the high-mass range, so that the overall FPR is also decreased (Fig. 2B). As one related example, we found that SESTAR incorrectly assigned a peptide (EKLCYVALDFEQEMATAASSSSLEK, about 3000 Da with chemical modifications) from ACTB as a selenopeptide, however, its MS/MS spectrum doesn't seem to contain selenium-encoded patterns. In SESTAR++, the peptide is filtered off by the larger "critical zone" with nine specific isotopic peaks (Fig. 2D).

We next compared the performance of SESTAR and SESTAR++ in identifying endogenous selenoproteins from proteome samples. A selenoprotein is regarded to be identified when a positive envelope recognized by SESTAR or SESTAR++ matches the mass of its selenopeptide. In selenopeptide-enriched samples from the HT1080 cellular proteomes, SESTAR++ identified the same set of selenoproteins as SESTAR did (Fig. 2E). These results suggested that SESTAR++ can achieve similar true discovery rate for selenopeptide identification but with a reduced FPR. Please note that neither SESTAR nor SESTAR++ can handle the identification of peptides containing multiple selenium atoms, but such cases should rarely exist due to the low abundance of selenium in living organisms.

4. Method

This section introduces how to employ SESTAR++ to proteomic profiling of selenoproteins, including the instruction of running the software, the application of it in targeted profiling modes, and the data processing procedures.

4.1 Instruction of running the software

The following procedures are demonstrated for SESTAR++. While this instruction might be also partially applicable to the original SESTAR, it should be noted that changing parameters in SESTAR requires revisions on the source code. SESTAR++ requires .Net Framework 4.7.2 and Microsoft Visual C++ 2015–2019 Redistributable (x86) for runtime environment.

Note: Microsoft Visual C++ must be x86 version but not x64 or other versions.

SESTAR++ download link: https://github.com/wangchulab/SESTARpp

1. Launch the software and select data:

   Click "open file" button to select the "MS1" file(s), which could be converted from the raw files by RawConverter (Woo et al., 2017). Selection of multiple files is allowed. The output file(s) will be saved to the same path as the source "MS1" file(s). The output files' names can be customized by double clicking "Output Name" column.

2. Select the parameters:

   Similarity score (S_S) and discrimination score (S_D) are the major parameters to set. S_S refers to the similarity between the observed envelope and the theoretically simulated envelope, and S_D refers to the uniqueness of the similarity to the selenopeptide's envelope with the non-selenium-containing envelope as the control. Lower cutoff values for S_S and S_D mean stricter thresholds, which will result in less FPR but risk losing potential candidates. We recommend the strict cutoff ($S_S < 10$, $S_D < 1/6$) for precise identification and the loose cutoff ($S_S < 15$, $S_D < 1/4$) for large-scale screening. The range of the numbers of peaks in an envelope can also be defined by users. The default is from 6 to 14 peaks. The maximum limit can be set as highest as 15. We suggest that the minimum limit should be set no lower than 5, as a selenopeptide's envelope should contain more peaks than that of a proteogenic one.

3. Match positive hits to target peptide sequences (optional):

   We offer the "peptide match" tool in SESTAR++ for convenient matching to the target selenopeptides. This tool matches each of the candidate envelopes by SESTAR++ to a list of selenopeptide sequences provided by the user and marks the peptide in the output file if they are the same or differ only by a neutron's mass. Users are suggested to provide a comprehensive list of selenopeptide sequences as many as possible, so that any peptide forms with missed cleavage could be considered. Our software by default provides two templates of selenopeptide sequences from the human and mouse proteomes, respectively, with at most two missed tryptic cleavage sites.

   This function can be activated by clicking "Peptide Match" in the "Tools" menu. Modifications and target sequences can be either loaded from existing templates or by manual entrance. The static modification refers to the difference of the molecular weights between the modified amino acids and the original amino acids. Data can be modified after template loading. Existing data can be modified by double clicking and new data can be entered by double clicking the next blank row. Selecting a row and pressing the "Delete" key can delete this row. We also offer the function of "template saving" that allows saving of the user-defined templates. While the "Peptide Match" tool is convenient, it can only provide suggestive evidences for matching to a potential selenopeptide candidate. Evidences from the MS/MS spectra are still required to confirm the target selenopeptide's sequence.

4. Generating the inclusion list (optional):

   An inclusion list of selenopeptide candidates is needed for the targeted profiling. SESTAR++ contains a tool that can merge positive hits with same *m/z* in the user-defined retention time window for convenient generation of an inclusion list. This function can be activated by clicking "Inclusion List" in the "Tools" menu. The number X in the "retention time tolerance" textbox indicates the envelopes with same *m/z* within +/− X min will be merged. An additional "InclusionList" file will be output. It contains some basic information about selenopeptide candidates including spectral count, *m/z*, charge, maximal and minimal retention time. Please note that retention time here refers to the retention time of the observed envelope. More buffering zones that flank the retention time window should be added for compiling the final inclusion list when the retention times of certain peptides are not stable under the experimental condition.

5. Check the output:

The output file of SESTAR++ contains the basic information about the positive hits. The *m/z* values and intensities of the observed and theoretical envelopes can facilitate manual confirmation. Retention time, charge and the m/z corresponding to the peak with the highest intensity can help assemble an inclusion list for targeted analysis. If "Peptide match" is activated, the matched protein and peptide is shown in the last column. "+1" or "−1" in the brackets indicates the difference between the observed and theoretical molecular weight. Candidates that are identified from multiple full MS scans are generally more reliable. Positive hits that are only identified in one full MS scan are filtered off when an inclusion list is generated.

4.2 SESTAR++ workflow

A standard SESTAR++ workflow contains two rounds of LC-MS/MS acquisitions (Fig. 3). Regular DDA data collection should be applied in the first round to acquire full MS scans. In the next step, SESTAR++ is applied to identify MS1 envelopes of potential selenopeptide candidates as long as their abundance are above the detection limit of the instrument. The output of SESTAR++ can then guide the targeted analysis by using an inclusion or running PRM (Bourmaud, Gallien, & Domon, 2016) to

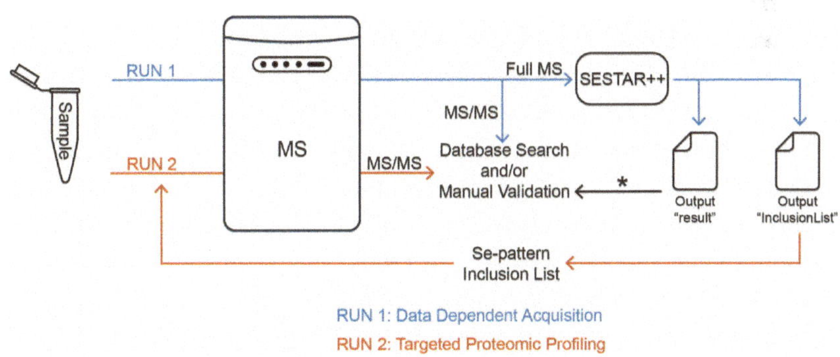

Fig. 3 Overall workflow of SESTAR++-guided targeted proteomic profiling. SESTAR++ workflow contains two rounds of LC-MS/MS acquisitions. Regular DDA data collection is applied in the first round to acquire full MS scans, and SESTAR++ computation is followed to identify MS1 envelopes of potential selenopeptide candidates. The output "InclusionList" file of SESTAR++ can guide targeted analysis such as PRM in the second round. *: a "result" file is generated that contains matches to known selenopeptide sequences for each positive hit if "peptide match" function is activated. It is helpful for manual validation of selenopeptide candidates.

improve the detection sensitivity of selenopeptides. SESTAR++ is independent of mass spectrometers in theory, but we have only tested it on the instruments from the Thermo Fisher Q-Exactive series and Fusion series. The detailed procedures are as following.

1. Sample preparation:

   For profiling of selenoproteins, SESTAR++ is compatible with regular bottom-up proteomic samples and don't require special sample processing. While not absolutely necessary, chemical enrichment of selenopeptides, e.g., by selenocysteine-selective labeling in acidic buffer (Bak et al., 2018; Guo et al., 2018), is recommended to further improve the overall profiling performance. Selenium-encoded patterns can be exogenously introduced to target peptides by labeling with chemical probes containing selenium. In this case, SESTAR++ is still effective at improving the detection sensitivity of such chemically labeled peptides, which is conceptually similar to IsoTaG (Woo et al., 2015). As an example, we prepared selenopeptide-enriched samples from mouse liver proteomes to illustrate how SESTAR++ works to enhance the detection of native selenoproteins.

2. Data-dependent acquisition (DDA):

   The first round of DDA analysis provides full MS scans for SESTAR++ to analyze. High-resolution full MS scans are required to extract high-quality envelopes for SESTAR++'s computation. On orbitrap instruments, we generally set the resolution of full MS above 70,000.

3. Running SESTAR++ to generate an inclusion list:

   Raw MS data files are firstly converted to the "MS1" format by transformation software such as RawConverter (He, Diedrich, Chu, & Yates 3rd, 2015). The "MS1" file(s) are then fed into SESTAR++ as described above to identify envelopes of selenopeptide candidates and to generate an inclusion list. "Inclusion list" function is highly recommended to be activated in this step to acquire the "InclusionList" file; otherwise, targets can only be selected from unmerged candidate spectra in basic output of SESTAR++. While not necessary, "peptide match" function is also recommended to be activated. The search parameters were set as loose cutoffs ($S_S < 15$, $S_D < 1/4$ and 6–14 peaks) in our example.

   Cautions should be taken when targets and parameters for the inclusion list are selected. Adding more target precursors into the inclusion list means on average less acquisition time for each precursor. Narrow retention time window may cause missing of the target peptide because a specific peptide's retention time might not be stable in different LC runs (Wichmann et al., 2019). It is therefore important to balance the number of target peptides added in the inclusion list with the width of each associated retention time window. The number of candidate selenopeptides in "InclusionList" file depends on the sample complexity and the cutoff strictness set in SESTAR++. In chemically enriched samples, there are usually about 100–1000 peptide hits to be targeted. We usually sort these target hits in "InclusionList" file by spectral counts, select top 1000 candidates at most for one inclusion list, and add additional window of +/− 3 min to the observed retention time range for each candidate. In our demonstrated example, we selected about 700 targets for further PRM analysis after filtering any positive hits that are only present in a single spectrum. Candidates whose molecular weight are matched to the target peptide sequences in "peptide match" function can also be added into the inclusion list individually, while we didn't do this in our example as a demonstration for seeking selenopeptides without pre-knowledge. Setting stricter (or looser) cutoff in SESTAR++ computation can also decrease (or increase) the number of target hits.

4. Targeted profiling:

Targeted methods such as using inclusion lists or running PRM is followed to generate a selenopeptide-specific profile. It's crucial to keep LC conditions unperturbed between the two rounds, otherwise some target peptides may be eluted out of the preset retention time window and elude the target analysis. In practice, columns, solvents, flow gradient and other materials/parameters used in liquid chromatography must remain unchanged. To acquire selenium-specific patterns in the MS/MS spectra which can provide further evidence of selenopeptides, isolation window must be wide enough to include enough number of selenium isotopes for fragmentation. However, the parameter should not be set too wide given the risk of being interfered by co-eluting peptides. Based on our experience, a total isolation window of 1.6–2.6 Th is recommended. As we mentioned in the previous step, an additional window should be added to the detected retention time range for each target to account for potential retention time shifts. For our demonstrated example, 3 min were added to each end of the detected retention time range for a 160-min gradient. Dynamic exclusion was disabled, and the other parameters for MS were set almost the same as the regular DDA analysis, including NCE (28), AGC target (full MS: 3e6, MS/MS: 1e5), etc.

4.3 Data processing

Common database search engines should be able to match MS/MS spectra to target selenopeptides while some adjustments might be needed. For example, ProLuCID (Xu et al., 2015) may not automatically recognize the one-letter code "U" for selenocysteine in the fasta file, and a tentative solution is to replace "U" with "C" in the protein database. As compensation, the mass difference of Se and S (47.94445 Da) should be added to match the modification on Sec, and such a modification should be set as a variable modification on Cys.

Although in most cases ProLuCID work as expected, we recently found that it sometimes failed to identify selenopeptides from the data of the targeted analysis. We reasoned that it's possibly because there are too few background spectra as negative examples so that the database search engine can't select positive hits by properly calculating the false discovery rate (FDR). A practical solution is to combine the MS/MS data of targeted profiling with those of the DDA analysis as a background and put them together in the database search engine. If a certain precursor's mass matches a target peptide sequence and gets no positive hit from the database search, or the database search reveals a possible selenopeptide but filters it off by FDR, it's necessary to manually validate the MS/MS spectra. In addition to the regular requirements such as the continuous fragmentation pattern on the MS/MS spectra, putative selenocysteine-containing fragment ions with high intensity need to show a clear selenium-encoded pattern. Fig. 4 shows an example for manual validation of selenopeptides. Manual validation reveals that the spectrum shows a good matching to theoretical fragments of a selenopeptide from GPX1 with clear selenium-specific patterns in most of the envelopes of selenocysteine-containing fragment ions. We are therefore convinced that this spectrum is a strong match to the selenopeptide of GPX1.

Following the data processing procedures above, we identified a total of seven native selenoproteins from the selenopeptide-enriched mouse liver proteome sample by traditional DDA analysis. As a comparison, three additional selenoproteins (SELT, SELS and MSRB1) were discovered by the SESTAR++ aided PRM (Fig. 5). These results demonstrated SESTAR++ is effective for enhancing selenoproteome profiling compared to the traditional DDA approaches, even in the highly enriched selenoproteome samples.

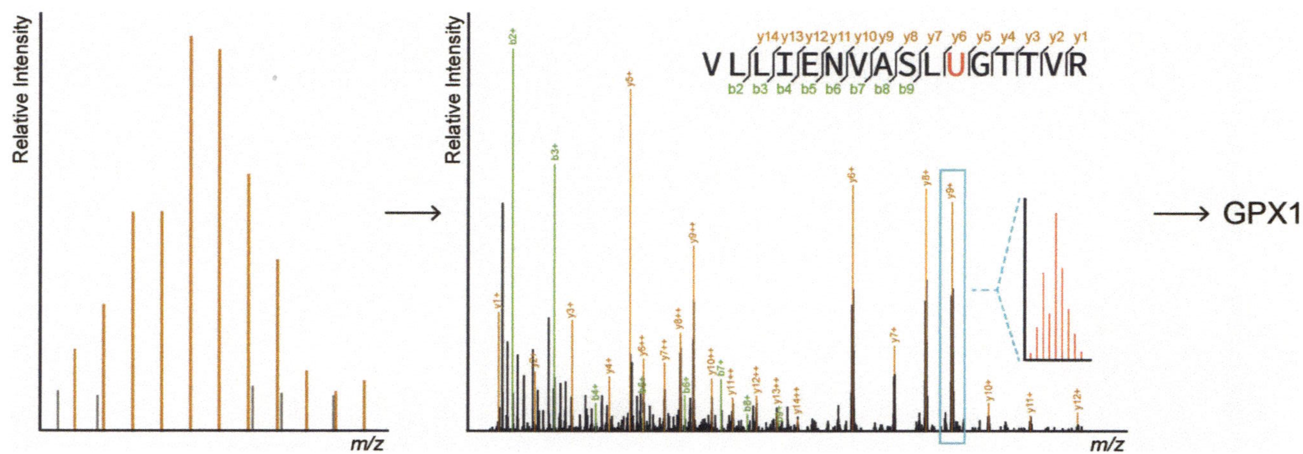

Fig. 4 Manual validation of the selenopeptide in GPX1. The selenium-encoded pattern is clear in the full MS scan (left). Continuous fragment ions can be matched to the peptide sequence and clear selenium-encoded patterns can be found in the MS/MS spectra (right).

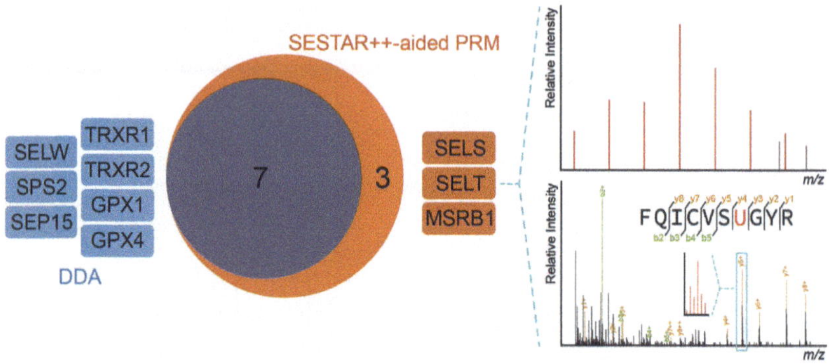

Fig. 5 Enhanced performance of SESTAR++-aided PRM. Three more selenoproteins (SELS, SELT and MSRB1) were discovered in the selenopeptide-enriched mouse liver proteome samples by the SESTAR++ aided PRM as compared to the traditional DDA analysis. Shown on the right are the isotopic envelope of the selenopeptide from SELT (top) and the MS/MS spectrum with corresponding b/y ions noted. Selenium-encode envelope patterns can be clearly observed in some of the y ions such as y5+.

5. Summary

Selenoproteins play crucial roles in organism, but direct profiling of this special class of proteins in proteomes remains challenging due to their extremely low abundance. By utilizing the unique isotopic distribution of selenium, SESTAR can enhance the sensitivity of detecting trace selenopeptides. We developed SESTAR++ as an accelerated version of SESTAR with similar performance of detecting native selenoproteins from cellular proteomic samples but with reduced false positive rates especially for peptides with higher masses. As one demonstrated example, SESTAR++ could guide targeted analysis to enable discovery of more selenoproteins from selenopeptide-enriched mouse liver proteome samples than the traditional DDA analysis. We envision SESTAR++ will become a convenient and powerful tool for selenoproteome analysis, which not only can be employed to profile known selenoproteins, but also can be applied to discover novel selenium-containing protein species in future.

Acknowledgment

We thank the Computing Platform of the Center for Life Science for supporting the proteomic data analysis. This work was supported by National Natural Science Foundation of China to C.W. (21778004, No. 21925701, No. 91953109).

References

Atkins, J. F., & Gesteland, R. F. (2000). The twenty-first amino acid. *Nature, 407*(463), 465.

Bak, D. W., Gao, J., Wang, C., & Weerapana, E. (2018). A quantitative Chemoproteomic platform to monitor Selenocysteine reactivity within a complex proteome. *Cell Chemical Biology, 25*, 1157–1167. (e1154).

Böck, A., Forchhammer, K., Heider, J., Leinfelder, W., Sawers, G., Veprek, B., et al. (1991). Selenocysteine: The 21st amino acid. *Molecular Microbiology, 5*, 515–520.

Bourmaud, A., Gallien, S., & Domon, B. (2016). Parallel reaction monitoring using quadrupole-Orbitrap mass spectrometer: Principle and applications. *Proteomics, 16*, 2146–2159.

Chen, J. (2012). An original discovery: Selenium deficiency and Keshan disease (an endemic heart disease). *Asia Pacific Journal of Clinical Nutrition, 21*, 320–326.

Flohé, L., Eisele, B., & Wendel, A. (1971). Glutathion peroxidase. I. Isolation and determinations of molecular weight. *Hoppe-Seyler's Zeitschrift fur Physiologische Chemie, 352*, 151–158.

Franke, K. W. (1934). A new toxicant occurring naturally in certain samples of plant foodstuffs I. results obtained in preliminary feeding trials. *Journal of Nutrition, 8*, 597–608.

Gao, J., Mfuh, A., Amako, Y., & Woo, C. M. (2018). Small molecule Interactome mapping by Photoaffinity labeling reveals binding site hotspots for the NSAIDs. *Journal of the American Chemical Society, 140*, 4259–4268.

Gao, J., Yang, F., Che, J., Han, Y., Wang, Y., Chen, N., et al. (2018). Selenium-encoded isotopic signature targeted profiling. *ACS Central Science, 4*, 960–970.

Goodlett, D. R., Bruce, J. E., Anderson, G. A., Rist, B., Pasa-Tolic, L., Fiehn, O., et al. (2000). Protein identification with a single accurate mass of a cysteine-containing peptide and constrained database searching. *Analytical Chemistry, 72*, 1112–1118.

Guo, L., Yang, W., Huang, Q., Qiang, J., Hart, J. R., Wang, W., et al. (2018). Selenocysteine-specific mass spectrometry reveals tissue-distinct Selenoproteomes and candidate Selenoproteins. *Cell Chemical Biology, 25*, 1380–1388 (e1384).

He, L., Diedrich, J., Chu, Y. Y., & Yates, J. R., 3rd. (2015). Extracting accurate precursor information for tandem mass spectra by RawConverter. *Analytical Chemistry, 87*, 11361–11367.

Ingold, I., Berndt, C., Schmitt, S., Doll, S., Poschmann, G., Buday, K., et al. (2018). Selenium utilization by GPX4 is required to prevent Hydroperoxide-induced Ferroptosis. *Cell, 172*, 409–422 (e421).

Kim, J. W., Lee, S., Park, K., Na, S., Paek, E., Park, H. S., et al. (2012). Monoisotopic mass determination algorithm for selenocysteine-containing polypeptides from mass spectrometric data based on theoretical modeling of isotopic peak intensity ratios. *Journal of Proteome Research, 11*, 4488–4498.

Kryukov, G. V., Castellano, S., Novoselov, S. V., Lobanov, A. V., Zehtab, O., Guigo, R., et al. (2003). Characterization of mammalian selenoproteomes. *Science, 300*, 1439–1443.

Labunskyy, V. M., Hatfield, D. L., & Gladyshev, V. N. (2014). Selenoproteins: Molecular pathways and physiological roles. *Physiological Reviews, 94*, 739–777.

McDonald, W. H., Tabb, D. L., Sadygov, R. G., MacCoss, M. J., Venable, J., Graumann, J., et al. (2004). MS1, MS2, and SQT-three unified, compact, and easily parsed file formats for the storage of shotgun proteomic spectra and identifications. *Rapid Communications in Mass Spectrometry, 18*, 2162–2168.

Palaniappan, K. K., Pitcher, A. A., Smart, B. P., Spiciarich, D. R., Iavarone, A. T., & Bertozzi, C. R. (2011). Isotopic signature transfer and mass pattern prediction (IsoStamp): An enabling technique for chemically-directed proteomics. *ACS Chemical Biology, 6*, 829–836.

Reich, H. J., & Hondal, R. J. (2016). Why nature chose selenium. *ACS Chemical Biology, 11*, 821–841.

Senko, M. W., Beu, S. C., & McLaffertycor, F. W. (1995). Determination of monoisotopic masses and ion populations for large biomolecules from resolved isotopic distributions. *Journal of the American Society for Mass Spectrometry*, *6*, 229–233.

Snider, G. W., Ruggles, E., Khan, N., & Hondal, R. J. (2013). Selenocysteine confers resistance to inactivation by oxidation in thioredoxin reductase: Comparison of selenium and sulfur enzymes. *Biochemistry*, *52*, 5472–5481.

Sokoloff, L. (1988). Kashin-Beck disease: Current status. *Nutrition Reviews*, *46*, 113–119.

Stanley, B. A., Sivakumaran, V., Shi, S., McDonald, I., Lloyd, D., Watson, W. H., et al. (2011). Thioredoxin reductase-2 is essential for keeping low levels of H(2)O(2) emission from isolated heart mitochondria. *Journal of Biological Chemistry*, *286*, 33669–33677.

Stergachis, A. B., MacLean, B., Lee, K., Stamatoyannopoulos, J. A., & MacCoss, M. J. (2011). Rapid empirical discovery of optimal peptides for targeted proteomics. *Nature Methods*, *8*, 1041–1043.

Sutherland, M., Shankaranarayanan, P., Schewe, T., & Nigam, S. (2001). Evidence for the presence of phospholipid hydroperoxide glutathione peroxidase in human platelets: Implications for its involvement in the regulatory network of the 12-lipoxygenase pathway of arachidonic acid metabolism. *Biochemical Journal*, *353*, 91–100.

Wichmann, C., Meier, F., Virreira Winter, S., Brunner, A. D., Cox, J., & Mann, M. (2019). MaxQuant.Live enables global targeting of more than 25,000 peptides. *Molecular & Cellular Proteomics*, *18*, 982–994.

Woo, C. M., Felix, A., Byrd, W. E., Zuegel, D. K., Ishihara, M., Azadi, P., et al. (2017). Development of IsoTaG, a chemical Glycoproteomics technique for profiling intact N- and O-Glycopeptides from whole cell proteomes. *Journal of Proteome Research*, *16*, 1706–1718.

Woo, C. M., Iavarone, A. T., Spiciarich, D. R., Palaniappan, K. K., & Bertozzi, C. R. (2015). Isotope-targeted glycoproteomics (IsoTaG): A mass-independent platform for intact N- and O-glycopeptide discovery and analysis. *Nature Methods*, *12*, 561–567.

Woo, C. M., Lund, P. J., Huang, A. C., Davis, M. M., Bertozzi, C. R., & Pitteri, S. J. (2018). Mapping and quantification of over 2000 O-linked Glycopeptides in activated human T cells with isotope-targeted Glycoproteomics (Isotag). *Molecular & Cellular Proteomics*, *17*, 764–775.

Xu, T., Park, S. K., Venable, J. D., Wohlschlegel, J. A., Diedrich, J. K., Cociorva, D., et al. (2015). ProLuCID: An improved SEQUEST-like algorithm with enhanced sensitivity and specificity. *Journal of Proteomics*, *129*, 16–24.

Zhang, Y., Fonslow, B. R., Shan, B., Baek, M. C., & Yates, J. R., 3rd. (2013). Protein analysis by shotgun/bottom-up proteomics. *Chemical Reviews*, *113*, 2343–2394.

CHAPTER ELEVEN

Assay of selenol species in biological samples by the fluorescent probe Sel-green

Baoxin Zhang and Jianguo Fang*

State Key Laboratory of Applied Organic Chemistry and College of Chemistry and Chemical Engineering, Lanzhou University, Lanzhou, China

*Corresponding author: e-mail address: fangjg@lzu.edu.cn

Contents

Abstract

Selenium (Se) is an essential trace element for diverse cellular functions. The biological significance of Se is predominantly dependent on its incorporation into the selenocysteine (Sec) for synthesis of selenoproteins (SePs), such as thioredoxin reductase family enzymes and glutathione peroxidase family enzymes. In general, the hyperactivity of the selenol group in Sec confers the Sec residue critical for functions of SePs. The Sec is much less abundant than its sulfur analog cysteine (Cys), and it remains a high challenge to detect Sec, especially in complex biological samples. We recently reported a selective fluorescent probe Sel-green for selenols and summarized the principles for design of selenol (and thiophenol) probes. Sel-green discriminates selenols from other biological species, especially thiols, under physiological conditions, and has been applied to detect both endogenous and exogenous selenol species in live cells. In this chapter, we describe a protocol and guideline for the selective detection of Sec by

Methods in Enzymology, Volume 662
ISSN 0076-6879
https://doi.org/10.1016/bs.mie.2021.10.021

applying the Sel-green. This protocol is also suitable for detection of other selenol species. This practical and convenient assay would assist scientists to better understand the pivotal roles of Sec as well as SePs.

1. Introduction

Selenium (Se) is an essential trace element involved in diverse physiological functions of humans (Duntas & Benvenga, 2015; Foster & Sumar, 1997; Rayman, 2000; Schwarz & Foltz, 1957; Thomson, 2004). Abnormal intake of Se has been associated with a number of diseases (Ballihaut, Kilpatrick, & Davis, 2011; Encinar, Schaumlöffel, Ogra, & Lobinski, 2004; Fairweather-Tait et al., 2010; Kucharzewski, Braziewicz, Majewska, & Góźdź, 2002; Rayman, 2012; Tanguy, Grauzam, de Leiris, & Boucher, 2012; Weekley & Harris, 2013). In mammalian tissues, the predominant chemical form of Se takes the form of selenocysteine (Sec), which is the 21st genetically encoded natural amino acid and an basic building block for selenoproteins (SePs) (Bindoli et al., 2009; Chen & Berry, 2003; Darras & Van Herck, 2012; Driscoll & Copeland, 2003; Hatfield, Tsuji, Carlson, & Gladyshev, 2014; Hondal, Nilsson, & Raines, 2001; Johansson, Gafvelin, & Arnér, 2005; Kanzok et al., 2001; Rocher, Lalanne, & ChaudiÈRe, 1992; Schomburg, Schweizer, & Köhrle, 2004; Weekley & Harris, 2013). SePs, such as thioredoxin reductase (TrxR) enzymes and glutathione peroxidase enzymes, have been implicated in redox state regulation and antioxidant defense (Arthur, 2001; Mustacich & Powis, 2000; Rotruck et al., 1973; Zhang et al., 2021; Zhang, Duan, Osama, & Fang, 2020; Zhong, Arnér, Ljung, Åslund, & Holmgren, 1998). As the Sec residues are essential for the biological activity of the various Se-containing species *in vivo*, there is an urgent need to develop reliable and rapid assays to detect the Sec and the related selenol species.

Fluorescent dyes have some key merits including high sensitivity, spatiotemporal resolution, and compatibility with biological systems. These advantages have enabled fluorescent dyes the powerful and popular tools to visualize the complicated biological processes (Gao, Yu, Lv, Choo, & Chen, 2017; Hu, Jia, Zhao, Cho, & Fang, 2019; Huang & Pu, 2021; Kobayashi & Choyke, 2011; Kong et al., 2015; Li et al., 2019; Li, Cheng, Huang, Li, & Zhu, 2015; Neto, Carvalho, & Correa, 2015; Yin et al., 2013; Zhang et al., 2013, 2014; Zhu et al., 2016). However, there is a big challenge to detect Sec by designing specific fluorescent probes because of the interference of high concentrations (millimolar levels) of biological thiols in cells. Generally, thiols have similar chemical properties to Sec. The *p*Ka value of selenol in Sec is 5.8, which is significantly lower than those of the most biological thiols (~8.3) (Huber & Criddle, 1967; Jacob, Giles, Giles, & Sies, 2003; Luo, Smith, & Anderson, 2005; Turner, Hager, & Shapiro, 1988). As a consequence, the Se in Sec is largely present as the nucleophilic selenolate (R-Se$^-$) form, while the majority of thiols present as a less reactive neutral form (R-SH) under physiological conditions (Hondal et al., 2001). Moreover, selenolate is more nucleophilic than thiolate (Singh & Whitesides, 1991). The above features provide novel insights for selective detection of Sec over biological thiols. In 2006, Maeda et al. developed a probe BESThio to determine the Sec with little suffering from the interference of its counterpart cysteine (Cys) under acidic conditions (pH 5.8) (Fig. 1A) (Maeda, Katayama, Matsuno, & Uno, 2006). In our previous work, we also utilized biotin conjugated iodoacetamide to selectively label the Sec residue in the TrxR at pH 6.5 (Fig. 1B) (Fang & Holmgren, 2006; Fang, Lu, & Holmgren, 2005). However, all these assays are not compatible with the neutral biological surroundings (pH ~ 7.4), which prevents their practical applications in live system.

Recently, our group has successfully developed the first selenol probe Sel-green that works under physiological conditions (Zhang et al., 2015). Sel-green has been successfully applied to study Sec by other groups (Chen et al., 2017, 2018; Tuladhar & Rein, 2018) and quantify selenol intermediates in reaction mixture by us (Li et al., 2020). This chapter is intended to provide readers with the information regarding the design of such a unique fluorescent probe and its applications for studying Sec and SePs. This chapter is outlined as follows. In Section 2, we present a guide of the design and storage of the probe. In Sections 3 and 4, we describe the *in vitro* characterization of Sel-green using fluorescence spectrophotometry and the details for imaging exogenous and endogenous Sec in live cells, respectively.

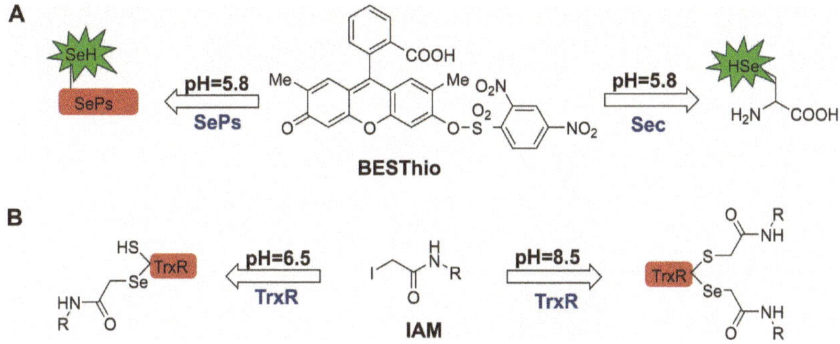

Fig. 1 Previous strategy for detection of Sec and SePs.

2. Fluorescent probe for Sec

2.1 Design of probe library

In past years, a wide variety of thiol-selective probes that possess a well-elucidated NAS (nucleophilic aromatic substitution) mechanism by selective cleavage of 2,4-dinitrobenzene functionality (a classical fluorescence quenching moiety) have been developed and applied for protein profiling and thiol sensing. Inspired by this recognition mechanism and the strong nucleophilicity of Sec in the neutral biological surroundings (pH $\sim$7.4), we have constructed a potential fluorescent probe library of selenols by adjusting the electron density of the fluorescence quenching moiety and the linker unit between the quenching moiety and the fluorophore. After a careful structure-activity relationship study, Sel-green was discovered as a highly selective fluorescent probe of selenols under physiological conditions. In addition, the structural determinants of such NAS type probes for selective recognition of selenols were recapitulated: (1) Two nitro-substitutions are essential for the quenching moiety; (2) The linker unit determines the selectivity, *i.e.*, molecules with either the sulfonamide linker or ether linker exhibit promising selectivity toward selenols while molecules with the sulfonate linker show promiscuous response to both thiols and selenols (Zhang et al., 2015).

2.2 Availability and synthesis of the probe

The probe (Sel-green) is commercially available from Glpbio (www.glpbio.com) or MCE (www.medchemexpress.cn).

The overall synthetic route is illustrated in Scheme 1. Compound **1** was prepared according to the literature from the commercially available 4-(diethylamino)salicylaldehyde (Li, Zhang, Yang, Yang, & Yang, 2014).

Synthesis of compound 7-Diethylamino-3-aminocoumarin (**2**). $SnCl_2{\cdot}2H_2O$ (19.4g, 85.81mmol) and 60mL of 37% HCl were slowly

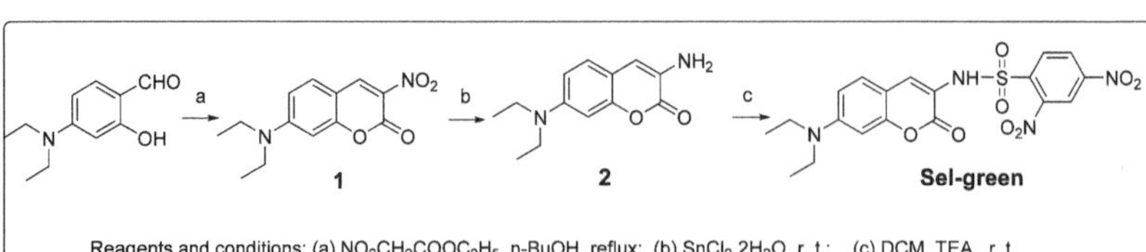

Scheme 1 Synthesis of Sel-green.

added into a 100 mL flask. Next, compound **1** (3.0 g, 11.45 mmol) was added portion-wisely, and the resulting solution was further stirred at room temperature for 6 h. After the reaction was completed, a solution of 5 M NaOH was employed to neutralize the excessive acid, and then extracted with ethyl acetate. The combined organic layer was dried with anhydrous Na_2SO_4. The crude product was purified by silica gel chromatography to afford desired product **2** (1.27 g, yield 48%). ^{1}H NMR (400 MHz, DMSO-d_6) δ: 7.19 (d, J=8.8 Hz, 1H), 6.68 (s,1H), 6.61 (dd, J=8.8, 2.4 Hz, 1H), 6.48 (d, J=2.4 Hz, 1H), 4.99 (s, 2H), 3.36 (m, 4H), 1.09 (t, 6H); ^{13}C NMR (100 MHz, DMSO-d_6) δ: 159.68, 150.84, 146.77, 129.14, 126.08, 111.55, 110.26, 109.66, 97.68, 44.19, 12.71; mp: 85–87 °C. ESI-MS (m/z): $[M+H]^+$ 233.4.

Synthesis of Sel-green. 2,4-Dinitrobenzenesulfonyl chloride (319 mg, 1.2 mmol) in 6 mL of CH_2Cl_2 was slowly added to the solution of **2** (232 mg, 1 mmol) and Et_3N (101 mg, 1 mmol) in 10 mL of CH_2Cl_2 under argon atmosphere. The mixture was stirred overnight at room temperature. After the reaction completed, the mixture was extracted with CH_2Cl_2 for three times and dried with anhydrous Na_2SO_4. After removing the solvent under reduced pressure, the residue was purified by flash column chromatography to give the title compound as a red solid (147 mg, yield 32%). ^{1}H NMR (400 MHz, DMSO-d_6) δ: 10.56 (s, 1H), 8.86 (d, J=2.0 Hz, 1H), 8.64 (dd, J=8.8, 2.0 Hz, 1H), 8.32 (d, J=8.4 Hz, 1H), 7.82 (s,1H), 7.51 (d, J=9.2 Hz, 1H), 6.75 (dd, J=8.8, 2.0 Hz, 1H), 6.50 (d, J=2.0 Hz, 1H), 3.45 (q, 4H), 1.27 (t, 6H); ^{13}C NMR (100 MHz, DMSO-d_6) δ: 159.70, 155.53, 151.15, 150.21, 147.83, 140.93, 138.30, 132.53, 130.25, 127.36, 120.12, 114.43, 109.86, 107.56, 96.74, 44.51, 12.69; mp: 172–174 °C; ESI-MS (m/z): $[M+H]^+$ 463.1; HRMS (m/z) $[M+H]^+$ calcd. For 463.0918, found 463.0907.

Tips: Purification (flash column chromatography) of the reaction product must be performed as soon as possible, because the Sel-green is prone to slow decomposition in solution. The product can be stored in a freezer at −20 °C.

2.3 Use and storage of the probe

To prepare the probe stock solution for imaging use, the probe powder is dissolved in the designated amount of DMSO to give a stock solution (50–100 mM). The stock is then diluted with DMSO to give a working concentration of 1 mM. The probe is most stable in powder form and can be stored for long time at −20 °C. Once dissolved in DMSO, the probe stocks only can be stored for several days at −20 °C. Thus, it is highly recommended to prepare the Sel-green solution freshly.

3. *In vitro* characterization of probes

3.1 Instrumentation and materials

All aqueous solutions in this protocol are prepared with double distilled water, and other reagents (analytical grade) are purchased from commercial supplies without further purification. All spectroscopic measurements are performed in 100 mM PBS buffer, pH 7.4. The UV–vis spectrometer evolution 200 (Thermo Scientific) is used to record absorption spectra. All fluorescence spectra are recorded on a LUMINA fluorescence spectrofluorometer (Thermo Scientific) at room temperature. Quartz cuvettes (1.0 cm × 1.0 cm) are used for emission measurements unless otherwise specified.

3.2 Spectroscopic properties

Before performing any biological experiments, the absorption and fluorescence spectra of Sel-green in responding to Sec were investigated. As Sec is not stable, Sec is freshly prepared by mixing equimolar of selenocystine $(Sec)_2$ with DTT in the same phosphate buffer. First, we determined the absorbance spectra of Sel-green toward Sec. As shown in Fig. 2A, the maximum absorbance of Sel-green is located at around 400 nm, which is blue-shifted to ~370 nm upon addition of Sec. Then, we further tested the fluorescence response of Sel-green to Sec, and the results are shown in Fig. 2B. As anticipated, Sel-green displays a marginal emission centered at 502 nm (λ_{ex} = 370 nm), while the emission intensity of the solution is remarkably enhanced after addition of Sec, and the fluorescence emissions is saturated nearly within 3 min. Finally, the emission spectra of Sel-green toward increasing concentrations of Sec are acquired (Fig. 2C). Fig. 2D shows the dose-dependent changes of the emission at 502 nm (λ_{ex} = 370 nm). All the results show that Sel-green possesses excellent response toward Sec.

3.3 Selectivity experiments

Next, the pH-dependent fluorescence change (Fig. 3A) and the kinetic behavior (Fig. 3B) of Sel-green were measured. The optimal pH for the probe is found to range from 6.0 to 7.5, and the probe displays much faster

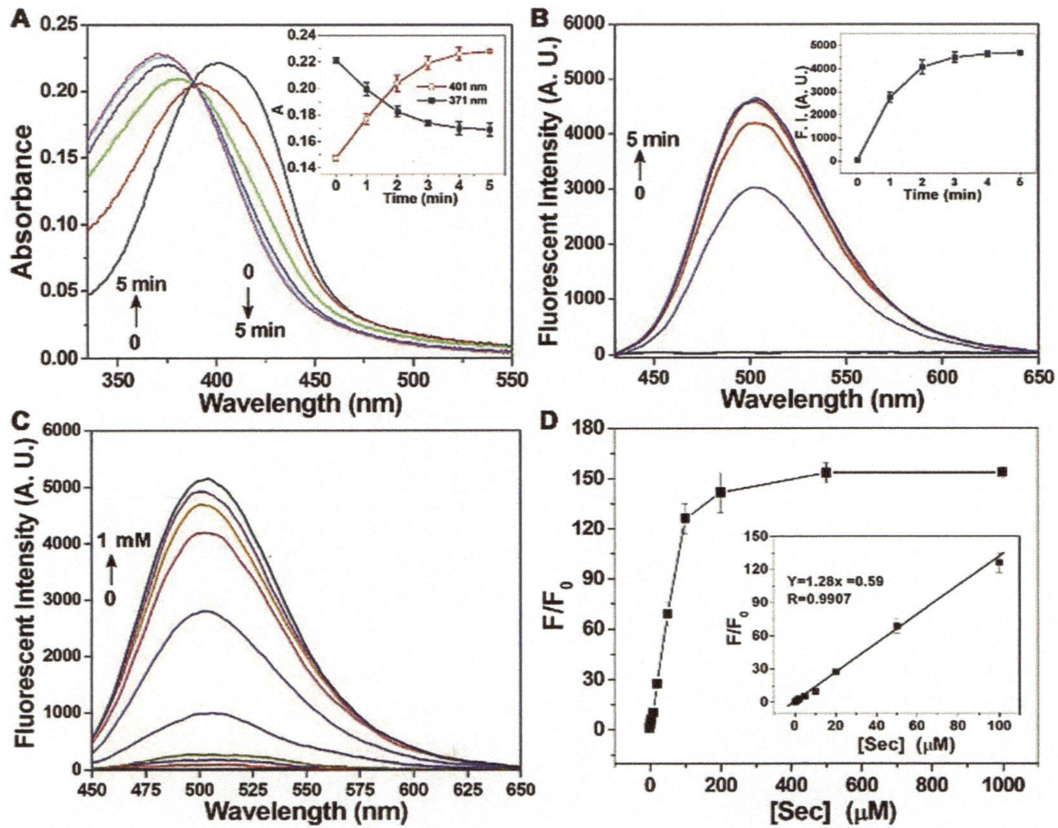

Fig. 2 Absorption and fluorescence spectra of Sel-green in responding to Sec. (A) Time-dependent absorbance spectra of Sel-green (10 μM) toward Sec (100 μM). The absorbance spectra were scanned every 1 min for 5 min. The inset shows the time-dependent changes of absorbance at 401 nm (red line) and 371 nm (black line). (B) Time-dependent emission spectra of Sel-green (10 μM) toward Sec (100 μM). (C) The emission spectra of Sel-green (10 μM) toward increasing concentrations of Sec. (D) Plotting the fold of fluorescence increase (F/F_0) at 502 nm *versus* the concentrations of Sec. *Reprinted with permission from Zhang, B., Ge, C., Yao, J., Liu, Y., Xie, H., & Fang, J. (2015). Selective selenol fluorescent probes: Design, synthesis, structural determinants, and biological applications. Journal of the American Chemical Society, 137, 757–769. Copyright 2015 American Chemical Society.*

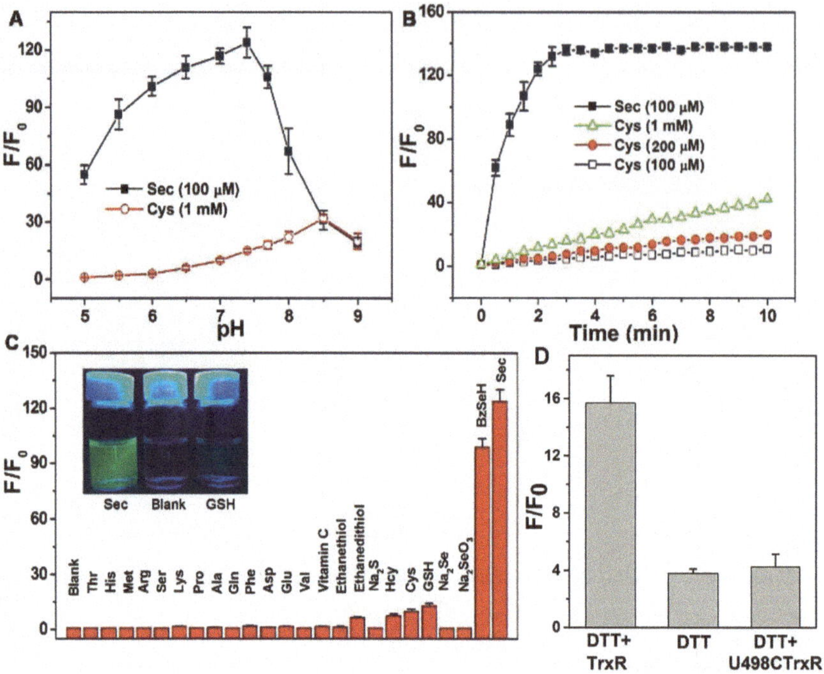

Fig. 3 (A) Influence of pH (from 5.0 to 9.0) on the response of Sec-green (10 μM) toward Sec (100 μM) and Cys (1 mM). The fluorescence at 502 nm was determined after 2 min of incubation. (B) Kinetics of the fluorescence change upon Sec-green (10 μM) responding to Sec (100 μM) and Cys (100 μM, 200 μM and 1 mM). The fluorescence at 502 nm was determined after 2 min of incubation in the PBS buffer (pH 7.4). (C) Fluorescence intensity changes of Sel-green (10 μM) toward Sec and various analytes. Concentrations of Sec and BzSeH are 100 μM, and others are 1 mM. The inserted photo was taken under UV illumination (365 nm). The fluorescence at 502 nm was determined after 2 min of incubation in the PBS buffer (pH 7.4). (D) Response of Sel-green to TrxR and U498CTrxR. Concentrations of the probe, DTT, and proteins are 50, 100 and 22 μM, respectively. The fluorescence intensity was read after incubation of the probe with the analytes at room temperature for 2 min. *Reprinted with permission from Zhang, B., Ge, C., Yao, J., Liu, Y., Xie, H., & Fang, J. (2015). Selective selenol fluorescent probes: Design, synthesis, structural determinants, and biological applications. Journal of the American Chemical Society, 137, 757–769. Copyright 2015 American Chemical Society.*

response to Sec than the counterpart Cys. We then further confirm the probe has high reactivity toward Sec over other relevant species which are present in biological system by adopting selectivity experiments (Fig. 3C). To this end, the probe Sel-green is incubated with a number of potentially interfering biomolecules such as common amino acids, vitamin C, inorganic sulfide and other selenocompounds at a final concentration of 1 mM. The results demonstrate that amino acids without a sulfhydryl group, vitamin C, inorganic sulfide and some selenocompounds exhibit negligible fluorescence changes, and other thiol species (Cys, homocysteine (Hcy), and glutathione (GSH)) give week response under the same conditions. Only the Sec and benzyl selenol (BzSeH, *in situ* generated from dibenzyl diselenide) remarkably switch on the fluorescence. Encouraged by the aforementioned results, we further applied the probe to measure the fluorescent response toward TrxR and U498C TrxR (Fig. 3D). Mammalian TrxR enzymes are essential Sec-containing flavoenzymes with a unique C-terminal -Gly-Cys-Sec-Gly active site (Zhang, Li, Han, Liu, & Fang, 2017). Compared to the wild type TrxR, the U498C TrxR is a mutant enzyme, where the Sec residue at the 498th position is replaced by Cys. The results show that TrxR could obviously switch on the fluorescence, while the U498C TrxR only gives a little fluorescence intensity changes. Thus, taken together, above results demonstrate that Sel-green can be employed to detect Sec selectively.

Tips: As shown in Fig. 3B, Sel-green also responds to GSH or Cys, yet the rate is slow. Thus to better discriminate selenols from thiols, short incubation time of the probe with analytes is recommended. Under our optimized conditions, 2 min of incubation is fine.

4. Detection of Sec in live cells

This section will provide detailed protocols for imaging exogenous and endogenous Sec in live cells. In Section 4.1, we describes control experiments imaging Sec generated from an exogenous donor $(Sec)_2$. In Section 4.2, we provide the experimental parameters to obtain reliable measurement results of endogenous Sec.

4.1 Imaging exogenous Sec

4.1.1 Instrumentation and materials

The HepG 2 cells were obtained from the Shanghai Institute of Biochemistry and Cell Biology, Chinese Academy of Sciences. Fluorescence microscopy studies were performed on a Floid cell imaging station (Life technology). Other compatible fluorescent microscopes should also work. The cells were cultured in Dulbecco's modified Eagle's medium (DMEM) supplemented with 10% FBS, 2 mM glutamine, and 100 units/mL penicillin/streptomycin and maintained in an atmosphere of 5% CO_2 at 37 °C. Probes and reagents: A 100 mM Sel-green stock solution in DMSO is made and the solution may be stored at −20 °C for several days. A 10 mM $(Sec)_2$ stock solution in PBS buffer (pH 8.0) is made and the solution may be stored at −20 °C for several days. A 100 mM BESThio stock solution in DMSO is made and the solution may be stored at −20 °C for several days. A 20 μM solution of Sel-green in complete medium should be prepared from the DMSO stock immediately before experiments. Cell culture supplies: Standard supplies for mammalian cell culture.

4.1.2 Cell culture

The HepG 2 cells were plated in 12-well plates at a density of 2×10^4 cells per well with 1 mL culture medium for 24 h. Then, the cells were incubated with $(Sec)_2$ (5 μM) for 1 or 12 h. After replacement with the fresh medium containing Sel-green (20 μM), the cells were allowed to continue incubation for 30 min. Then the cells were washed with PBS three times, and the fluorescence images were immediately obtained with the Floid cell imaging station. In the BESThio-treated group, the cells were incubated with the thiol probe BESThio (20 μM) for 30 min.

4.1.3 Imaging and results

To further investigate the potential application of Sec-green, we first determine if Sel-green could respond to the exogenous Sec. In this experiment, the cells were preincubated with $(Sec)_2$ for 1 or 12 h, and then treated with Sel-green. As shown in Fig. 4, the control cells (cells without any treatment, Fig. 4A) and the cells treated with only Sec-green (20 μM, Fig. 4C) give no detectable fluorescence signal, while the cells with additional $(Sec)_2$ display a

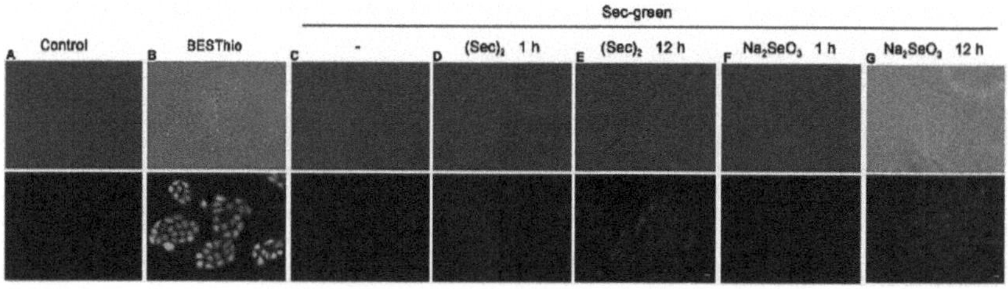

Fig. 4 Imaging Sec in cells. HepG 2 cells were treated with vehicle (A), the thiol probe BESThio (20 μM, B) and Sec-green (20 μM, C) for 30 min. HepG 2 cells were pretreated with $(Sec)_2$ (5 μM) for 1 h (D) or 12 h (E), and Na_2SeO_3 (5 μM) for 1 h (F) or 12 h (G), followed by incubation with Sec-green (20 μM) for additional 30 min. *Reprinted with permission from Zhang, B., Ge, C., Yao, J., Liu, Y., Xie, H., & Fang, J. (2015). Selective selenol fluorescent probes: Design, synthesis, structural determinants, and biological applications. Journal of the American Chemical Society, 137, 757–769. Copyright 2015 American Chemical Society.*

obivious green fluorescence signal (Fig. 4D and E). Another control group which was added the thiol probe BESThio also shows bright fluorescence signal (Fig. 4B). These results indicate that Sel-green could recognize Sec without suffering from the interference of the cellular thiols.

4.2 Imaging endogenous Sec

4.2.1 Instrumentation and materials

A 100 mM Na_2SeO_3 stock solution in double distilled water is made and could be stored at −20 °C for several weeks. Cell culture supplies: Standard supplies for mammalian cell culture. Fluorescence microscopy studies were performed on the Floid cell imaging station (Life technology). Other compatible fluorescent microscopes should also work.

4.2.2 Cell culture

The HepG 2 cells were plated in 12-well plates at a density of 2×10^4 cells per well with 1 mL cultured medium for 24 h. To increase the endogenous level of Sec, the cells were preincubated with sodium selenite (Na_2SeO_3, 5 μM) for 1 or 12 h. After replacement with the fresh medium containing Sel-green (20 μM), the cells were allowed to continue incubation for 30 min. The cells were washed with PBS three times, and the fluorescence images were then immediately obtained with a Floid cell imaging station (life technology).

4.2.3 Imaging results

Sodium selenite is a precursor of Sec biosynthesis, and supplement of cells with Na_2SeO_3 could significantly increases the Sec level in cells (Kryukov et al., 2003). In this experiment, the cells were preincubated with Na_2SeO_3, and then were exposed to Sel-green. After washing the cells with PBS three times, the fluorescence images were acquired. When the cells were incubated with sodium selenite for 12 h, the notable fluorescence appears (Fig. 4G). A short time treatment (1 h) gives weak but reproducible signal (Fig. 4F). These results demonstrate that the probe Sel-green is suitable for imaging endogenous Sec in live cells.

5. Conclusions

As Sec is an essential building block for SePs which have been implicated in various cellular functions and linked to several human diseases, there is an urgent need of reliable and rapid assays to monitor Sec. In order to address the needs in this field, our group has developed the Sec fluorescent probe Sel-green, which can respond to Sec and other selenols with more than 100-fold enhancement of emission in neutral aqueous solution. The content in this chapter would help readers to understand the design principles, optical properties and applications of Sel-green. We expect that the Sel-green probe would be a useful tool to dissect the biological functions of Sec and SePs.

Acknowledgments

The financial supports from the National Natural Science Foundation of China (21778028 and 22077055), the Natural Science Foundation of Gansu Province (20JR5RA311 and 20JR10RA612) are greatly acknowledged.

References

Arthur, J. R. (2001). The glutathione peroxidases. *Cellular and Molecular Life Sciences: CMLS, 57*, 1825–1835.

Ballihaut, G., Kilpatrick, L. E., & Davis, W. C. (2011). Detection, identification, and quantification of selenoproteins in a candidate human plasma standard reference material. *Analytical Chemistry, 83*, 8667–8674.

Bindoli, A., Rigobello, M. P., Scutari, G., Gabbiani, C., Casini, A., & Messori, L. (2009). Thioredoxin reductase: A target for gold compounds acting as potential anticancer drugs. *Coordination Chemistry Reviews, 253*, 1692–1707.

Chen, J., & Berry, M. J. (2003). Selenium and selenoproteins in the brain and brain diseases. *Journal of Neurochemistry, 86*, 1–12.

Chen, W., Colon, R., Louda, J. W., del Rey, F. R., Durham, M., & Rein, K. S. (2018). Brevetoxin (PbTx-2) influences the redox status and NPQ of Karenia brevis by way of thioredoxin reductase. *Harmful Algae, 71*, 29–39.

Chen, W., Tuladhar, A., Rolle, S., Lai, Y., Rodriguez del Rey, F., Zavala, C. E., et al. (2017). Brevetoxin-2, is a unique inhibitor of the C-terminal redox center of mammalian thioredoxin reductase-1. *Toxicology and Applied Pharmacology, 329*, 58–66.

Darras, V. M., & Van Herck, S. L. (2012). Iodothyronine deiodinase structure and function: From ascidians to humans. *The Journal of Endocrinology, 215*, 189–206.

Driscoll, D. M., & Copeland, P. R. (2003). Mechanism and regulation of selenoprotein synthesis. *Annual Review of Nutrition, 23*, 17–40.

Duntas, L. H., & Benvenga, S. (2015). Selenium: An element for life. *Endocrine, 48*, 756–775.

Encinar, J. R., Schaumlöffel, D., Ogra, Y., & Lobinski, R. (2004). Determination of selenomethionine and selenocysteine in human serum using speciated isotope dilution-capillary HPLC–Inductively coupled plasma collision cell mass spectrometry. *Analytical Chemistry, 76*, 6635–6642.

Fairweather-Tait, S. J., Bao, Y., Broadley, M. R., Collings, R., Ford, D., Hesketh, J. E., et al. (2010). Selenium in human health and disease. *Antioxidants & Redox Signaling, 14*, 1337–1383.

Fang, J., & Holmgren, A. (2006). Inhibition of thioredoxin and thioredoxin reductase by 4-hydroxy-2-nonenal in vitro and in vivo. *Journal of the American Chemical Society, 128*, 1879–1885.

Fang, J., Lu, J., & Holmgren, A. (2005). Thioredoxin reductase is irreversibly modified by curcumin: A novel molecular mechanism for its anticancer activity. *Journal of Biological Chemistry, 280*, 25284–25290.

Foster, L. H., & Sumar, S. (1997). Selenium in health and disease: A review. *Critical Reviews in Food Science and Nutrition*, *37*, 211–228.

Gao, M., Yu, F., Lv, C., Choo, J., & Chen, L. (2017). Fluorescent chemical probes for accurate tumor diagnosis and targeting therapy. *Chemical Society Reviews*, *46*, 2237–2271.

Hatfield, D. L., Tsuji, P. A., Carlson, B. A., & Gladyshev, V. N. (2014). Selenium and selenocysteine: Roles in cancer, health, and development. *Trends in Biochemical Sciences*, *39*, 112–120.

Hondal, R. J., Nilsson, B. L., & Raines, R. T. (2001). Selenocysteine in native chemical ligation and expressed protein ligation. *Journal of the American Chemical Society*, *123*, 5140–5141.

Hu, G., Jia, H., Zhao, L., Cho, D.-H., & Fang, J. (2019). Small molecule fluorescent probes of protein vicinal dithiols. *Chinese Chemical Letters*, *30*, 1704–1716.

Huang, J., & Pu, K. (2021). Near-infrared fluorescent molecular probes for imaging and diagnosis of nephro-urological diseases. *Chemical Science*, *12*, 3379–3392.

Huber, R. E., & Criddle, R. S. (1967). Comparison of the chemical properties of selenocysteine and selenocystine with their sulfur analogs. *Archives of Biochemistry and Biophysics*, *122*, 164–173.

Jacob, C., Giles, G. I., Giles, N. M., & Sies, H. (2003). Sulfur and selenium: The role of oxidation state in protein structure and function. *Angewandte Chemie International Edition*, *42*, 4742–4758.

Johansson, L., Gafvelin, G., & Arnér, E. S. J. (2005). Selenocysteine in proteins—Properties and biotechnological use. *Biochimica et Biophysica Acta (BBA) - General Subjects*, *1726*, 1–13.

Kanzok, S. M., Fechner, A., Bauer, H., Ulschmid, J. K., Müller, H.-M., Botella-Munoz, J., et al. (2001). Substitution of the thioredoxin system for glutathione reductase in *Drosophila melanogaster*. *Science*, *291*, 643–646.

Kobayashi, H., & Choyke, P. L. (2011). Target-cancer-cell-specific activatable fluorescence imaging probes: Rational design and in vivo applications. *Accounts of Chemical Research*, *44*, 83–90.

Kong, F., Hu, B., Gao, Y., Xu, K., Pan, X., Huang, F., et al. (2015). Fluorescence imaging of selenol in HepG2 cell apoptosis induced by Na2SeO3. *Chemical Communications*, *51*, 3102–3105.

Kryukov, G. V., Castellano, S., Novoselov, S. V., Lobanov, A. V., Zehtab, O., Guigó, R., et al. (2003). Characterization of mammalian selenoproteomes. *Science*, *300*, 1439–1443.

Kucharzewski, M., Braziewicz, J., Majewska, U., & Góźdź, S. (2002). Concentration of selenium in the whole blood and the thyroid tissue of patients with various thyroid diseases. *Biological Trace Element Research*, *88*, 25–30.

Li, J., Cheng, F., Huang, H., Li, L., & Zhu, J.-J. (2015). Nanomaterial-based activatable imaging probes: From design to biological applications. *Chemical Society Reviews*, *44*, 7855–7880.

Li, X., Hou, Y., Zhao, J., Li, J., Wang, S., & Fang, J. (2020). Combination of chemotherapy and oxidative stress to enhance cancer cell apoptosis. *Chemical Science*, *11*, 3215–3222.

Li, X., Zhang, B., Yan, C., Li, J., Wang, S., Wei, X., et al. (2019). A fast and specific fluorescent probe for thioredoxin reductase that works via disulphide bond cleavage. *Nature Communications*, *10*, 2745.

Li, J., Zhang, C.-F., Yang, S.-H., Yang, W.-C., & Yang, G.-F. (2014). A Coumarin-based fluorescent probe for selective and sensitive detection of thiophenols and its application. *Analytical Chemistry*, *86*, 3037–3042.

Luo, D., Smith, S. W., & Anderson, B. D. (2005). Kinetics and mechanism of the reaction of cysteine and hydrogen peroxide in aqueous solution. *Journal of Pharmaceutical Sciences*, *94*, 304–316.

Maeda, H., Katayama, K., Matsuno, H., & Uno, T. (2006). 3′-(2,4-Dinitrobenzenesulfonyl)-2′,7′-dimethylfluorescein as a fluorescent probe for selenols. *Angewandte Chemie International Edition, 45*, 1810–1813.

Mustacich, D., & Powis, G. (2000). Thioredoxin reductase. *The Biochemical Journal, 346*(Pt. 1), 1–8.

Neto, B. A. D., Carvalho, P. H. P. R., & Correa, J. R. (2015). Benzothiadiazole derivatives as fluorescence imaging probes: Beyond classical scaffolds. *Accounts of Chemical Research, 48*, 1560–1569.

Rayman, M. P. (2000). The importance of selenium to human health. *The Lancet, 356*, 233–241.

Rayman, M. P. (2012). Selenium and human health. *The Lancet, 379*, 1256–1268.

Rocher, C., Lalanne, J.-L., & ChaudiÈRe, J. (1992). Purification and properties of a recombinant sulfur analog of murine selenium-glutathione peroxidase. *European Journal of Biochemistry, 205*, 955–960.

Rotruck, J. T., Pope, A. L., Ganther, H. E., Swanson, A. B., Hafeman, D. G., & Hoekstra, W. G. (1973). Selenium: Biochemical role as a component of glutathione peroxidase. *Science, 179*, 588–590.

Schomburg, L., Schweizer, U., & Köhrle, J. (2004). Selenium and selenoproteins in mammals: Extraordinary, essential, enigmatic. *Cellular and Molecular Life Sciences: CMLS, 61*, 1988–1995.

Schwarz, K., & Foltz, C. M. (1957). Selenium as an integral part of factor 3 against dietary necrotic liver degeneration. *Journal of the American Chemical Society, 79*, 3292–3293.

Singh, R., & Whitesides, G. M. (1991). Selenols catalyze the interchange reactions of dithiols and disulfides in water. *The Journal of Organic Chemistry, 56*, 6931–6933.

Tanguy, S., Grauzam, S., de Leiris, J., & Boucher, F. (2012). Impact of dietary selenium intake on cardiac health: Experimental approaches and human studies. *Molecular Nutrition & Food Research, 56*, 1106–1121.

Thomson, C. D. (2004). Assessment of requirements for selenium and adequacy of selenium status: A review. *European Journal of Clinical Nutrition, 58*, 391–402.

Tuladhar, A., & Rein, K. S. (2018). Manumycin a is a potent inhibitor of mammalian thioredoxin reductase-1 (TrxR-1). *ACS Medicinal Chemistry Letters, 9*, 318–322.

Turner, E., Hager, L. J., & Shapiro, B. M. (1988). Ovothiol replaces glutathione peroxidase as a hydrogen peroxide scavenger in sea urchin eggs. *Science, 242*, 939–941.

Weekley, C. M., & Harris, H. H. (2013). Which form is that? The importance of selenium speciation and metabolism in the prevention and treatment of disease. *Chemical Society Reviews, 42*, 8870–8894.

Yin, C., Huo, F., Zhang, J., Martínez-Máñez, R., Yang, Y., Lv, H., et al. (2013). Thiol-addition reactions and their applications in thiol recognition. *Chemical Society Reviews, 42*, 6032–6059.

Zhang, L., Duan, D., Liu, Y., Ge, C., Cui, X., Sun, J., et al. (2014). Highly selective off–on fluorescent probe for imaging thioredoxin reductase in living cells. *Journal of the American Chemical Society, 136*, 226–233.

Zhang, J., Duan, D., Osama, A., & Fang, J. (2020). Natural molecules targeting thioredoxin system and their therapeutic potential. *Antioxidants & Redox Signaling, 34*, 1083–1107.

Zhang, J., Duan, D., Song, Z.-L., Liu, T., Hou, Y., & Fang, J. (2021). Small molecules regulating reactive oxygen species homeostasis for cancer therapy. *Medicinal Research Reviews, 41*, 342–394.

Zhang, B., Ge, C., Yao, J., Liu, Y., Xie, H., & Fang, J. (2015). Selective selenol fluorescent probes: Design, synthesis, structural determinants, and biological applications. *Journal of the American Chemical Society, 137*, 757–769.

Zhang, J., Li, X., Han, X., Liu, R., & Fang, J. (2017). Targeting the thioredoxin system for cancer therapy. *Trends in Pharmacological Sciences, 38*, 794–808.

Zhang, X., Tian, Y., Li, Z., Tian, X., Sun, H., Liu, H., et al. (2013). Design and synthesis of curcumin analogues for in vivo fluorescence imaging and inhibiting copper-induced cross-linking of amyloid beta species in Alzheimer's disease. *Journal of the American Chemical Society*, *135*, 16397–16409.

Zhong, L., Arnér, E. S. J., Ljung, J., Åslund, F., & Holmgren, A. (1998). Rat and calf thioredoxin reductase are homologous to glutathione reductase with a carboxyl-terminal elongation containing a conserved catalytically active penultimate selenocysteine residue. *Journal of Biological Chemistry*, *273*, 8581–8591.

Zhu, B., Li, P., Shu, W., Wang, X., Liu, C., Wang, Y., et al. (2016). Highly specific and ultrasensitive two-photon fluorescence imaging of native HOCl in lysosomes and tissues based on thiocarbamate derivatives. *Analytical Chemistry*, *88*, 12532–12538.

CHAPTER TWELVE

Identification of selenoprotein O substrates using a biotinylated ATP analog

Meghomukta Mukherjee[a] and Anju Sreelatha[a,b,*]
[a]Department of Physiology, University of Texas Southwestern Medical Center, Dallas, TX, United States
[b]Center for Mineral Metabolism and Clinical Research, University of Texas Southwestern Medical Center, Dallas, TX, United States
*Corresponding author: e-mail address: anju.sreelatha@utsouthwestern.edu

Contents

Abstract

Selenoprotein O is one of 25 human selenoproteins that incorporate the 21st amino acid selenocysteine. Recent studies have revealed a previously undocumented mechanism of redox regulation by which SelO protects cells from oxidative damage.

Methods in Enzymology, Volume 662
ISSN 0076-6879
https://doi.org/10.1016/bs.mie.2021.10.001

SelO catalyzes the covalent addition of AMP from ATP to the hydroxyl side chain of protein substrates in a post translational modification known as AMPylation. Although AMPylation was discovered over 50 years ago, methods to detect and enrich substrates are limited. Here, we describe protocols to clone, purify, and identify the substrates of bacterial SelO using a biotinylated ATP analog. Identification of SelO substrates and the functional consequences of AMPylation will illuminate the significance of this evolutionarily conserved selenoprotein.

1. Introduction

Selenium is an essential micronutrient that acts as a major cellular antioxidant. In particular, selenoproteins incorporate selenium in the form of the 21st amino acid, selenocysteine (Stadtman, 1996). In comparison to the standard amino acid cysteine, selenocysteine contains a selenol group with a lower p*Ka* resulting in higher nucleophilicity and oxidoreductase efficiency (Labunskyy, Hatfield, & Gladyshev, 2014). Thus, several selenoproteins regulate redox homeostasis by neutralizing reactive oxygen species (ROS) such as hydrogen peroxide and lipid hydroperoxide (Steinbrenner & Sies, 2009). Recent studies have underscored the importance of selenoproteins in development as well as several pathologies including neurodegeneration, cardiovascular disorders and cancer (Bellinger, Raman, Reeves, & Berry, 2009; Rocca, Pasqua, Boukhzar, Anouar, & Angelone, 2019; Wu et al., 2021; Zhang & Song, 2021). Despite their physiological importance, the biochemical activity of nearly half of the selenoproteins remain a mystery.

Mitochondria are the primary source of ROS within the cell; the mitochondrial electron transport chain produces ATP and subsequently yields ROS as a byproduct (Zhao, Jiang, Zhang, & Yu, 2019). Hence, the mitochondria rely on antioxidant enzymes, including selenoproteins to reduce oxidative burden and maintain redox homeostasis (Sheu, Nauduri, & Anders, 2006). The human genome encodes 25 selenoproteins including the three mitochondrial selenoproteins, thioredoxin reductase 2 (TrxR2), selenoprotein O (SelO) and glutathione peroxidase 4 (Gpx4) (Kryukov et al., 2003). TrxR2 catalyzes the reduction of thioredoxin using reducing equivalents from NADPH as part of the cellular disulfide reductase system (Conrad et al., 2004). Similarly, Gpx4 utilizes glutathione as an electron donor to catalyze the reduction of lipid hydroperoxides and prevent oxidative damage of cellular membranes and lipoproteins (Maiorino, Conrad, & Ursini, 2018). While TrxR2 and Gpx4 harbor the selenocysteine residue within the active site, SelO contains a C-terminal selenocysteine (Kryukov et al., 2003). Interestingly, SelO is the largest selenoprotein and is conserved from bacteria to humans (Dudkiewicz, Szczepinska, Grynberg, & Pawlowski, 2012). In bacteria and lower eukaryotes, the selenocysteine residue of SelO is replaced with an invariant cysteine (Dudkiewicz et al., 2012). The cysteine containing bacterial and yeast homologs provide a unique advantage for biochemical characterization by avoiding the complications of expression and purification of selenocysteine-containing recombinant protein. Furthermore, bacterial and yeast genetics facilitate robust functional analysis of SelO in cells. For simplicity, we will refer to the *Escherichia coli* SelO as *E. coli* SelO, although the bacterial homolog does not contain selenocysteine.

SelO harbors a mitochondrial targeting sequence at the N-terminus followed by a protein kinase-like domain (Dudkiewicz et al., 2012). However, SelO was predicted to be a pseudokinase because it lacked the catalytic aspartate that acts as a general base to deprotonate the hydroxyl group of the substrate. Structural and biochemical analysis of the SelO active site revealed ATP bound in a flipped orientation relative to canonical kinases (Fig. 1A) (Sreelatha et al., 2018). The mode of ATP binding allows for the transfer of AMP from ATP onto serine, threonine, and tyrosine residues on protein substrates in a post-translational modification known as AMPylation (Fig. 1B) (Sreelatha et al., 2018). SelO is one of only two mammalian enzymes that catalyze AMPylation (Chatterjee & Truttmann, 2021). The other mammalian AMPylator is the ER-resident protein HYPE, which AMPylates heat shock chaperone, BiP, to regulate the unfolded protein response (Ham et al., 2014). The identification of SelO, along with HYPE, suggests that AMPylation may be a more widespread post-translational modification than previously appreciated.

Given the central role of the mitochondria in cellular metabolism and redox homeostasis, identification of SelO substrates will shed light into the mechanisms by which AMPylation modulates mitochondrial signaling. Since the discovery of AMPylation in 1967, research has been hampered by the lack of enrichment techniques for AMPylated proteins. Methods to detect AMPylation include autoradiography of radiolabeled $[\alpha\text{-}^{32}P]$ATP, immunoblotting with AMPylation specific antibodies, mass spectrometry analysis, and click reactions with alkynyl tagged ATP that allow for chemical handles such as fluorophores or affinity tags (Casey & Orth, 2018). Among these methods, immunoblotting and ATP analogs with chemical handles have been used for enrichment with limited success.

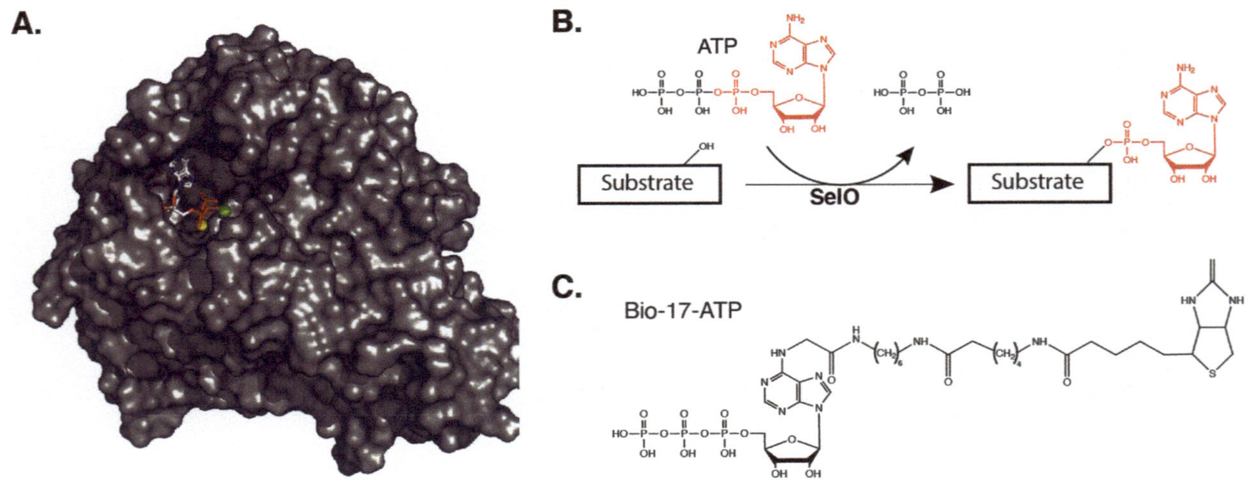

Fig. 1 *E. coli* SelO binds ATP in the active site to AMPylate protein substrates. (A) Surface rendering of *Pseudomonas syringae* SelO demonstrates the orientation of ATP bound in the active site of SelO. (B) *E. coli* SelO catalyzes the transfer of AMP from ATP to serine, threonine and tyrosine residues of protein substrates. (C) Structure of Bio-17-ATP, which allows for biotin labeling of substrates in an *in vitro* AMPylation reaction with *E. coli* SelO. *Panels (B) and (C) Figure adapted from Sreelatha, A., Yee, S.S., Lopez, V.A., Park, B.C., Kinch, L.N., Pilch, S., et al. (2018). Protein AMPylation by an evolutionarily conserved pseudokinase.* Cell 175, *809–821.e819.*

To determine the molecular consequences of protein AMPylation, we devised a strategy to enrich and identify targets of SelO. In the active site of SelO, the γ-phosphate of ATP is held in a charged pocket while the adenine moiety faces outward. This orientation of ATP facilitates the use of a biotinylated ATP analog in an AMPylation reaction in which SelO catalyzes the transfer of biotinylated AMP to substrates that can be enriched using biotin-streptavidin affinity (Fig. 1C). The high affinity binding of biotin to streptavidin is resistant to harsh conditions such as detergents and denaturants. Therefore, this strategy offers a unique advantage by increasing the sensitivity of detection and reducing background binding for protein identification by mass spectrometry. Here, we describe methods to identify the targets of *E. coli* SelO using biotinylated ATP to label and enrich the substrates.

2. Key resources table

Reagent or resource	Source	Identifier
Antibodies		
High sensitivity streptavidin HRP	Pierce	21130
Bacterial Strains		
E. coli DH5α	Dixon lab	N/A
E. coli Rosetta DE3	Novagen	70954
Chemicals, Peptides, and Recombinant Proteins		
Adenosine 5′-triphosphate disodium salt	Sigma	A2383
Biotin-17-adenosine 5′-triphosphate	Enzo life sciences	ENZ 42817
[α-^{32}P]ATP-3000 Ci/mmol 10 mCi/mL	Perkin Elmer	BLU003H250UC
High capacity streptavidin agarose resin	Pierce	20357
cOmplete™, Protease Inhibitor Cocktail	Roche	11873580001
HisPur™ Ni-NTA agarose beads	Pierce	88221
Q5® High-Fidelity 2 × Master mix	NEB	M0492
*Bam*HI-HF®	NEB	R3136
*Hin*dIII-HF®	NEB	R3104

T4 DNA Ligase	NEB	M0202
6 × loading dye	NEB	B7024
Coomassie Brilliant Blue R	Sigma	B0149
Critical Commercial Assays		
QuikChange II site mutagenesis kit	Agilent	200523
GeneJet Plasmid Miniprep kit	ThermoFisher	K0503
DNA Clean & Concentrator™-25	Zymo	D4006
Zymoclean™ Gel DNA recovery kit	Zymo	D4002
Supersignal West Pico Chemiluminescent Substrate	Fisher	PI34578
Oligonucleotides		
aaaaggatccatgaccctgtcttttattac ccgc	Sigma	N/A
aaaaaagcttttaactcgagcagctgactt cc	Sigma	N/A
caaacggcccgtaagcaagcgtcagcccc	Sigma	N/A
ggggctgacgcttgcttacgggccgtttg	Sigma	N/A
Recombinant DNA		
ppSumo	(Nguyen et al., 2016)	N/A

3. Materials and equipment

3.1 Materials and reagents

- 1% agarose gel
- 12% SDS-PAGE gel (precast or handcast)
- 50 mg/mL kanamycin. Store aliquots at −20 °C
- 34 mg/mL chloramphenicol. Store aliquots at −20 °C
- Luria Bertani (LB) broth
- LB Agar plates with antibiotics. Store plates at 4 °C
- 0.4 M Isopropyl β-D-thiogalactoside (IPTG). Store aliquots at −20 °C
- 1 M Tris-HCl pH 8
- 1 M $MgCl_2$

- 2.5 M NaCl
- 1 M DTT
- 8 M Urea
- 0.5 M EDTA pH 8
- 10% TX-100 (w/v)
- 10 mM ATP: dissolve ATP in 25 mM Tris-HCl pH 7.5 and adjust the pH with NaOH if necessary. Measure the absorbance at 259 nm to calculate ATP concentration ($\varepsilon = 15.4 \times 10^3\ M^{-1}\ cm^{-1}$). Store aliquots at −80 °C
- 5 × SDS loading buffer: 62.5 mM Tris-PO_4 pH 6.8, 50% glycerol, 6.25% SDS, 0.1% Bromophenol blue. Add 5% β-mercaptoethanol just before use
- Buffer A: 50 mM Tris-HCl pH 8, 300 mM NaCl
- Buffer B: 10 mM Tris-HCl pH 8, 150 mM NaCl
- Ulp1 deSUMOylase: can be purified by Ni affinity from 6 × His Ulp plasmid that is available upon request or purchased from Invitrogen 12588018
- Wash buffer 1: 2% SDS (w/v)
- Wash buffer 2: 50 mM HEPES pH 7.5, 500 mM NaCl, 1 mM EDTA, 1% TX-100 (w/v), 0.1% (w/v) deoxycholic acid
- Wash buffer 3: 10 mM Tris-HCl pH 7.5, 250 mM LiCl, 1 mM EDTA, 0.5% (w/v) NP-40, 0.5% (w/v) deoxycholic acid
- Wash buffer 4: 50 mM Tris-HCl pH 7.5
- Coomassie staining solution: dissolve 3 g of Coomassie Brilliant Blue in 50% methanol and 10% acetic acid
- Destaining solution: 40% methanol, 10% acetic acid
- Gel drying solution: 30% methanol, 3% glycerol

3.2 Equipment

- PCR cycler
- Water bath incubator
- Dry bath incubator
- Refrigerated floor centrifuge
- 37 °C shaking Incubator
- 4 °C cold room
- Probe Sonicator
- DNA gel electrophoresis apparatus
- Protein gel electrophoresis apparatus

- Electroblotting/western transfer apparatus
- BioRad Econo-Pac chromatography column
- Amicon centrifugal filter unit (30 kDa MWCO)
- Fast protein liquid chromatography (FPLC) system
- HiLoad Superdex 200 pg column
- Imager to visualize DNA bands for molecular cloning
- Spectrophotometer to determine DNA and protein concentrations
- Rocking platform shaker
- Gel drying apparatus
- X-ray film processor

4. Step-by-step method details

4.1 Expression and purification of recombinant *E. coli* SelO

4.1.1 Construction of E. coli *SelO (YdiU) ppSumo plasmid*

Timing: 1 week

Note: We find that 6 × His-SUMO-SelO fusion protein has high expression and stability that allows for high yield purification. The 6 × His will be used for the nickel affinity purification, and the SUMO tag increases the expression level and solubility of SelO. Here we briefly describe a strategy to clone *E. coli* SelO into ppSumo. Requests for plasmids can be directed to the lead contact.

Day 1

(a) PCR amplify *E. coli* SelO from genomic *E. coli* DNA using the forward primer: AAAA GGATCC ATGACCCTGTCTTTTATTAC CCGC and the reverse primer: AAAA AAGCTT TTAACTCGAG CAGCTGACTTCC with Q5 DNA polymerase according to the NEB protocol.

Note: A bacterial DNA template for PCR can be prepared by heating 5 μL of an overnight Rosetta *E. coli* culture in 45 μL of water for 10 min at 95 °C.

(b) Upon completion of the PCR, add DNA loading dye to 2 μL of PCR product and load on a 1% agarose gel along with a 1 kb plus ladder.

(c) After electrophoresis, image the agarose gel to verify the PCR product. The expected band for *E. coli* SelO migrates at approximately 1450 bp.

(d) If the correct size if observed, then proceed to purification of the PCR product.

(e) Digest ppSumo (plasmid backbone) and the PCR cleaned *E. coli* SelO (insert) using *Bam*HI and *Hin*dIII according to the NEB protocols.
(f) Add DNA loading dye to the digested *E. coli* SelO and ppSumo DNA and resolve on 1% agarose gel along with a 1 kb plus ladder.
(g) After electrophoresis, image the gel and excise the digested insert (~1450 bp) and the plasmid backbone (~5600 bp).
(h) Gel extract the digested DNA and measure the concentration.
(i) Ligate *E. coli* SelO and ppSumo using T4 DNA ligase according to the NEB protocol.
(j) Transform the ligation into chemically competent DH5α *E. coli* cells.
(k) Plate the cells on LB agar plates with 50 μg/mL kanamycin and incubate overnight at 37 °C.

Day 2

(a) Pick individual colonies from the agar plates and inoculate 5 mL cultures of LB broth containing 50 μg/mL kanamycin.
(b) Grow overnight at 37 °C shaking at 225 rpm.

Day 3

(a) Miniprep the culture and measure the concentration.
(b) Verify the insert by restriction digest followed by Sanger sequencing using forward primer: ATTAGAATCCAAGCTGATCAGACCC and reverse primer: GCTAGTTATTGCTCAGCGG.

Day 4–6

To construct the plasmid coding for the inactive mutant of *E. coli* SelO, the aspartate at amino acid 256 is mutated to alanine (D256A). Use the forward primer: caaacggcccgtaagcaagcgtcagcccc and the reverse primer: ggggctgacgcttgcttacgggccgtttg with the Agilent QuickChange Site Mutagenesis kit to make *E. coli* SelO D256A ppSumo.

4.1.2 Expression of E. coli *SelO ppSumo in Rosetta* E. coli *cells*

Timing: 3 days

The following describes a protocol for the expression and purification of *E. coli* SelO ppSumo. The same procedure can be adapted for *E. coli* SelO D256A ppSumo.

Critical Step: Take caution to prevent cross contamination of the inactive D256A protein with wildtype *E. coli* SelO protein during expression and purification.

Day 1

(a) Transform the sequence verified *E. coli* SelO ppSumo plasmid into chemically competent Rosetta *E. coli* cells.

(b) Plate the cells on LB agar plates with 50 μg/mL kanamycin and 34 μg/mL chloramphenicol.
(c) Incubate overnight at 37 °C.

Day 2

(a) Inoculate 10 mL of LB broth with 2–3 colonies of *E. coli* SelO ppSumo Rosetta, 50 μg/mL kanamycin, and 34 μg/mL chloramphenicol.
(b) Place the starter culture in a 37 °C incubator shaking at 225 rpm for 4 h or until OD_{600} reaches 0.8–1.
(c) Inoculate 500 mL of LB with the 10 mL starter culture and grow at 37 °C until OD_{600} reaches 0.6.

Note: In order to analyze the expression and purification of *E. coli* SelO, we suggest taking samples as demonstrated in Fig. 2. Sample 1, Pre-induction: Spin down 0.5 OD of cells and resuspend in 100 μL of 1 × SDS sample buffer. Boil for 5 min and store at −20 °C.

(d) Induce protein expression with the addition of 0.4 mM IPTG and continue shaking for 16–18 h at 25 °C.

Day 3

(a) Spin down the induced 500 mL culture at 5000 × rcf for 15 min.

Note: Sample 2, Post-induction: Spin down 0.5 OD of cells and resuspend in 100 μL of 1 × SDS sample buffer. Boil for 5 min and store at −20 °C.

(b) Discard the supernatant and resuspend the pellet from the 500 mL bacterial culture in 50 mL of pre-chilled Buffer A supplemented with 1 mM DTT and 1 mM PMSF.

Pause point: bacterial cell pellet can be stored at −80 °C.

4.1.3 Purification of E. coli *SelO ppSumo*

Timing: 2 days

Critical Step: Pre-cool buffers and equipment. Perform all protein purification steps at 4 °C to avoid heating the samples.

Day 1

(a) Place the bacterial resuspension in an ice bath and sonicate at 35% amplitude on a 20 s on/off cycle for 9 cycles.
(b) Centrifuge the lysate for 25 min at 25,000 × *g* at 4 °C.
(c) Transfer the supernatant to a pre-chilled 50 mL conical.

Note: Sample 3, soluble: Transfer 10 μL of supernatant to 10 μL of 2 × SDS sample buffer. Boil for 5 min and store at −20 °C.

Sample 4, insoluble and unbroken cells: Resuspend pellet in 50 mL of 8 M urea. Transfer 10 μL of resuspension to 10 μL of 2 × SDS sample buffer. Boil for 5 min and store at −20 °C.

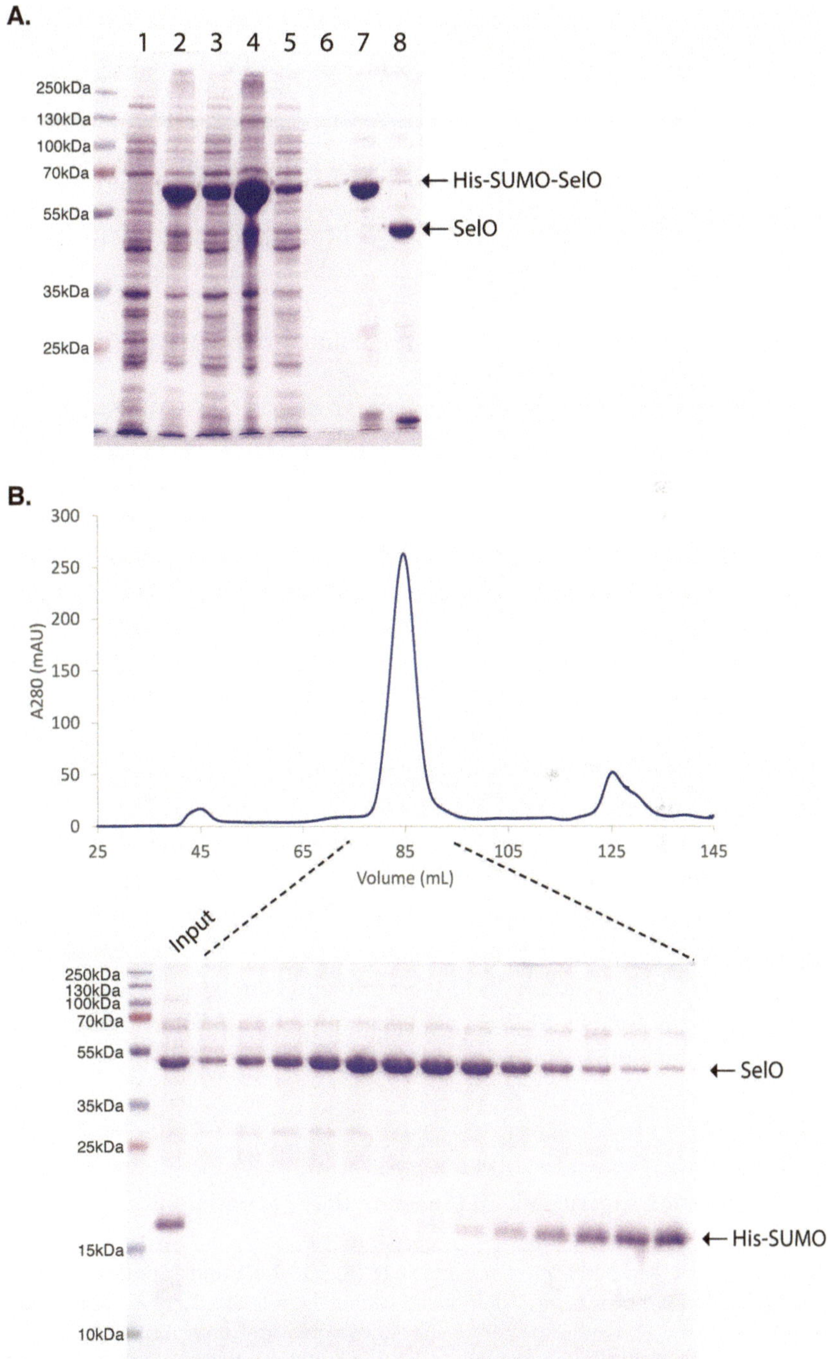

Fig. 2 See figure legend on next page.

(d) Add 500 μL of pre-equilibrated Ni-NTA agarose to the clarified supernatant.

(e) Nutate at 4 °C for 1 h to enable optimal binding of *E. coli* SelO to Ni-NTA beads.

(f) Transfer the lysate slowly to a Econo-column and allow the column to drain by gravity.

Note: Sample 5, unbound/flow-through: Transfer 10 μL of the lysate flow through to 10 μL of 2 × SDS sample buffer. Boil for 5 min and store at −20 °C.

Critical step: Make sure that the column does not run dry during the binding or following wash steps.

(g) Gently wash the column three times with 5 mL of buffer A supplemented with 1 mM DTT and 10 mM Imidazole without disturbing the agarose bed.

Note: Sample 6, wash: Transfer 10 μL of the wash flow through to 10 μL of 2 × SDS sample buffer. Boil for 5 min and store at −20 °C.

Note: We recommend a minimum wash volume of 20 column volumes.

(h) Cap the bottom of the column and add 500 μL of Buffer A supplemented with 1 mM DTT and 300 mM Imidazole to the column.

(i) Incubate on ice for 10 min.

(j) Elute the protein into a pre-chilled conical vial.

(k) Repeat the elution steps for a total elution volume of 2 mL.

Note: We recommend a minimum elution volume of 5 column volumes.

(l) Measure the concentration of eluted protein using Bradford reagent.

Note: Sample 7, elution: Prepare 5 μg of eluted protein in 20 μL of 1 × SDS sample buffer. Boil for 5 min and store at −20 °C.

(m) Cleave the 6 × His-SUMO tag by adding Ulp1 deSUMOylase as per the manufacturer's instructions.

(n) Incubate overnight at 4 °C.

Day 2

(a) Spin down protein at 21,000 × *g* for 10 min.

(b) Transfer the supernatant and measure the concentration.

Fig. 2 Expression and purification of *E. coli* SelO. (A) SDS-PAGE and Coomassie staining of the following samples. Lane 1: uninduced bacterial cell lysate. Lane 2: induced bacterial lysate. Lane 3: soluble protein. Lane 4: insoluble and unbroken cells. Lane 5: unbound protein/flowthrough. Lane 6: wash. Lane 7: eluted protein. Lane 8: eluted protein cleaved with Ulp1. (B) Size exclusion chromatography trace (upper) and SDS-PAGE and Coomassie staining (lower) of *E. coli* SelO after treatment with Ulp1.

Note: Sample 8, post-Ulp1 elution: Prepare 5 μg of cleaved protein in 20 μL of 1 × SDS sample buffer. Boil for 5 min and store at −20 °C.

(c) Resolve 5 μL of samples 1–8 by SDS-PAGE and visualize proteins with Coomassie blue staining. If protein yield and quality are sufficient post Ulp1, proceed to the next step. We recommend proceeding to the next purification step with greater than 4 mg of *E. coli* SelO.

(d) Separate the 6 × His-SUMO tag from *E. coli* SelO by size exclusion chromatography using a HiLoad Superdex 200 pg column connected to a Fast Protein Liquid Chromatography (FPLC) system pre-equilibrated with Buffer B supplemented with 1 mM DTT.

(e) Transfer 10 μL of eluted fraction to 10 μL of 2 × SDS sample buffer, resolve fractions by SDS-PAGE, and visualize the proteins with Coomassie blue staining.

(f) Combine and concentrate the fractions with *E. coli* SelO that are void of 6 ×-His SUMO tag using an Amicon centrifugal filter unit (30 kDa MWCO).

(g) Measure the concentration and adjust the protein concentration to 5 mg/mL.

(h) Aliquot and flash freeze protein in liquid nitrogen for storage at −80 °C.

4.2 Preparation of *E. coli* lysate

Timing: 1 day

1. Inoculate 2–3 colonies of Rosetta in 20 mL of LB broth.
2. Grow at 37 °C shaking at 225 rpm until OD_{600} reaches 1.

Note: Cultures may be harvested from various growth conditions, such as log *vs* stationary phase, to analyze *E. coli* SelO-mediated AMPylation.

3. Centrifuge the bacterial culture at 5000 × *g* for 15 min.
4. Wash the cell pellet once with 500 μL of Buffer B.
5. Resuspend the cells in 500 μL of Buffer B supplemented with 0.1% TX-100, 1 mM EDTA, and 1 × PIC.
6. Sonicate the cell resuspension on a 5 s on/10 s off cycle for 6 cycles.
7. Centrifuge the lysate for 21,000 × *g* for 10 min.
8. Transfer the supernatant and measure the concentration.

Note: Concentrations between 5 and 10 mg/mL work well for AMPylation reactions with biotinylated ATP.

Pause Point: Supernatant may be aliquoted and frozen at −80 °C. However, multiple freeze thaw cycles may affect the integrity of the proteins in the lysate.

4.3 *In vitro* AMPylation reaction

Timing: 1 day

Prepare *E. coli* SelO AMPylation reaction mixture to determine the optimal concentration of lysate for the AMPylation reaction including the proper controls (*E. coli* SelO inactive mutant and no lysate controls). See Fig. 3.

1. Each reaction of 20 µL contains 50 mM Tris-HCl pH 8, 5 mM $MgCl_2$, 5 mM DTT, 0.20 mg/mL *E. coli* SelO, 0.5 mM Bio-17-ATP, and 0–40 µg of lysate.

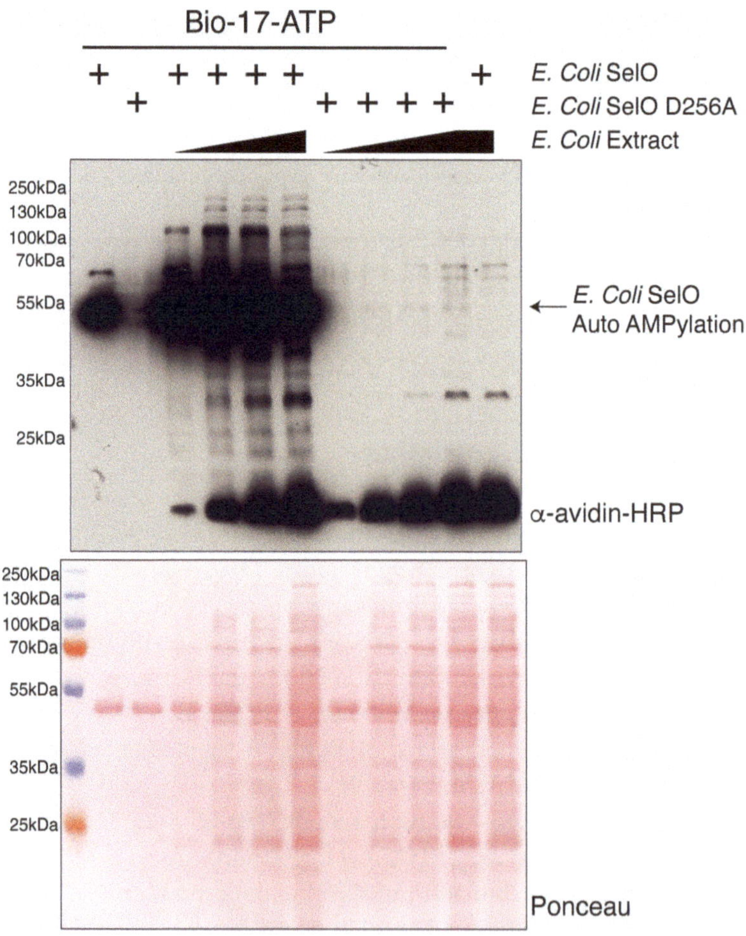

Fig. 3 Optimization of *in vitro* AMPylation assay using Bio-17-ATP. Varying concentrations of bacterial lysate were used in an AMPylation reaction with recombinant *E. coli* SelO or the inactive mutant, D256A. Biotinylated reaction products were detected usingstreptavidin-HRP following SDS-PAGE and transfer to nitrocellulose membrane. The Ponceau stained membrane is shown.

Note: Initiate the reaction by adding Bio-17-ATP.

2. Incubate the reactions at 37 °C for 30 min.
3. Terminate the reaction by adding 0.6 μL of 500 mM EDTA.
4. Add 20 μL of 2 × SDS sample loading buffer.
5. Heat samples at 95 °C for 3 min.
6. Resolve 5 μL of reaction products by SDS-PAGE.
7. Following electrophoresis, transfer the proteins onto nitrocellulose membranes and perform Streptavidin HRP blot according to Roux, Kim, and Burke (2013).

4.4 Enrichment and identification of putative substrates

Timing: 2 days

Day 1

1. Prepare a 50 μL AMPylation reaction mixture with 50 mM Tris-HCl pH 8, 5 mM MgCl2, 5 mM DTT, 0.20 mg/mL *E. coli* SelO (wildtype or inactive mutant), 0.5 mM Bio-17-ATP and the optimal concentration of lysate. See Fig. 4.

   Note: Initiate the reaction by adding the Bio-17-ATP.
2. Incubate the reactions at 37 °C for 30 min.
3. Terminate the reactions by adding 1.5 μL of 500 mM EDTA.
4. Transfer 5 μL of terminated reaction to 5 μL of 2 × SDS samples buffer. Heat the 10 μL samples at 95 °C for 3 min and store at −20 °C as input samples
5. Add 455 μL of 50 mM Tris-HCl pH 7.4, 150 mM NaCl, 1% TX-100, 1 × protease inhibitor, 1 mM EDTA, and 1 mM DTT buffer to the remaining 45 μL and mix by inverting
6. Transfer 20 μL of pre-equilibrated high affinity streptavidin beads to each reaction
7. Nutate the samples overnight at 4 °C

Day 2

The following briefly describes the Streptavidin pulldown protocol. Refer to Roux et al. (2013) for a detailed protocol.

1. Spin down the samples at 5000 × *g* for 30 s
2. Remove the supernatant and wash with 500 μL of wash buffer 1
3. Repeat the wash step with wash buffer 2, wash buffer 3, and wash buffer 4
4. Resuspend the beads in 500 μL of wash buffer 4

Optional: 50 μL of bead suspension can be processed separately for SDS-PAGE and verification of biotinylated proteins by a streptavidin blot.

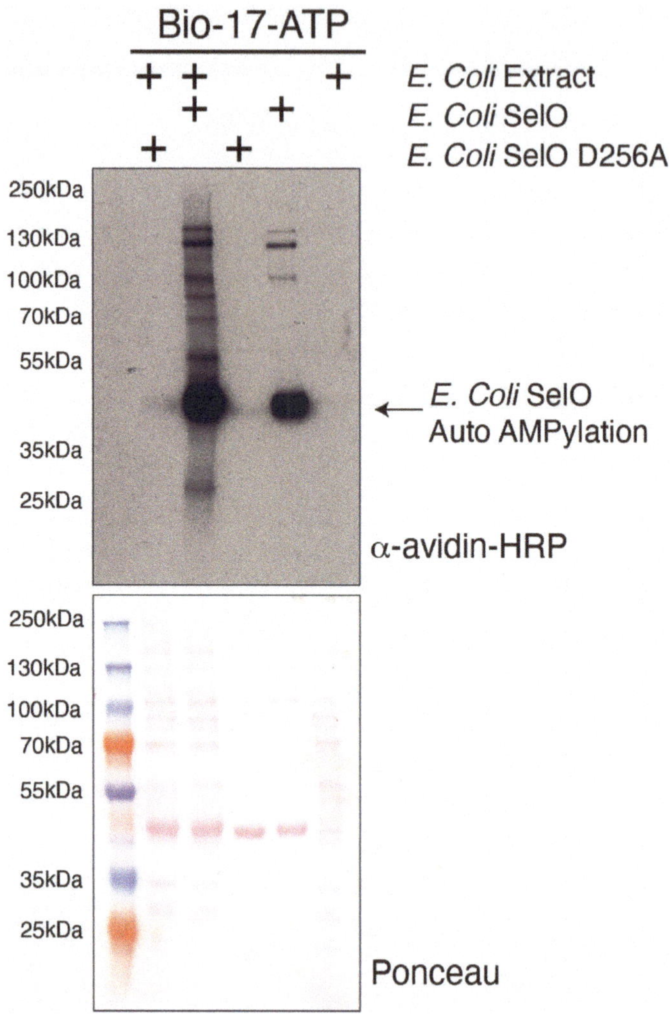

Fig. 4 *E. coli* SelO AMPylates proteins in bacterial lysate. *In vitro* AMPylation assay with bacterial lysate, Bio-17-ATP and recombinant *E. coli* SelO or the inactive mutant, D256A. Biotinylated reaction products were detected using streptavidin-HRP following SDS-PAGE and transfer to nitrocellulose membrane. The Ponceau stained membrane is shown. *Figure adapted from Sreelatha, A., Yee, S.S., Lopez, V.A., Park, B.C., Kinch, L.N., Pilch, S., et al. (2018). Protein AMPylation by an evolutionarily conserved pseudokinase.* Cell 175, *809–821.e819.).*

However, biotinylated proteins bound to the streptavidin beads may not elute with SDS sample buffer and/or heating. Therefore, Streptavidin HRP signal from the samples may be misrepresentative of the labeling and pulldown efficiency.

5. Submit samples for on-bead trypsin digestion and protein identification by mass spectrometry. Determine relative amount of proteins by label free quantification based on peak area using Proteome discoverer software. In order to prioritize substrates for validation, compare protein abundances in *E. coli* SelO treated samples relative to the inactive mutant

4.5 Validation of substrates using radiolabeled ATP

Timing: 1 day

Putative substrates of *E. coli* SelO identified by mass spectrometry analysis may be validated using recombinant proteins in an *in vitro* AMPylation reaction with radiolabeled [α-^{32}P]ATP.

1. Clone, express and purify recombinant substrates as described in Section 4.1.

Note: The protocol described above has been optimized for *E. coli* SelO. Successful recombinant protein production of putative substrates might require optimization of tags, expression conditions, and purification scheme.

2. Assemble each reaction of 20 μL with 50 mM Tris-HCl pH 8, 5 mM $MgCl_2$, 5 mM DTT, 0.05 mg/mL *E. coli* SelO, and 0.25 mg/mL recombinant substrate
3. Initiate the reaction by adding 100 μM [α-^{32}P] ATP (specific radioactivity of 1000 cpm/pmol)
4. Incubate the reactions at 37 °C for 30 min
5. Terminate the reaction by adding 0.6 μL of 500 mM EDTA
6. Add 5 μL of 5 × SDS sample loading buffer
7. Heat samples at 95 °C for 3 min
8. Resolve 10 μL of the reaction products by SDS-PAGE
9. After electrophoresis, stain the gel with Coomassie blue staining solution for 1 h with gentle shaking
10. Destain the gel for 4 h or overnight to remove background staining
11. Wash the gel twice in gel drying solution for 10 min
12. Dry the gel on a filter paper using a gel drying apparatus
13. In a dark room, place an autoradiography film on the dried gel with an autorad marker in a light proof cassette
14. Exposure times vary depending on signal strength, but we recommend 1 day at −80 °C. Develop the film in an X-ray film processor

5. Expected outcomes

In this chapter, we describe the overexpression and purification of bacterial SelO which can be used for *in vitro* AMPylation assays to identify putative substrates. The average yield after Ni-NTA affinity is approximately 15–20 mg of *E. coli* SelO for a 500 mL culture. In the event that the purification yield is low, the expression and purification can be optimized based on the SDS-PAGE analysis of the samples collected at each step as shown in Fig. 2. As a measure of protein quality, the auto-AMPylation activity of the purified enzyme can be assayed using Bio-17-ATP or radiolabeled [α-^{32}P]ATP as detailed in Section 8.

Another critical factor in determining the outcome of the AMPylation reaction is the lysate. Adding large volumes of lysate to the AMPylation reaction will inhibit SelO due to presence of EDTA and NaCl. Furthermore, adding high concentrations of lysate will inhibit the reaction since endogenous proteins in the *E. coli* lysate will compete for the biotinylated ATP. The inactive mutant, *E. coli* SelO D256A, is an important control that will distinguish AMPylated proteins from the endogenous proteins that bind Bio-17-ATP and/or background signal from the streptavidin blot and affinity purification. We have also successfully used lysates from highly enriched mitochondria with Bio-17-ATP to identify the targets of the yeast and mammalian homologs of SelO. In preparing lysates from mitochondria, as well as bacteria, an important consideration is the growth conditions prior to cell harvest as this dictates the protein expression profile. Successful completion of this unbiased screening method will identify the substrates and shed light on the molecular processes regulated by SelO.

6. Advantages

The main advantage of working with bacterial SelO is the ease of expression and purification. In our hands, the bacterial SelO protein has the highest yield and activity. However, these methods can be adapted to purify the mammalian homolog of SelO with a cysteine substituted for the selenocysteine, U667C. The average yield after Ni-NTA affinity is approximately 10 mg of human SelO U667C for 500 mL culture. Caution must be used when analyzing substrates of human SelO as the reactivity of selenocysteine may alter the repertoire of substrates in the mitochondria.

Similar to the Bio-17-ATP, the click reaction has been used to conjugate chemical handles, such as biotin, to ATP derivatives with an alkynyl tag at N6 position (N6pATP) (Grammel, Luong, Orth, & Hang, 2011). However, the use of Bio-17-ATP circumvents the need for a click reaction. The advantage of using biotinylated ATP analogs stems from the binding affinity of biotin to streptavidin which enables robust purification. Roux et al. (2013) has detailed an effective protocol to enrich for biotinylated proteins using stringent conditions to wash out contaminating proteins. In this chapter, we combine the utility of Bio-17-ATP with established biotin affinity methods to lower background binding and detect AMPylated substrates with high sensitivity in comparison to immunoprecipitation with AMPylation-specific antibodies.

7. Limitations

While this strategy has been utilized to successfully label substrates of *E. coli* SelO (Sreelatha et al., 2018; Yang et al., 2020), it may not be generally applicable to all AMPylators due to the bulky biotin handle on the ATP. Even with SelO, *in vitro* AMPylation with Bio-17-ATP may miss substrates that are of low abundance in a cell lysate or whose expression is regulated such as stress responsive proteins. Furthermore, the biotinylated ATP is membrane impermeable rendering the probe impracticable for *in vivo* labeling.

False positives are common in the *in vitro* AMPylation reaction with lysates. Therefore, the inactive SelO mutant and the no lysate controls are important to eliminate background. Additionally, thorough validation of substrates identified by mass spectrometry using immunoblotting and autoradiography is recommended. For detailed methods on the detection of AMPylation, refer to Casey and Orth (2018). Although radiolabeled ATP offers high sensitivity, working with ^{32}P may be cumbersome. We suggest validating the results with radiolabeled ATP which can detect low levels of AMPylation on substrates in comparison to other methods such as immunoblotting.

8. Optimization and troubleshooting

8.1 Low or no protein expression

Confirm that the *E. coli* SelO ppSumo plasmid is sequence verified, in frame, and with no mutations. Some factors that affect *E. coli* SelO expression are temperature, time of expression, and OD of induction. Also, we recommend only filling 20% of the flask volume with LB broth for proper aeration.

8.2 Low protein yield after purification

One major reason for low yield of *E. coli* SelO is insufficient cell lysis. This is the case when the majority of *E. coli* SelO is present in Sample 4 which contains insoluble and unbroken cells. To increase lysis, sonication can be combined with freeze-thaw cycles. Alternatively, we have used high-pressure homogenizers, such as the EmulsiFlex C3, to attain more complete lysis in comparison to sonication. Analysis of the samples, as shown in Fig. 2, will greatly aid in troubleshooting the purification scheme.

Another cause for low protein yield is improper wash and/or elution conditions. Ensure the concentration of imidazole used for wash and elution are accurate. Buffers with DTT in excess of 1 mM will also reduce the Ni-NTA agarose and prevent *E. coli* SelO binding. To monitor the binding and elution efficiency, add 10 μL of beads pre- and post-elution to 10 μL of 2 × SDS sample buffer and analyze as shown in Fig. 2.

8.3 High background/non-specific labeling in Streptavidin blot

The protocol outlined by Roux et al. (2013) is very useful in performing the streptavidin blot and affinity purification. In the Bio-17-ATP assay, the ratio of lysate to *E. coli* SelO is critical in detecting AMPylated proteins. The addition of too much lysate will compete for the biotinylated ATP and increase non-specific signal from the lysate. Similar to the optimization of the lysate outlined in Section 6, one could also optimize the concentration of *E. coli* SelO to increase signal to noise ratio.

8.4 Low signal in streptavidin blot

A reason for suboptimal signal is the addition of too little lysate in the AMPylation reaction, which may fade the signal from low abundance proteins. Low quality/activity of the purified *E. coli* SelO protein could also contribute to the absence of signal in the AMPylation assay. Active *E. coli* SelO, but not the inactive D256A mutant, should autoAMPylate with Bio-17-ATP as seen in Figs. 3 and 4.

9. Safety considerations and standards

Caution must be exercised when working with radioactive ^{32}P. Reduce exposure by using plexiglass shielding and wearing personal protective equipment. Follow institutional standard operating procedure for proper storage and disposal of radioactive samples.

10. Alternative methods/procedures

While Bio-17-ATP has been demonstrated to work well with *E. coli* SelO, alternatives such as N6pATP may also allow robust detection of

AMPylated substrates. In contrast to Bio-17-ATP, the N6pATP is a less bulky analog that can be conjugated to a variety of chemical handles to profile substrates (Grammel et al., 2011). ATP analogs have poor cellular uptake due to the charged phosphate residues. Recent work has aimed to overcome this limitation by the use of a pronucleotide, pro-N6pA, which has enhanced cellular uptake (Kielkowski et al., 2020). The pronucleotide is metabolized to N6pATP and labels AMPylated proteins within cells to provide spatiotemporal resolution.

Another exciting recent development is the production of monoclonal antibodies against AMPylation. Hopfner et al. (2020) have generated antibodies from mice that recognize AMP-serine, threonine and tyrosine residues. The antibodies are a valuable tool for the detection of AMPylated proteins by immunoblot as well as enrichment of AMPylated proteins by immunoprecipitations.

Acknowledgments

We would like to thank Victor Lopez for providing feedback on the manuscript. We are grateful for the support of NIH Grant K01DK123194 (A.S.), Welch Foundation Grant I-2046 (A.S.) and Cancer Prevention Research Institute of Texas Grant RR190106 (A.S.). is a W.W. Caruth, Jr. Scholar in Biomedical Research.

References

Bellinger, F. P., Raman, A. V., Reeves, M. A., & Berry, M. J. (2009). Regulation and function of selenoproteins in human disease. *The Biochemical Journal*, *422*, 11–22.

Casey, A. K., & Orth, K. (2018). Enzymes involved in AMPylation and deAMPylation. *Chemical Reviews*, *118*, 1199–1215.

Chatterjee, B. K., & Truttmann, M. C. (2021). Fic and non-Fic AMPylases: Protein AMPylation in metazoans. *Open Biology*, *11*, 210009.

Conrad, M., Jakupoglu, C., Moreno, S. G., Lippl, S., Banjac, A., Schneider, M., et al. (2004). Essential role for mitochondrial thioredoxin reductase in hematopoiesis, heart development, and heart function. *Molecular and Cellular Biology*, *24*, 9414–9423.

Dudkiewicz, M., Szczepinska, T., Grynberg, M., & Pawlowski, K. (2012). A novel protein kinase-like domain in a selenoprotein, widespread in the tree of life. *PLoS One*, *7*, e32138.

Grammel, M., Luong, P., Orth, K., & Hang, H. C. (2011). A chemical reporter for protein AMPylation. *Journal of the American Chemical Society*, *133*, 17103–17105.

Ham, H., Woolery, A. R., Tracy, C., Stenesen, D., Kramer, H., & Orth, K. (2014). Unfolded protein response-regulated Drosophila Fic (dFic) protein reversibly AMPylates BiP chaperone during endoplasmic reticulum homeostasis. *The Journal of Biological Chemistry*, *289*, 36059–36069.

Hopfner, D., Fauser, J., Kaspers, M. S., Pett, C., Hedberg, C., & Itzen, A. (2020). Monoclonal anti-AMP antibodies are sensitive and valuable tools for detecting patterns of AMPylation. *iScience*, *23*, 101800.

Kielkowski, P., Buchsbaum, I. Y., Kirsch, V. C., Bach, N. C., Drukker, M., Cappello, S., et al. (2020). FICD activity and AMPylation remodelling modulate human neurogenesis. *Nature Communications*, *11*, 517.

Kryukov, G. V., Castellano, S., Novoselov, S. V., Lobanov, A. V., Zehtab, O., Guigo, R., et al. (2003). Characterization of mammalian selenoproteomes. *Science, 300*, 1439–1443.

Labunskyy, V. M., Hatfield, D. L., & Gladyshev, V. N. (2014). Selenoproteins: Molecular pathways and physiological roles. *Physiological Reviews, 94*, 739–777.

Maiorino, M., Conrad, M., & Ursini, F. (2018). GPx4, lipid peroxidation, and cell death: Discoveries, rediscoveries, and open issues. *Antioxidants & Redox Signaling, 29*, 61–74.

Nguyen, K. B., Sreelatha, A., Durrant, E. S., Lopez-Garrido, J., Muszewska, A., Dudkiewicz, M., et al. (2016). Phosphorylation of spore coat proteins by a family of atypical protein kinases. *Proceedings of the National Academy of Sciences of the United States of America, 113*, E3482–E3491.

Rocca, C., Pasqua, T., Boukhzar, L., Anouar, Y., & Angelone, T. (2019). Progress in the emerging role of selenoproteins in cardiovascular disease: Focus on endoplasmic reticulum-resident selenoproteins. *Cellular and Molecular Life Sciences, 76*, 3969–3985.

Roux, K. J., Kim, D. I., & Burke, B. (2013). BioID: A screen for protein-protein interactions. *Current Protocols in Protein Science, 74*. 19.23.11-19.23.14.

Sheu, S. S., Nauduri, D., & Anders, M. W. (2006). Targeting antioxidants to mitochondria: A new therapeutic direction. *Biochimica et Biophysica Acta, 1762*, 256–265.

Sreelatha, A., Yee, S. S., Lopez, V. A., Park, B. C., Kinch, L. N., Pilch, S., et al. (2018). Protein AMPylation by an evolutionarily conserved pseudokinase. *Cell, 175*(809–821), e819.

Stadtman, T. C. (1996). Selenocysteine. *Annual Review of Biochemistry, 65*, 83–100.

Steinbrenner, H., & Sies, H. (2009). Protection against reactive oxygen species by selenoproteins. *Biochimica et Biophysica Acta, 1790*, 1478–1485.

Wu, W., Li, D., Feng, X., Zhao, F., Li, C., Zheng, S., et al. (2021). A pan-cancer study of selenoprotein genes as promising targets for cancer therapy. *BMC Medical Genomics, 14*, 78.

Yang, Y., Yue, Y., Song, N., Li, C., Yuan, Z., Wang, Y., et al. (2020). The YdiU domain modulates bacterial stress signaling through Mn(2+)-dependent UMPylation. *Cell Reports, 32*, 108161.

Zhang, Z. H., & Song, G. L. (2021). Roles of selenoproteins in brain function and the potential mechanism of selenium in Alzheimer's disease. *Frontiers in Neuroscience, 15*, 646518.

Zhao, R. Z., Jiang, S., Zhang, L., & Yu, Z. B. (2019). Mitochondrial electron transport chain, ROS generation and uncoupling (review). *International Journal of Molecular Medicine, 44*, 3–15.

CHAPTER THIRTEEN

Application of alpha-methyl selenocysteine as a tool for the study of selenoproteins

Emma J. Ste. Marie[a] and Robert J. Hondal[b,*]
[a]Department of Chemistry and Biochemistry, Norwich University, Northfield, VT, United States
[b]Department of Biochemistry, University of Vermont, Burlington, VT, United States
*Corresponding author: e-mail address: robert.hondal@uvm.edu

Contents

Abstract

Selenocysteine (Sec) is the 21st proteogenic amino acid and it is now widely accepted that Sec is involved in redox biochemistry when incorporated in proteins. However, many of the chemical mechanisms for Sec bioactivity remain unknown. Herein, we describe a derivative of Sec, alpha-methyl Sec ((αMe)Sec), that is a useful chemical tool to study selenoenzyme mechanisms. (αMe)Sec is identical to Sec except the C^{α}-H is replaced with a C^{α}-methyl group, which prevents this derivative from undergoing oxygen-mediated β-syn elimination to dehydroalanine, which is a common problem with Sec-containing peptides and proteins. Thus, since (αMe)Sec-containing peptides and proteins cannot lose the side-chain selenium atom when oxidized, mechanistic

Methods in Enzymology, Volume 662
ISSN 0076-6879
https://doi.org/10.1016/bs.mie.2021.10.016

studies can be performed that are not always possible with Sec. In this chapter, we provide detailed methods for the incorporation of (αMe)Sec into peptides using solid phase peptide synthesis and subsequent incorporation into mammalian thioredoxin reductase using protein semisynthesis. We then provide two examples of how (αMe) Sec has been used as a chemical tool to study selenoenzyme mechanism. Finally, we discuss future applications where we envision (αMe)Sec will be useful.

Abbreviations

(αMe)AA	alpha-methyl amino acid
(αMe)Sec	alpha-methyl selenocysteine
5-Npys	2-thio(5nitropyridyl)
AA	amino acid
Arg, R	arginine
Asp, D	aspartate
Aib	α-aminoisobutyric acid
A340	absorbance at 340 nm
Boc	tert-butyloxycarbonyl
Boc-Cys(Mob)-OH	*N*-(tert-Butoxycarbonyl)-*S*-(4-methyoxybenzyl)-L-cysteine
Boc-Cys(Mob)-OPfp	*N*-(tert-Butoxycarbonyl)-*S*-(4-methyoxybenzyl)-L-cysteine pentafluorophenyl ester
CBD	chitin binding domain
Cys, C	cysteine
DBG	$C^{\alpha,\alpha}$-diisobutylglycine
DCC	*N,N'*-dicyclohexyl carbodiimide
DCM	dichloromethane
ddI H_2O	distilled and deionized water
DHA	dehydroalanine
DIC	*N,N'*-diisopropylcarbodiimide
DMF	dimethylformamide
DmTrxR-CCS	*Drosophila melanogaster* thioredoxin reductase with C-terminal sequence: Cys-Cys-Ser
DPG	$C^{\alpha,\alpha}$-dipropylglycine
DTNP	2,2′-dithiobis(5-nitropyridine)
E. coli	*Escherichia coli*
EtOAc	ethyl acetate
EDTA	ethylenediaminetetraacetic acid
Fmoc	fluorenylmethoxycarbonyl
Fmoc-Sec(Mob)-OH	Fmoc-Se-4-methoxybenzyl selenocysteine
Fmoc-(αMe)Sec(But)-OH	N^{α}-[(9H-fluoren-9-ylmethoxy)carbonyl]-Se-(tert-butyl)-L-alpha-methyl selenocysteine
Gln, Q	glutamine
GSSG	glutathione oxidized
GPX	glutathione peroxidase
GSH	glutathione reduced
GR	glutathione reductase

Gly, G	glycine
h	hour
HATU	hexafluorophosphate azabenzotriazole tetramethyl uranium
HOAt	1-hydroxy-7-azabenzotriazole
HIS	histidine
H_2O_2	hydrogen peroxide
IC_{50}	half-maximal inhibitory concentration
KCl	potassium chloride
KP	potassium phosphate
mTrxR	mouse thioredoxin reductase
mTrxR-C(αMe)UG	mouse thioredoxin reductase with C-terminal sequence: Cys-(αMe)Sec-Gly
mTrxR-CUG	mouse thioredoxin reductase with C-terminal sequence: Cys-Sec-Gly
NADPH	nicotinamide adenine dinucleotide phosphate
NaOAc	sodium acetate
NCL	native chemical ligation
NMM	*N*-methylmorpholine
PyAOP	7-azobenzotriazolyoxytris(pyrrolidino)phosphonium hexafluorophosphate
Pfp	pentafluorophenol
RES	reactive electrophilic species
ROS	reactive oxygen species
rt	room temperature
Sec, U	selenocysteine
Ser, S	serine
SPPS	solid phase peptide synthesis
TFA	trifluoroacetic acid
Thr, T	threonine
TIS	triisopropylsilane
Trx	thioredoxin
Tyr, Y	tyrosine
Val, V	valine
WT	wild type

1. Introduction

Selenocysteine (Sec, U) is the 21st amino acid (AA) in the genetic code where this AA is primarily involved in redox biochemistry. Beyond proteins, Sec has also found wide utility in synthetic peptides, especially as a replacement for cysteine. One limitation of Sec in both proteins and synthetic peptides is that it can undergo a β-syn elimination reaction after

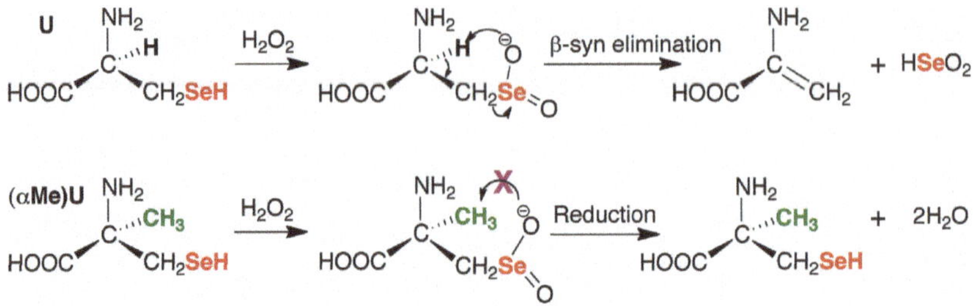

Fig. 1 Avoidance of β-syn elimination in (αMe)Sec. (Top panel): Sec or Sec-containing peptides can undergo deselenization due to β-syn elimination as a result of overoxidation of the Sec residue. (Bottom panel): The (αMe)Sec analog avoids β-syn elimination as it lacks hydrogen on Cα. The overoxidized form of (αMe)Sec can be reduced back to the parent form by the use of mild reducing agents such as ascorbic acid.

oxidation forming dehydroalanine (DHA), rendering the Sec-containing peptide or protein inactive due to loss of the selenium atom (Fig. 1).

Formation of dehydroalanine can be avoided by the use of alpha-methyl selenocysteine ((αMe)Sec), an analogue of Sec. This derivative replaces the α-hydrogen with a methyl group, removing the possibility of β-syn elimination (Fig. 1, bottom panel). This analog is advantageous because overoxidation no longer becomes a problem. It can be used in glutathione peroxidase (GPX) mimetic peptides since resistance to oxidative inactivation is greatly increased and it can also be used to study the mechanism of naturally occurring selenoenzymes.

In this chapter, we provide detailed methods for the incorporation of (αMe)Sec into peptides using solid phase peptide synthesis (SPPS), and subsequent incorporation into mammalian thioredoxin reductase (mTrxR) using protein semisynthesis. It has been previously noted in literature that the incorporation of α-methyl AA's into peptides is challenging due to increased steric hinderance of the N-terminal amine that the α-methylene causes (Carpino, Abdel-Maksoud, Mansour, & Zewail, 2007; Frérot, Coste, Pantaloni, Dufour, & Jouin, 1991; Ste Marie et al., 2020). Herein, we discuss the unique challenges associated with α-methyl AA's, and describe a novel method for the coupling of an Fmoc-protected AA onto the amine of α-methyl AA's on resin.

In addition, we describe two applications of (αMe)Sec: (i) Incorporation into a peptide that acts as a GPX mimic (Wehrle, Ste Marie, Hondal, & Masterson, 2019), and (ii) Incorporation into the active site of mTrxR as a replacement for Sec (Ste Marie et al., 2020). In both applications, (αMe)Sec is utilized as a chemical tool to study selenium redox biochemistry. In the case of the first application, we found that an (αMe)Sec-containing peptide was able to catalytically reduce hydrogen peroxide (H_2O_2) and showed significantly enhanced stability to oxidative inactivation compared to its Sec-containing analog. This study provides potential mechanistic insight into the oxidative inactivation of GPX and other selenoproteins. In the case of the second application, we found that an (αMe)Sec-containing mTrxR allowed for enhanced rescue from inactivation by reactive electrophilic species (RES). Studies over the past three decades have led to the idea that Sec evolved in part to help resist reactive oxygen species (ROS). A recent study by the authors extends this hypothesis to include resistance to RES. Resistance to ROS and RES by Sec could be a significant biological and chemical advantage compared to the use of Cys in enzymes. Finally, we discuss other potential biotechnological applications of (αMe)Sec and provide future direction for this unique chemical tool.

2. Materials and methods

2.1 Materials

Solvents including dichloromethane (DCM), dimethylformamide (DMF), methanol, anhydrous diethyl ether, trifluoroacetic acid (TFA), and ethyl acetate (EtOAc), along with Corning™ Costar™ Spin-X™ centrifuge tube filters to remove catalase agarose, and hydrogen peroxide (30%, ACS grade) were purchased from Fisher Scientific (Waltham, MA). *N*-Fmoc amino acids and Boc-Cys(Mob)-OH were purchased from RSsynthesis (Louisville, KY). 2-Chlorotrityl chloride resin SS (100–200 mesh), and 1-[bis(dimethylamino)methylene]-1H-1,2,3-triazolo[4,5-b]pyridinium 3-oxide hexafluorophosphate (HATU) for solid-phase peptide synthesis (SPPS) were purchased from Advanced ChemTech (Louisville, KY). Chitin resin for TrxR semisynthesis was purchased from New England Biolabs (Ipswich, MA). Immobilized catalase beads were obtained from BioVision Incorporated (San Francisco, CA). Acrolein solution (100 μg/mL or 5000 μg/mL) in methanol was purchased from Ultra Scientific (Santa Clara, CA). NADPH for enzyme assays was purchased from AppliChem (Darmstadt, Germany). Imidazole (reagent grade), bovine serum albumin (lyophilized powder), 2,2′-dithiobis(5-nitropyridine) (DTNP), pentafluorophenol (>99%), Amicon® ultra centrifugal filter units, L-glutathione reduced (≥98.0%), and silica gel for column chromatography were purchased from MilliporeSigma (Burlington, MA). Triisopropylsilane (TIS, 98%), piperidine, ninhydrin dye,

N-methylmorpholine (99%), and *N*, *N*′-dicyclohexylcarbodiimide (DCC, 99%), were purchased from ACROS Organics (Pittsburgh, PA). All other chemicals were purchased from MilliporeSigma (Burlington, MA), Fisher Scientific (Waltham, MA), or ACROS Organics (Pittsburgh, PA). Enzyme kinetic assays were performed on a Cary50 UV-Vis spectrophotometer (Walnut Creek, CA).

2.2 Synthesis of (αMe)Sec

Briefly, this synthesis starts by alkylation of an achiral methyl malonate through the use of a selenium-containing alkylating agent synthesized in the presence of dichloromethane. The seleno-malonate was then subjected to an enzymatic hydrolysis utilizing pig liver esterase followed by a Curtius rearrangement producing a protected derivative of (αMe)Sec that could be used in solid phase peptide synthesis. By utilizing this methodology, access to both enantiomers, as well as other relevant analogues, such as α-AAs, are possible. Full details of the synthesis of (αMe)Sec are described by Wehrle et al. (2019).

2.3 Incorporation of (αMe)Sec into a peptide

The unique challenges associated with incorporating (αMe)AAs into peptides using traditional SPPS have been noted in the literature, and are discussed in greater detail in the Section 3.3.1. However, in Section 2.3.1, (αMe)Sec is the N-terminal AA in the peptide sequence and therefore does not require special procedures since there was no AA coupled onto its sterically hindered amine. Conversely, in Section 2.3.2, there is a cysteine (Cys) residue coupled onto the N-terminal amino of (αMe)Sec, and this coupling required extra steps. Our novel method for coupling an AA onto the sterically hindered amine of an (αMe)AA is described in Section 2.3.2 and 2.3.3.

2.3.1 Synthesis of GPX peptide mimic containing (αMe)Sec

Either Fmoc-(αMe)Sec(Bu^t)-OH (synthesized as described in Wehrle et al., 2019), or Fmoc-Sec(Mob)-OH (Hondal & Raines, 2002) were used to synthesize peptides of sequence H-(αMe)Sec-Gly-Thr-Thr-Val-Arg-Asp-Tyr-Thr-Gln-OH (H-(αMe)UGTTVRDYTQ-OH), or H-Sec-Gly-Thr-Thr-Val-Arg-Asp-Tyr-Thr-Gln-OH (H-UGTTVRDYTQ-OH). Both peptides were synthesized on a 0.1 mmol scale using a glass vessel that was shaken with a model 75 Burrell wrist action shaker. For each batch,

300 mg of 2-chlorotrityl chloride resin SS (100–200 mesh), was swelled in DCM for 30 min. The first AA was directly coupled to the resin using 2% *N*-methylmorpholine (NMM) in DCM, shaking for 1 h. The resin was then capped using 10 mL of 8:1:1 DCM:methanol:NMM. Subsequent AAs were coupled using 0.2 mmol of Fmoc-protected AA, 0.2 mmol HATU, and 2% NMM in dimethylformamide (DMF), shaking for 1 h, rt. Fmoc-Sec(Mob)-OH or Fmoc-(αMe)Sec(Bu^t)-OH were coupled using a solution of 0.2 mmol AA and 0.2 mmol 1-hydroxy-7-azabenzotriazole (HOAt) in DMF with 1% *N,N*′-diisopropylcarbodiimide (DIC), shaking for 2 h, rt. Coupling did not require preactivation of any AA. Between AA couplings, the Fmoc protecting group was removed via two 10 min agitations with 10 mL of 20% piperidine in DMF. Success of Fmoc removal steps and AA couplings were monitored qualitatively using a ninhydrin test (Kaiser, Colescott, Bossinger, & Cook, 1970). Removal of the final Fmoc protecting group completed the peptide synthesis. Peptides were cleaved from the resin via a 4 h reaction with a cleavage cocktail consisting of 0.4 mmol of 2,2′-dithiobis(5-nitropyridine) (DTNP), in 10 mL of 96:2:2 TFA/TIS/H_2O. Following cleavage and side chain deprotection, the resin was washed with TFA and DCM and the volume of the cleavage solution was reduced by evaporation with nitrogen gas. The peptide was precipitated from solution by addition of cold, anhydrous diethyl ether. Centrifugation at 3000 rpm on a clinical centrifuge (International Equipment Co., Boston, MA) for 10 min pelleted the peptide. To wash the peptides, peptide pellets were dried under nitrogen gas, dissolved in minimal 96:2:2 TFA/TIS/H_2O, then precipitated and pelleted as described previously. This wash process was repeated five times for each peptide. After washing, the final peptide pellet was dried, dissolved in a minimal amount of water, lyophilized, and used without further purification.

2.3.2 Synthesis of Boc-Cys(Mob)-OPfp

Boc-Cys(Mob)-OH (6.0 mmol, 2.05 g) was dissolved in 10 mL of anhydrous DCM and cooled to 0 °C via ice bath, under nitrogen gas. To this, dicyclohexyl carbodiimide (DCC) (3.0 mmol, 619.0 mg) dissolved in 10 mL of anhydrous DCM was added. Reaction was allowed to proceed ~15 min with constant stirring. A white precipitate was observed to form after ~5 min. Pentafluorophenol (6.0 mmol, 1.1 g) dissolved in 20 mL anhydrous ethyl acetate was added via syringe and the reaction was immediately removed from ice and allowed to warm to rt (~22 °C), and stirred an additional 1.5 h. The white urea precipitate was removed from solution via filtration, and the filtrate was placed on ice for 15 min to prompt formation of more precipitate and subsequently filtered again. The final filtrate was reduced to a yellow oil via rotary evaporation. Column purification using ~25 g silica gel and ~300 mL of 80:20 hexanes: ethyl acetate as the solvent afforded pure Boc-Cys(Mob)-OPfp with an $R_f = 0.44$. The fractions containing desired product were concentrated using rotary evaporation and hi-vac producing a white, crystalline solid in 89% yield (4.3 g).

2.3.3 Synthesis of tripeptide: H-Cys-(αMe)Sec-Gly-OH

The preparation of 2-chlorotrityl chloride resin, coupling and subsequent deprotection of Fmoc-Gly-OH was completed as described in Section 2.3.1 on a 0.1 mmol scale. Next, Fmoc-(αMe)Sec(Bu^t)-OH, synthesized as described previously (Wehrle et al., 2019), was coupled to *N*-deprotected Gly using a solution of 0.2 mmol AA and 0.2 mmol HOAt in DMF with 1% DIC, shaking for 2 h, rt. The Fmoc group was then removed from (αMe)Sec via six, 45 min agitations with 10 mL of 20% piperidine in DMF. Notably, this deprotection step is extensive as we found it difficult to remove Fmoc from the (αMe)Sec-amine to be incomplete under standard conditions. In addition, ninhydrin test was not used to monitor complete Fmoc removal as we found the (αMe)Sec-amine to be too hindered to react with the ninhydrin dye. To couple Cys to (αMe)Sec, the resin was washed extensively with DCM to remove any DMF, then swelled in a minimal volume of DCM and shaken with 2% NMM for 15 min. A solution of 0.22 mmol Boc-Cys(Mob)-Opfp (synthesized as described in Section 2.3.2) in 5 mL DCM was then added to the resin, and the resin shaken for 1.5 h at rt. After 1.5 h shaking, 0.15 mmol HATU in 5 mL of DMF was added and allowed to react for an additional 1 h. The resin was washed 4x with DMF, then 4x with DCM and the Boc-Cys(Mob)-OPfp coupling step was repeated 12–15 times. Note: we were able to scale this tripeptide synthesis up to a 0.3 mmol scale, but for best results typically made the tripeptide on a 0.1 mmol scale.

Peptides were cleaved from the resin via a 1.5 h reaction with a cleavage cocktail consisting of four-fold excess of DTNP, in 10 mL of cleavage cocktail (96:2:2 TFA/TIS/H_2O). Following cleavage and side chain deprotection, the resin was washed with TFA and the volume of the cleavage solution was reduced by evaporation with nitrogen gas. The peptide was precipitated from solution by addition of cold, anhydrous diethyl ether. Centrifugation at 3000 rpm on a clinical centrifuge (International Equipment Co., Boston, MA) for 10 min pelleted the peptide. To wash the peptides, peptide pellets were dried under nitrogen gas, dissolved in minimal 96:2:2 TFA/TIS/H_2O, then precipitated and pelleted as described previously. This wash process was repeated five times for each peptide. After washing, the final peptide pellet was dried, dissolved in a minimal amount of water, lyophilized, and used without further purification.

2.4 Measurement of peroxidase activity of peptides via glutathione reductase coupled assay

A coupled glutathione reductase (GR) assay was used to test the abilities of the peptides to function as GPX mimics. A 0.5 mL solution containing 100 mM potassium phosphate (KP) buffer, pH 8.0, 1 mM ethylenediaminetetraacetic acid (EDTA), 100 μM glutathione reduced (GSH), 200 μM nicotinamide adenine dinucleotide phosphate (NADPH), 2 units of GR, and 10 μM of ebselen, H-UGTTVRDYTQ-OH or H-(αMe)UGTTVRDYTQ-OH (synthesized as described in Section 2.3.1), were added to a quartz cuvette and mixed thoroughly. This was used to blank the spectrophotometer at 340 nm. Assays were initiated by addition of 200 μM H_2O_2. Activity was followed by the decrease of absorption at 340 nm using an extinction coefficient of 6220 M^{-1} cm^{-1}, at 25 °C, and each initial rate was measured at least three times. Background NADPH consumption was corrected for by subtracting control experiment activities in which no peptide was present as well as an additional control in which no H_2O_2 was present.

2.5 Stability test of Sec- and (αMe)Sec—Containing peptides

To test the stability of peptides in the presence of oxygenated buffer, H-UGTTVRDYTQ-OH or H-(αMe)UGTTVRDYTQ-OH (synthesized in Section 2.3.1) were incubated in 100 mM KP buffer, pH 8.0, at 25 °C. Aliquots were taken and centrifuged to pellet precipitated selenium after 24, 48, and 96 h. GPX-like activity of the aliquots were monitored as described in Section 2.3.

2.6 Semi-synthesis of TrxR enzymes containing (αMe)Sec

The semi-synthesis of the mTrxR WT enzyme has been previously reported (Eckenroth et al., 2006; Eckenroth, Lacey, Lothrop, Harris, & Hondal, 2007). In order to replace the active site Sec of mTrxR (Sec489) with (αMe)Sec, we used a mTrxR-intein fusion protein in which the final three AAs (Cys-Sec-Gly) are missing. A tripeptide of sequence H-Cys-(αMe)Sec-Gly-OH (synthesized as described in Section 2.3.3) was then ligated to the enzyme using the technique of protein semisynthesis (Hondal & Raines, 2002). Procedures for construction of this fusion protein and the conditions needed to carry out peptide ligation in order to create mutants of mTrxR by semisynthesis have previously been reported (Eckenroth et al., 2006), and we used analogous procedures in this work to obtain mTrxR-C(αMe)UG. The concentrations of the final, purified mTrxR-C(αMe)UG was determined based on the extinction coefficient of flavin at 460nm ($\varepsilon 460 = 22.6\,mM^{-1}\,cm^{-1}$ for the dimer) (Zhong & Holmgren, 2000). Please note that since the last three AAs of wild type mTrxR are Cys-Sec-Gly, we abbreviate the enzyme as mTrxR-CUG, where "m" stands for mouse (we used the mouse enzyme), and CUG are the one-letter abbreviations for each amino acid. Similarly, we abbreviated the mutant as m mTrxR-C(αMe)UG.

2.6.1 Characterization of selenoenzymes

All activity assays were performed on a Cary 50 UV/VIS spectrophotometer from Varian (Walnut Creek, CA) at rt, in 100mM KP, 1mM EDTA, pH 7.0. Activity was monitored over 2min and V_o was determined from the linear slope. Plots of V_o/total moles of enzyme vs substrate concentration were fit by the Michaelis-Menten equation using KaleidaGraph 4.02 from Synergy Software (Reading, PA) and activities reported as moles of NADPH consumed per min per mole of TrxR.

2.6.2 Activity assays with Trx, H_2O_2, and selenocysteamine substrates

Activity assays for Trx, H_2O_2, and selenocysteamine substrates were performed by monitoring NADPH consumption via decrease in absorbance at 340nm (A340) using an extinction coefficient of $6220\,M^{-1}\,cm^{-1}$. Stock solutions of H_2O_2 were freshly prepared in ddI H_2O before each experiment and the concentration of the solution was determined spectrophotometrically using an extinction coefficient of $43.6\,M^{-1}\,cm^{-1}$ at 240nm. Stock solutions of selenocysteamine were freshly prepared in ddI H_2O before each experiment. A stock solution of 20mM NADPH in ddI H_2O was prepared prior to experiments and frozen in aliquots. Thioredoxin (Trx) reductase activity was measured via a 0.5mL reaction containing 200μM NADPH and varying concentrations of *E. coli* Trx that ranged from 0 to 180μM. All assays were initiated by addition of 260nM mTrxR-C(αMe)UG or 2nM mTrxR-CUG. H_2O_2-peroxidase activity was measured via a 0.5mL reaction containing 200μM NADPH and varying concentrations of H_2O_2 that ranged from 0 to 400mM. All assays were initiated by addition of the following concentrations of enzyme: 260 nM mTrxR-C(αMe)UG and 6 nM mTrxR-CUG. Selenocysteamine reductase activity was measured via a 0.5 mL reaction containing 200 μM NADPH and varying concentrations of selenocysteamine that ranged from 0 to 500 μM. All assays were initiated by addition of the following concentrations of enzyme: 104 nM mTrxR-C(αMe)UG and 2 nM mTrxR-CUG. Background NADPH consumption was corrected for by subtracting the activity of control assays in which either TrxR enzymes or substrates were omitted from the reaction. All assays were repeated in triplicate for each enzyme.

2.7 Inhibition of (αMe)Sec—Containing protein with reactive electrophilic species (RES)

To determine the half-maximal inhibitory concentration (IC_{50}) of acrolein, inhibition of TrxR enzymes (both mTrxR-C(αMe)UG and mTrxR-CUG) by acrolein was monitored via a two-part assay in which TrxR enzymes were first exposed to increasing concentrations of acrolein, and second, any remaining enzyme activity was assessed relative to a control reaction where acrolein was omitted. Prior to experimentation, a fresh stock solution of acrolein was prepared in ethanol. To inhibit enzymes with acrolein, mTrxR-CUG (10 nM) or mTrxR-C(αMe)UG (780 nM), were pre-reduced by 5 min incubation with 200 μM NADPH in a 150 μL assay in 100 mM KP, 1 mM EDTA, pH 7 buffer at rt. Next, TrxR enzymes were incubated with increasing amounts of acrolein (0–18 nM) for 10 min followed by a second addition of 200 μM NADPH. Next, activity assays were initiated by addition of 150 μM Trx and Trx-reductase activity was measured by monitoring the consumption of NADPH as a decrease in A340 for three minutes. The change in absorbance was converted to percentage of activity remaining relative to the acrolein-untreated (ethanol only) control. Additionally, for each assay, a "no substrate" background was performed. In each of these control experiments, no activity was detected.

2.7.1 Reversal of acrolein TrxR inhibition using imidazole and hydrogen peroxide

Inhibition and subsequent reversal of TrxR enzymes by acrolein was completed by a three-part assay in which TrxR enzymes were first inhibited by acrolein, then acrolein inhibition was reversed via imidazole/ H_2O_2 combinations, and third, enzyme activity remaining was assessed relative to a control reaction where acrolein was omitted. Prior to experimentation, fresh stock solutions of acrolein were prepared in ethanol and stock solutions of H_2O_2 were prepared in ddI H_2O and the concentration was determined spectrophotometrically using an extinction coefficient of $43.6\,M^{-1}\,cm^{-1}$ at 240 nm. Reversal assays utilized a combination of three buffers to facilitate pH changes. For all enzymes, "Buffer A" was composed of 10 mM KP, 10 mM sodium acetate (NaOAc), 5 mM potassium chloride (KCl), pH 7.0; "Buffer C + imidazole" was composed of 200 mM KP, 1 mM EDTA, 210 μM BSA, 21 mM imidazole, pH 7.25; and "Buffer C—imidazole" was composed of 200 mM KP, 1 mM EDTA, 210 μM BSA, pH 7.25. For reactions with mTrxR-CUG, "Buffer B" was composed of 50 mM NaOAc, pH 4.5. For reactions with mTrxR-C(αMe)UG, "Buffer B" was composed of 50 mM NaOAc, pH 5.3.

To begin, either mTrxR-CUG (30 nM) or mTrxR-C(αMe)UG (780 nM), were incubated with 400 μM NADPH in 55 μL of Buffer A for 5 min at rt. To this, 35 μL of Buffer B was added (adjusting the reaction pH to 4.6 for mTrxR-CUG; and to pH 5.5 for mTrxR-C(αMe)UG), followed immediately by addition of either acrolein solution, or ethanol for control experiments. To achieve 75–85% inhibition: for mTrxR-CUG, 2.5 μM acrolein was used; for mTrxR-C(αMe)UG, 800 nM acrolein was used. Enzymes could react with acrolein or ethanol for 10 min at rt.

Next, to reverse acrolein adduction, 40 μL of Buffer C (+/-) imidazole was added to the reaction (i.e. for reversal conditions with imidazole, Buffer C + imidazole was used and for reversal conditions with no imidazole, Buffer C – imidazole was used). For reversal conditions involving H_2O_2, 2 mM H_2O_2 was added immediately after addition of Buffer C (+/-) imidazole. Addition of Buffer C (+/-) imidazole adjusted reaction pH to ~7.1. Reversal reactions were incubated for 10 min in a 37 °C water bath. After incubation, to all reactions, 50 μL catalase agarose was added and incubated for 15 min at rt. To remove catalase agarose, reactions were transferred to a Corning™ Costar™ Spin-X™ Centrifuge Tube Filter and centrifuged for 2 min at 3000 RPM. Flow through was collected and an additional 200 μM NADPH was added, then enzymes were assayed with Trx (150 μM Trx for assays with mTrxR-CUG and 220 μM Trx for assays with mTrxR-C(αMe)UG). For all enzymes, final assays were 150 μL and Trx reductase activity was measured by monitoring the consumption of NADPH as a decrease in A340 for 3 min. For each assay, a "no substrate" background was performed. In each of these control experiments, no activity was detected. For all substrates, the change in absorbance was converted to percentage of activity remaining relative to the acrolein-untreated (ethanol only) control.

3. Results and discussion

3.1 Efficient chemical synthesis of (αMe)Sec

While syntheses of (αMe)Sec have been previously reported (Iwaoka, Ito, Miyazaki, & Michibata, 2016; Makiyama, Komatsu, Iwaoka, & Yatagai, 2010; Reich & Wollowitz, 1993), limitations existed that prevented (αMe)Sec from being used as a chemical tool. To overcome these limitations, we developed a synthetic method that allows for quick synthesis, with moderate yields while also obtaining high enantiomeric excess (Wehrle et al., 2019). We provide a scheme showing the synthetic steps as Scheme 1, but note that we do not go into great detail of the chemical synthesis since it is beyond the scope of this chapter and is previously reported by Wehrle et al. (2019). This method allows access to both enantiomers and could also be used to access other (αMe)AAs.

3.2 (αMe)Sec-containing peptide as a novel GPX mimic

GPX is a selenoenzyme found in many organisms throughout all kingdoms of life, although Cys-containing forms also exist, including in humans (Toppo, Vanin, Bosello, & Tosatto, 2008). GPX catalyzes the reduction

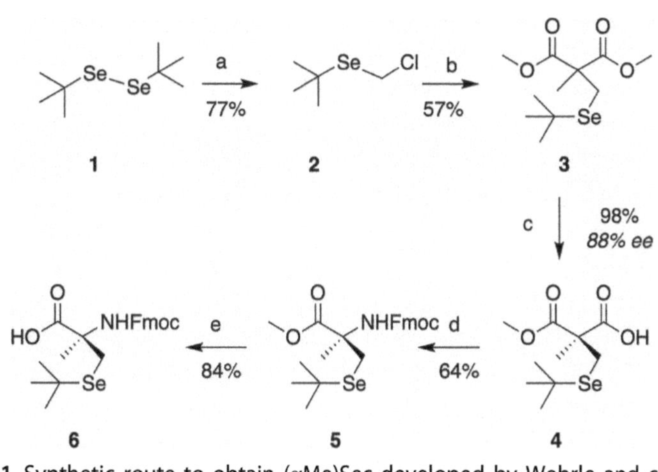

Scheme 1 Synthetic route to obtain (αMe)Sec developed by Wehrle and coworkers. (Wehrle et al., 2019) Reagents and conditions: (A) Sodium bis(2-methoxyethoxy)aluminium hydride, DCM, rt; (B) dimethyl methylmalonate, sodium hydride, tetrahydrofuran, reflux; (C) Pig liver esterase, KP pH 7.4, rt; (D) (i) diphenylphosphoryl azide, triethylamine, dichloroethane, reflux; (ii) 9-fluorenemethanol, titanium (IV) tert-butoxide, benzene, rt; (E) hydrochloric acid, 1,4-dioxane, reflux.

of H_2O_2 to water and lipid hydroperoxides to their corresponding alcohols (Flohe, Günzler, & Schock, 1973; Rocher, Lalanne, & Chauidere, 1992). When concentrations of H_2O_2 are elevated, such as during reoxygenation after a period of ischemia, GPX cannot reduce H_2O_2 fast enough to prevent reperfusion injury (Maulik, Yoshida, & Das, 1999). This may be due in part to overoxidation of the Sec residue with concomitant conversion to DHA by β-syn elimination and loss of enzyme activity (Cho et al., 2010; Ma, Caprioli, Hill, & Burk, 2003; Rhee & Cho, 2010). Small molecule GPX mimics such as ebselen that cannot undergo β-syn elimination may have therapeutic value in times of high oxidative stress.

We envisioned that (αMe)Sec could be used in two capacities as a: (i) small molecule GPX mimic, and, (ii) chemical tool to study the chemical mechanism of Sec β-syn elimination to DHA. For this, we synthesized the peptide of sequence: H-(αMe)Sec-Gly-Thr-Thr-Val-Arg-Asp-Tyr-Thr-Gln-OH, and the control peptide of sequence: H-Sec-Gly-Thr-Thr-Val-Arg-Asp-Tyr-Thr-Gln-OH as described in Section 2.3.1. Both peptides correspond to AAs 49-58 of the human GPX active site. During cleavage of both peptides from the 2-cholortrityl resin, DTNP was included in the cleavage cocktail (TFA/TIS/H_2O) to remove the Mob protecting groups (Harris, Flemer, & Hondal, 2007), and this deprotection afforded a mixture of three different forms of the peptide: (i) peptide diselenide, (ii) peptide selenol, and (iii) Sec(5-Npys) or (αMe)Sec(5-Npys) protected peptide, where (5-Npys) is the 2-thio(5nitropyridyl) adduct formed from treatment with DTNP during cleavage (Wehrle et al., 2019). This crude peptide mixture containing these three redox states of the peptide were used without further purification for the study.

3.2.1 GPX like peroxidase activity of peptide mimics

A coupled GR assay was used to test the abilities of the peptides to function as GPX mimics (Wehrle et al., 2019; Wilson, Zucker, Huang, & Spector, 1989). Both peptides were able to reduce H_2O_2 significantly higher than the background rate when used in an enzyme-like assay described in Section 2.3. The rate for the (αMe)Sec-containing peptide with 200 μM H_2O_2 was 23.50 ± 1.88 μM/min, while the rate for the Sec-containing peptide was 17.96 ± 0.37 μM/min based on the slopes of the plot of A_{340} vs time shown in Fig. 2A. These values compare favorably to that of ebselen, a known GPX mimic (Kil, Pierce, Tran, Gu, & Lynch, 2007; Lynch & Kil, 2009; Maiorino, Roveri, Coassin, & Ursini, 1988) (in our hands, 38.84 ± 1.80 μM/min, data not shown). While the (αMe)Sec-containing

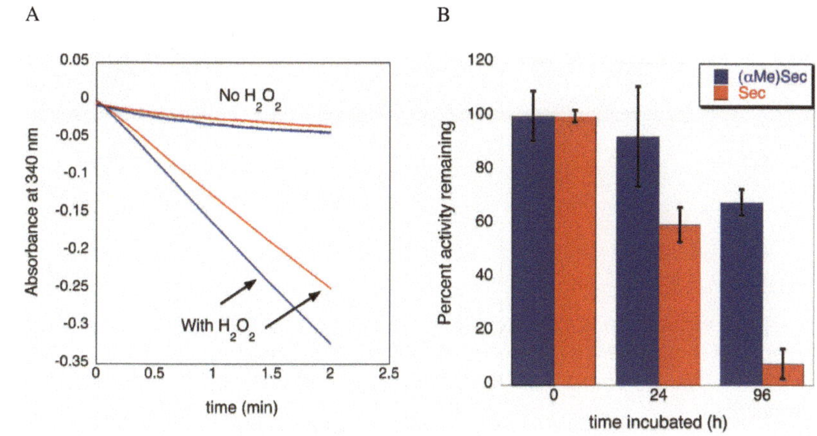

Fig. 2 GPX-like activity and stability of Sec- and (αMe)Sec-containing peptides. (A) GPX-like activity of (αMe)Sec-peptide (blue) and Sec-peptide (red) assayed with H_2O_2. Oxidized glutathione disulfide (GSSG) is formed as a by-product of the peptide catalytic cycle. GR reduces GSSG back to GSH using NADPH, and the consumption of NADPH can be measured via absorbance at 340 nm. (B) Percent activity remaining after incubation of peptides in 50 mM potassium phosphate buffer, pH 8.0, after 24 h and 96 h, at rt.

peptide had only a slight kinetic advantage, a much larger difference in stabilities of the two peptides was observed as shown in Fig. 2B, which shows the activity remaining after incubation in oxygenated KP buffer, pH 8.0, as a function of time (described in Section 2.4).

3.2.2 Mechanisms of loss in activity of GPX-peptide mimics

The loss in activity of the Sec-containing peptide in oxygenated buffer quantified in Fig. 2B is most likely due to oxidation of the Sec residue followed by deselenization, which can occur by various mechanisms including β-syn elimination. We provide further evidence of this in our earlier work, but the detailed data is beyond the scope of this chapter (Wehrle et al., 2019). Deselenization results in the appearance of a red precipitate due to the insolubility of elemental red selenium. Visual evidence for this phenomenon can be seen in Fig. 3, which shows a comparison of deselenization of peptides.

As can be seen, a red precipitate clearly forms after 96 h in the tube with the Sec-containing peptide, but not in the (αMe)Sec-containing peptide. As is evident in Fig. 3, no color change was observed in the reaction containing our (αMe)Sec-peptide even after 96 h of incubation. Fitting with these

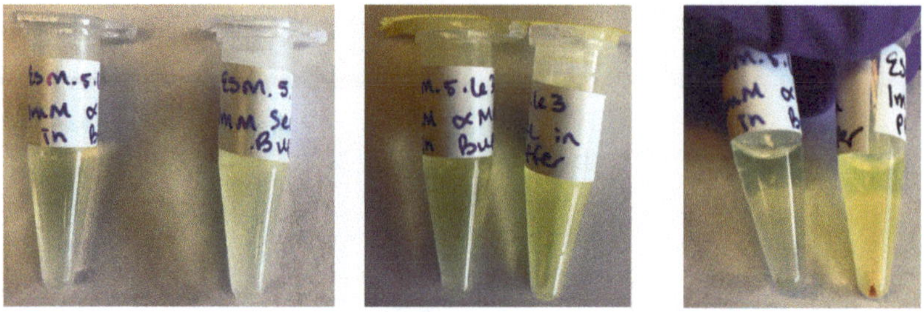

Initial (0 hr) 24 hr 96 hr

Fig. 3 β-syn elimination in a Sec-containing peptide exposed to oxygenated buffer, pH 8.0: Peptides (1 mM) were incubated in KP buffer, pH 8.0, for the indicated times above. After 96 h β-syn elimination is clearly visible in the Sec-peptide (tube at right in each photo).

observations, the (αMe)Sec-containing peptide maintained >65% activity (Fig. 2B). The (αMe)Sec-containing peptide cannot lose selenium through an oxidation/β-syn elimination pathway and thus this cannot be the reason for the modest loss in activity after 96 h in oxygenated buffer. Some possible explanations for the observed decrease in activity of the Sec-peptide are photo-degradation of the peptide due to the presence of selenium or oxidation of other AAs.

3.2.3 (αMe)Sec in a peptide: Conclusions

In conclusion, we described the synthesis of an (αMe)Sec-containing peptide that functions as a GPX mimic to convert H_2O_2 into water. This peptide demonstrated peroxidase activity that was 60% of that of ebselen, a known small molecule GPX mimic. We also found that replacement of Sec with (αMe)Sec confers superior stability to the peptide as the oxidized form of (αMe)Sec cannot undergo β-syn elimination. This peptide has potential to function as a peptide therapeutic since it is a stable GPX mimic. Furthermore, this study provides mechanistic evidence supporting the hypothesis that a major reason for observed loss of activity of selenoproteins if due to the loss of selenium via a β-syn elimination mechanism.

3.3 Tool to study β-elimination reactions of naturally occurring Sec-enzymes

Another application for which (αMe)Sec has proven its utility is for the study of the role of selenium in nature. It is now accepted that one advantage that selenoenzymes have is that they can resist inactivation by oxidation much better than their Cys-orthologs (Snider, Ruggles, Khan, & Hondal, 2013). Recently, we were able to provide evidence to expand this hypothesis to include RES using a mutant version of mTrxR that contains (αMe)Sec in place of Sec in the active site. Therefore, using (αMe)Sec as a chemical tool, we provided mechanistic insight into β-elimination reactions that allow selenoenzymes to resist permanent inactivation by RES. This work is described in Sections 3.3.1–3.3.4.

3.3.1 Incorporation of (αMe)Sec into a tripeptide

Importantly, in the work described in Section 3.2, (αMe)Sec was the N-terminal AA in the peptide sequence and did not require coupling of another AA onto the amine of (αMe)Sec during peptide synthesis. However, for the incorporation of (αMe)Sec into a protein, we synthesized the peptide of sequence H-Cys-(αMe)Sec-Gly-OH for the semisynthesis of mutant mTrxR, and therefore had to couple Fmoc-Cys-OH onto the amine of (αMe)Sec during SPPS. This coupling is much more difficult compared to the coupling of AAs to α-aminoisobutyric acid (Aib), which is described in great detail in the peptide literature.

Previous reports have highlighted the synthetic challenges of making peptides that contain sterically hindered and bulky (αMe)AAs (Fu, 2002; Fu et al., 2001; Fu & Hammer, 2002; Humphrey & Chamberlin, 1997; Katritzky, Todadze, Angrish, & Draghici, 2007). Specifically, it has been noted that forming an amide bond between an (αMe)AA and the carboxylic acid of the next AA in the peptide sequence is difficult due to the steric hinderance of the amine group, which is adjacent to the disubstituted α-carbon, of an (αMe)AA (Fu, 2002; Fu et al., 2001). Incorporation of (αMe)AAs into peptides is exceptionally challenging since the reactivities of both (i) the amino groups and (ii) the carboxyl groups of sterically hindered α,α-disubstituted AAs are significantly lower than those of typical AAs (Katritzky et al., 2007). Traditional activation methods employed for amide bond formation, such as carbodiimide protocol, are usually inefficient for (αMe)AA coupling, suffering from low yields, slow reaction rates, and a necessity for large excess of reagents (Fu, 2002; Fu et al., 2001; Humphrey & Chamberlin, 1997; Katritzky et al., 2007). Other coupling reagents, such as hexafluorophosphate azabenzotriazole tetramethyl uronium (HATU) or HOAt (both commercially available), have also proven to be ineffective in the synthesis of peptides containing (αMe)AAs. Albericio and coworkers previously reported synthesis of peptides that consist of difficult sequences with 7-azobenzotriazolyoxytris(pyrrolidino)phosphonium hexafluorophosphate (PyAOP) (Albericio et al., 1997). PyAOP has been successfully applied for the preparation of a range of peptides containing hindered AAs, such as Aib (Frérot et al., 1991; Katritzky et al., 2007). PyAOP and the related PyBrOP were also used for the synthesis of peptides containing the disubstituted $C^{\alpha,\alpha}$-dipropylglycine (DPG) and $C^{\alpha,\alpha}$-diisobutylglycine (DBG) (Fu, 2002), and α-methylcysteine (Masterson, Kedrowski, & Blair, 2010). The symmetric anhydride method has also been previously employed to facilitate the acylation of Aib and other bulky α-disubstituted AA's (Fu & Hammer, 2002). This method involves the formation of the symmetric anhydride of the incoming AA, then coupling this directly to the resin in the absence of other coupling reagents (Atherton, Cameron, & Sheppard, 1988; Fu & Hammer, 2002). Additionally, Carpino and coworkers previously showed that including a proton sponge greatly improved the yield of an acylation reaction between resin-bound Aib and an incoming 5-mer peptide without causing racemization (Carpino et al., 2007).

While we attempted many of the above mentioned methods, we did not achieve successful synthesis of the H-Cys-(αMe)Sec-Gly-OH tripeptide using any of them: our syntheses suffered from low yields, numerous side products, or truncated peptide. Therefore, to overcome these challenges, we developed a new method that utilizes Boc-Cys(Mob)-OPfp (Ste Marie et al., 2020), and this method is described in Sections 2.3.2 and 2.3.3, and displayed in Scheme 2. Using this new method, we were able to obtain the desired H-Cys-(αMe)Sec-Gly-OH tripeptide in sufficient yield and purity for protein semisynthesis. While this method requires an extra synthetic step to synthesize Boc-Cys(Mob)-OPfp since this derivative is not commercially available, we envision that it would be a useful method for incorporating other large (αMe)AAs into peptides, including DPG and DBG.

The method shown in Scheme 2 is specific for the tripeptide H-Cys-(αMe)Sec-Gly-OH, but could be adapted to a more general procedure for other (αMe)Sec- or (αMe)AA-containing peptides. The generalized procedure involves synthesis of tripeptides containing the (αMe)AA such that the C-terminal AA is coupled to the solid support, followed by coupling of the (αMe)AA using standard procedures. The (αMe)AA to be coupled would first be converted to the Boc(αMe)AA-OPfp. The Boc(αMe)AA-OPfp derivative can be stored stably for months at –20 °C, which is an advantage over the use of the symmetrical anhydride method.

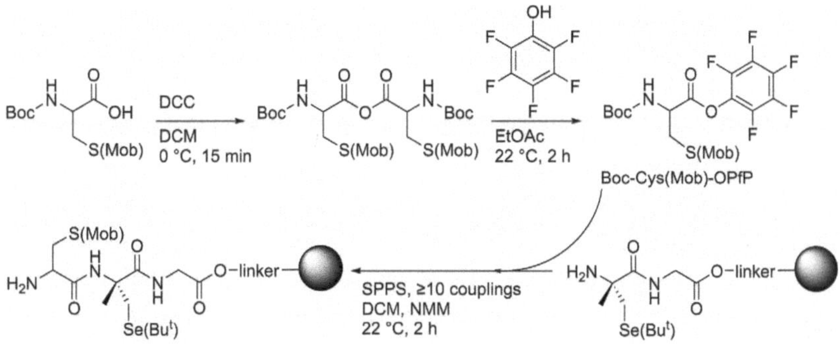

Scheme 2 Synthetic route to obtain Cys-(αMe)Sec-Gly tripeptide. (Top) Starting with Boc-Cys(Mob)-OH, *N,N'*-dicyclohexylcarbodiimide (DCC) is used to make a symmetric anhydride intermediate in dichloromethane (DCM), which is then opened up using pentafluorophenol in ethyl acetate (EtOAc). The resulting Boc-Cys(Mob)-OPfP is added directly to the resin (grey sphere) with H-(αMe)Sec(Bu^t)-Gly-OH dipeptide attached. SPPS procedures are used to perform ≥10 Boc-Cys(Mob)-OPfp couplings with NMM and no coupling reagents.

Symmetrical anhydrides of AAs must be made right before the coupling step and cannot be stored (Fu & Hammer, 2002). The reason for the use of a (αMe)AA-OPfp derivative for coupling is two-fold: (i) The OPfp ester is a very electrophilic acylating reagent for synthesis of amide bonds, having higher reactivity than a symmetrical anhydride (Atherton et al., 1988), but providing higher stability (as mentioned above), and (ii) The use of Boc to protect the α-amino group allows for many coupling reactions to be performed to drive amide bond formation to completion. This is necessary because base-mediated coupling reactions involving Fmoc protected AAss result in "over coupling" of the incoming AA. This is due to premature removal of the Fmoc group on the (αMe)AA due to the multiple coupling steps needed to form the amid bond between the (αMe)AA and the incoming AA. Normal coupling reactions not involving (αMe)AAs do not require the extensive coupling reactions needed for amide bond formation, so short term exposure to base is not typically a problem.

Once the Boc-AA_1(αMe)AA_2AA_3-OH tripeptide is synthesized, the tripeptide can be cleaved from the solid support with TFA without scavenger so that side-chain protecting groups remain on the peptide (Ste Marie & Hondal, 2018). The use of TFA results in cleavage from the solid support with concomitant removal of the Boc group. In our case, we used the tripeptide directly for protein semisynthesis. For longer peptides, the isolated H-AA_1(αMe)AA_2AA_3-OH tripeptide (with side-chains protected) could be then be treated with Fmoc-O-Succinimide to form Fmoc-AA_1(αMe)AA_2AA_3-OH. This tripeptide unit would then be used as a unit to couple to the desired AA or peptide pre-attached to a solid support. Once this tripeptide is coupled, further coupling would occur using standard procedures. This method provides a solution to the difficult coupling of an AA to an (αMe)AA. This method may seem a bit cumbersome, but we think this is the best method to achieve a (αMe)AA with high yield and purity until a new direct coupling method is developed that overcomes the problems of coupling an AA to the sterically hindered amine of an (αMe)AA.

3.3.2 Incorporation of (αMe)Sec-containing tripeptide into a protein

As described in Section 2.5, we incorporated (αMe)Sec into mTrxR using protein semi-synthesis (Hondal & Raines, 2002). Protein semisynthesis, championed by the Muir group in 2000 (Huse, Holford, Kuriyan, & Muir, 2000), is defined as the assembly of a protein from a combination of synthetic and recombinant fragments (Thompson & Muir, 2020). Semisynthesis works particularly well for Sec-containing mTrxR (Hondal & Raines, 2002). In mTrxR, Sec is the penultimate residue of the AA sequence, and it is adjacent to a Cys residue, which can undergo ligation with the expressed portion of the enzyme, as depicted in Scheme 3. To obtain mTrxR and its derivatives, only the short peptide H-Cys-Sec-Gly-OH, which corresponds to the C-terminus of the enzyme, needs to be synthesized using SPPS while a majority of the protein (487 residues) can be expressed and purified in *E. coli*. The Hondal research group has used this method to make mTrxR-CUG wild type (WT), along with a myriad of mTrxR mutants where the C-terminal synthetic peptide is mutated (Barber & Hondal, 2019; Eckenroth et al., 2006, 2007; Hondal & Ruggles, 2011; Lacey, Eckenroth, Flemer Jr, & Hondal, 2008; Lothrop et al., 2014; Lothrop, Ruggles, & Hondal, 2009; Snider et al., 2013).

With the tripeptide H-Cys-(αMe)Sec-Gly-OH in hand, we used semisynthesis to obtain mutant mTrxR-C(αMe)UG where the Sec active site residue of native mTrxR is replaced with (αMe)Sec. MS analysis of mTrxR enzymes confirmed the presence of the full-length WT and mutant proteins, and enzyme kinetics for several known mTrxR substrates (including Trx, H_2O_2, selenocysteamine, and DTNB), showed k_{cat} and K_m parameters that are similar to previously reported values for the WT. This data is not included in this chapter, but can be found in our recent publication (Ste Marie et al., 2020).

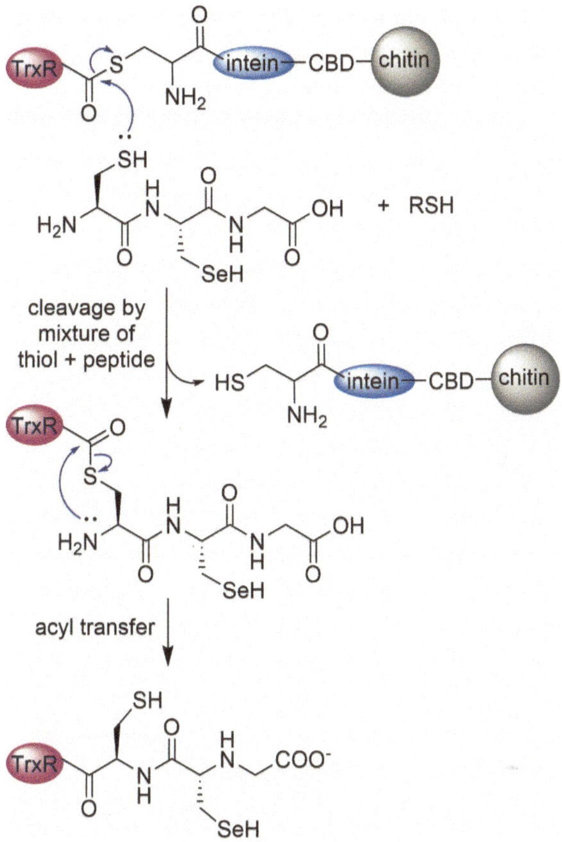

Scheme 3 Cleavage of mTrxR-intein fusion construct by addition of tripeptide and thiol. Previous work from Hondal and coworkers(Eckenroth et al., 2006, 2007) constructed a fusion protein that consists of mouse thioredoxin reductase-2 (mTrxR-2, pink sphere), intein, and a chitin binding domain peptide (CBD). The CBD binds tightly to chitin-agarose resin (grey sphere), and this affinity is used for purification of the fusion protein. The fusion protein was made such that glycine 487 of mTrxR-2 is the AA at the junction between mTrxR-2 and the intein (blue sphere) through a thioester linkage. Addition of a cleavage buffer containing thiol and H-CUG-OH (which are the final three AAs in the protein sequence) initiates an *N*→*S* acyl transfer, with subsequent cleavage of full length mTrxR-CUG from the column. *Adapted from Eckenroth, B., Harris, K., Turanov, A. A., Gladyshev, V. N., Raines, R. T., & Hondal, R. J. (2006). Semisynthesis and characterization of mammalian thioredoxin reductase.* Biochemistry, 45, *5158–5170.*

3.3.3 Inhibition of (αMe)Sec-containing mTrxR with RES

The adduction of TrxR enzymes by RES has been well documented (Cai et al., 2012; Gan et al., 2013; Saccoccia et al., 2014; Zhang et al., 2019). Many RES that are found to inhibit TrxR are biological Michael acceptors as they contain an α, β-unsaturated carbonyl functional group. Acrolein is the simplest and most reactive of the α, β-unsaturated aldehydes, and rapidly adducts protein nucleophiles. The high reactivity of acrolein is the basis for the cytotoxicity, genotoxicity, and mutagenic activity observed in cells exposed to high concentrations of acrolein (Kehrer & Biswal, 2000). Acrolein is universally present in the environment, cooked foods, and cigarette smoke (Stevens & Maier, 2008). Acrolein can also be released *in vivo* via biochemical reactions such as lipid peroxidation of polyunsaturated fatty acids and degradation of spermine and spermidine in situations of oxidative stress and inflammation (Stevens & Maier, 2008).

We began by studying the effect of acrolein on our semisynthetic mTrxRs. Pre-reduced mTrxR enzymes (both the WT that contains Sec in the active site, and the mutant that contains (αMe)Sec in the active site) were incubated with various concentrations of acrolein in buffer. The percent enzyme activity remaining was measured spectrophotometrically and used to determine the IC_{50} value for each acrolein/enzyme combination While we did test other RES (including curcumin and 4-hydroxynonenal, detailed in Ste Marie et al., 2020) we found that acrolein was the most potent RES, as only nM concentrations of acrolein was needed to totally inhibit enzymes (Fig. 4) (Ste Marie et al., 2020). Therefore, acrolein was used for subsequent reversal studies. Please note that our enzyme abbreviations are explained in Section 2.5.

3.3.4 Reversal of acrolein adduct from (αMe)Sec-TrxR using oxidative conditions

We wanted to test the hypothesis that Sec-enzymes can resist RES by undergoing a reversal of the Michael adduction via a β-syn selenoxide elimination, a non-enzymatic mechanism. Notably, such a reversal would be very slow for orthologous Cys-TrxR enzymes as sulfur carbon bonds are much less labile compared to selenium carbon bonds. We used the Cys-TrxR from *D. melanogaster*, whose last three amino acids are Cys-Cys-Ser. As such we abbreviate the enzyme as DmTrxR-CCS. Our hypothesized β-syn elimination mechanism for the removal of acrolein or other RES adducts in the presence of oxidant is shown in Fig. 5.

One limitation in testing the reversibility of RES adducts using WT selenoenzymes is that there are two possible ways in which the selenoxide can undergo β-elimination (Fig. 5). It can eliminate through the desired *pathway 1* to remove the RES adduct, but it could also abstract the α-proton of the

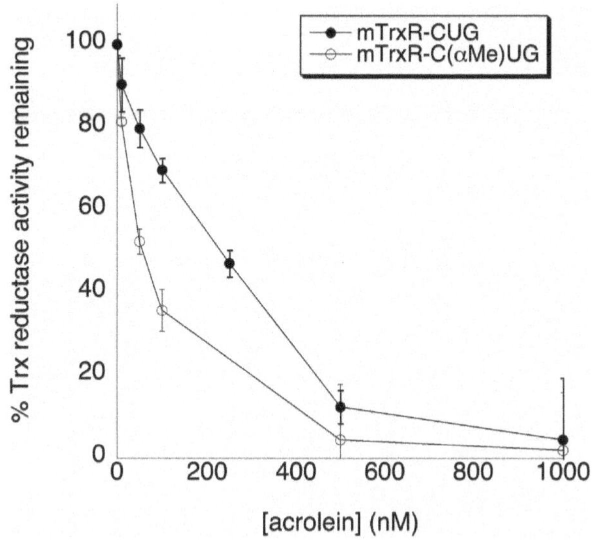

Fig. 4 Inhibition of TrxR enzymes by acrolein. Inhibition of mTrxR-CUG (WT) and mTrxR-C(αMe)UG (mutant) by acrolein after 10 min incubation in 100 mM potassium phosphate, 1 mM EDTA, pH 7 buffer at rt.

enzyme backbone resulting in the formation of DHA in place of the catalytic Sec residue, as shown in Fig. 5A, *pathway 2*. Without Sec, the enzyme would be inactive.

Therefore, if we were unable to restore Trx-reductase activity under conditions in which the adduct could be reversed, we would not be able to distinguish if formation of DHA or the inability to reverse acrolein adduction was the reason for our observed negative result. Our mutant mTrxR-C(αMe)UG enzyme negates this ambiguity since it lacks the α-proton necessary for β-syn elimination with the protein backbone. The formation of DHA is not possible, and therefore, it forces reversibility to occur through the mechanism depicted in Fig. 5B.

To test for reversal of the Michael adduct, we inhibited pre-reduced mTrxR-CUG and mTrxR-C(αMe)UG with acrolein at low pH of (pH 4.6 for WT and pH 5.5 for mutant) for 10 min at rt. The greater acidity of Sec ($pK_a \sim 5.3$) vs Cys ($pK_a \sim 8.3$) or His ($pK_a \sim 9.2$) causes it to be deprotonated at lower pH, and this property allowed us to selectively adduct Sec instead of other AA residues. We attempted to adduct the (αMe) Sec-mutant at pH 4.6 (as we did the WT), however the enzyme precipitated so we chose pH 5.5, which we found to be the lowest pH the mutant enzyme could tolerate for the 10 min acrolein incubation step. As a negative

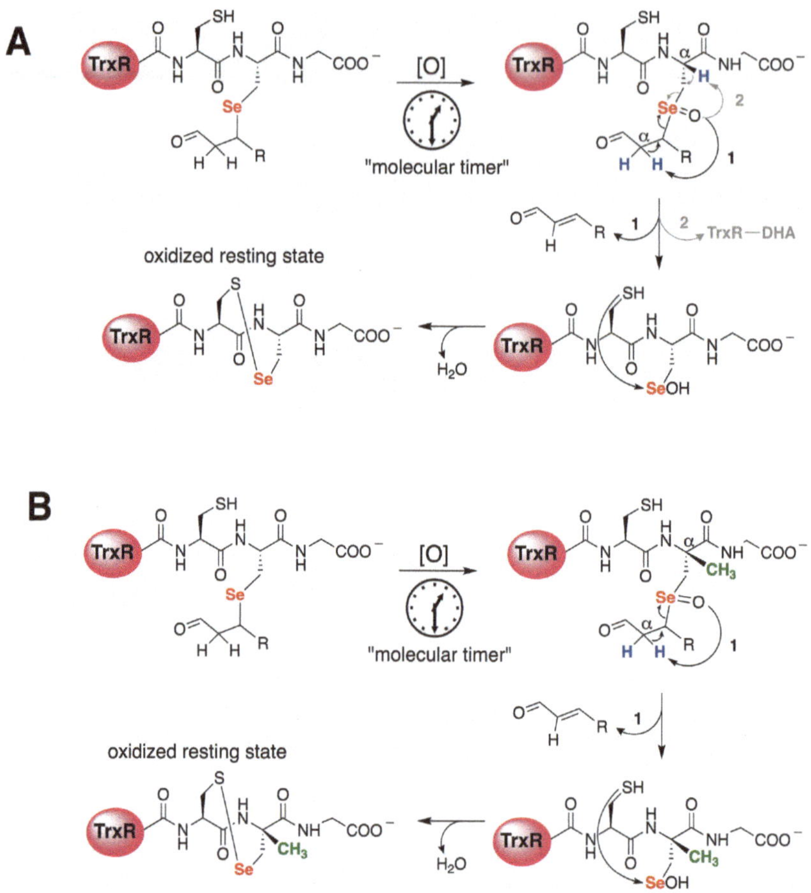

Fig. 5 Selenoxide elimination mechanism for reversing alkylation on Sec of TrxR.(Ste Marie et al., 2020) (A) Once Sec is alkylated by an electrophile, mTrx will produce H_2O_2, which can act as a type of "molecular timer". This H_2O_2 will convert the selenide to a selenoxide which can undergo rapid β-syn elimination via pathway 1 (bold pathway). Abstraction of the Cα-proton of the electrophile leads to elimination of the electrophile and restoration of TrxR back to its oxidized selenosulfide state. However, abstraction of the Cα-H from the enzyme backbone via pathway 2 (gray) results in elimination of selenium from the enzyme and the formation of dehydoalanine (DHA). Without Sec in the active site, enzyme activity cannot be restored. (B) The mutant mTrxR can only utilize pathway 1 in the β-syn elimination mechanism. Thus, the mutant should be better able to reverse the electrophilic alkylation and restore Trx-reductase activity.

control (i.e. we are confident reversal would not be obtained), we also adducted DmTrxR-CCS with acrolein under similar conditions. To ensure the Cys residues were primarily in the reactive thiol form, we conducted the adduction reaction at pH 7 instead of pH 4.5 or 5.5.

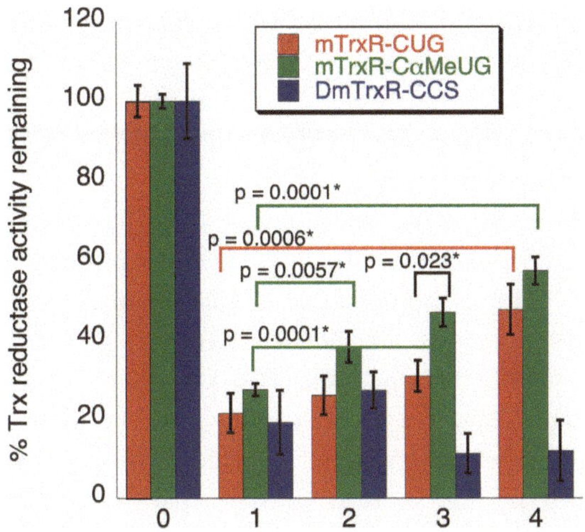

Fig. 6 Inhibition of TrxR by acrolein and subsequent reversal of acrolein inhibition using H_2O_2 and imidazole.(Ste Marie et al., 2020) For all, Trx reductase activity was monitored by depletion of NADPH spectrophotometrically at A340. mTrxR-CUG (red), mTrxR-C (αMe)UG (green), and DmTrxR-CCS (blue) were inhibited by acrolein (condition 1), and no reversal reaction was attempted for this condition. Next, acrolein adduction was reversed via one of the following: (condition 2) 5 mM imidazole; (condition 3) 2 mM H_2O_2; or (condition 4) 2 mM H_2O_2 and 5 mM imidazole. All "reversal" reactions were conducted in 200 mM potassium phosphate, 1 mM EDTA, pH 7, at 37 °C for 10 min. Condition 0 (control) represents the normalized Trx reductase activity of all enzymes before acrolein inhibition or reversal, and all other bars represent the amount of activity remaining relative to the acrolein-untreated control. The calculated *P*-values between relevant sets of data are indicated in the figure and the * indicates significance at the 95% confidence level.

For all acrolein adduction experiments, we chose an acrolein concentration that provided 70–80% inhibition of Trx-reductase activity (see Fig. 6, *condition 1*). At this level of reduced activity, we are confident that only enough acrolein was used to adduct the most reactive residue, which should be the penultimate Sec at low pH. Using a larger excess of acrolein would undoubtably inhibit all enzyme activity, but we aimed to use minimal concentrations of acrolein to avoid the unwanted adduction of other AAs. We next explored conditions by which we could restore Trx reductase activity in the acrolein adducted enzymes. Since alkylated mTrxR enzymes produce their own biological oxidants (Anestål, Prast-Nielsen, Cenas, & Arnér, 2008), and loss of their antioxidant activity leads to increased cellular oxidative stress, we postulated that reversal might be accomplished by adding oxidant. We predict that this accumulation of surrounding oxidant acts as a "molecular timer" to oxidize the selenide to a selenoxide catalyzing the β-elimination reaction (Fig. 5). We also anticipated needing a scavenger to bind to acrolein to prevent it from reattaching to the Sec (or other nucleophilic AAs) once removed. We found that imidazole, which is the functional group found in His residues, worked well as a scavenger, and did not cause other side reactions.

Using the above rationale, we incubated acrolein adducted enzymes with 2 mM H_2O_2 and 5 mM imidazole in 200 mM potassium phosphate buffer, pH 7.15, at 37 °C for 10 min (Fig. 6, *condition 4*) and were able to restore 30% Trx-reductase activity for the mTrxRC(αMe)UG mutant, and 25% Trx-reductase activity for the mTrxR-CUG WT. The ability for our (αMe)Sec-mutant to restore more activity than the WT is likely due to the inability of the mutant to undergo β-elimination with the peptide backbone. As shown in Fig. 5, it is possible that when acrolein-adducted mTrxR-CUG WT was oxidized, it underwent β-syn elimination with both the Cα-H of the protein backbone (forming inactive DHA), and the Cα-H of the acrolein adduct. Since both pathways are possible for the WT enzyme, less activity is expected to be recovered, which is what we observed experimentally.

To further explore the mechanism of the reversal reaction, we attempted to reverse acrolein inhibition using either imidazole alone (Fig. 6, *condition 2*) or H_2O_2 alone (Fig. 6, *condition 3*). Notably, some Trx reductase-activity is restored for both the mTrxR WT and mutant with H_2O_2 alone. This strongly supports a β-syn selenoxide mechanism as the way in which activity is regained.

In contrast to the above results, we were not able to restore any activity under the above discussed conditions for the Cys-TrxR, DmTrxR-CCS. In fact, when H_2O_2 was included in the reversal experiment, DmTrxR-CCS lost more Trx-reductase activity than the non-reversed condition (Fig. 6) (Ste Marie et al., 2020). This is most likely due to oxidative inactivation as DmTrxR-CCS is known to lose activity readily in the presence of oxidants (Snider et al., 2013; Ste Marie et al., 2020).

As described in Section 2.6.1, the methods for performing these inhibition and reversibility assay requires many steps and is very precise. Scheme 4 demonstrates how this assay was performed. Notably, the pH indicated at each step is the resulting pH from the mixing of all components of the system. Since it is impossible to measure the pH of a solution of such small volume (the final volume of each assay was just 150 μL), the pH at each step was determined via mixing equal volumes of buffers and components at a larger scale (i.e. trial and error). This provided the appropriate ratio of buffers

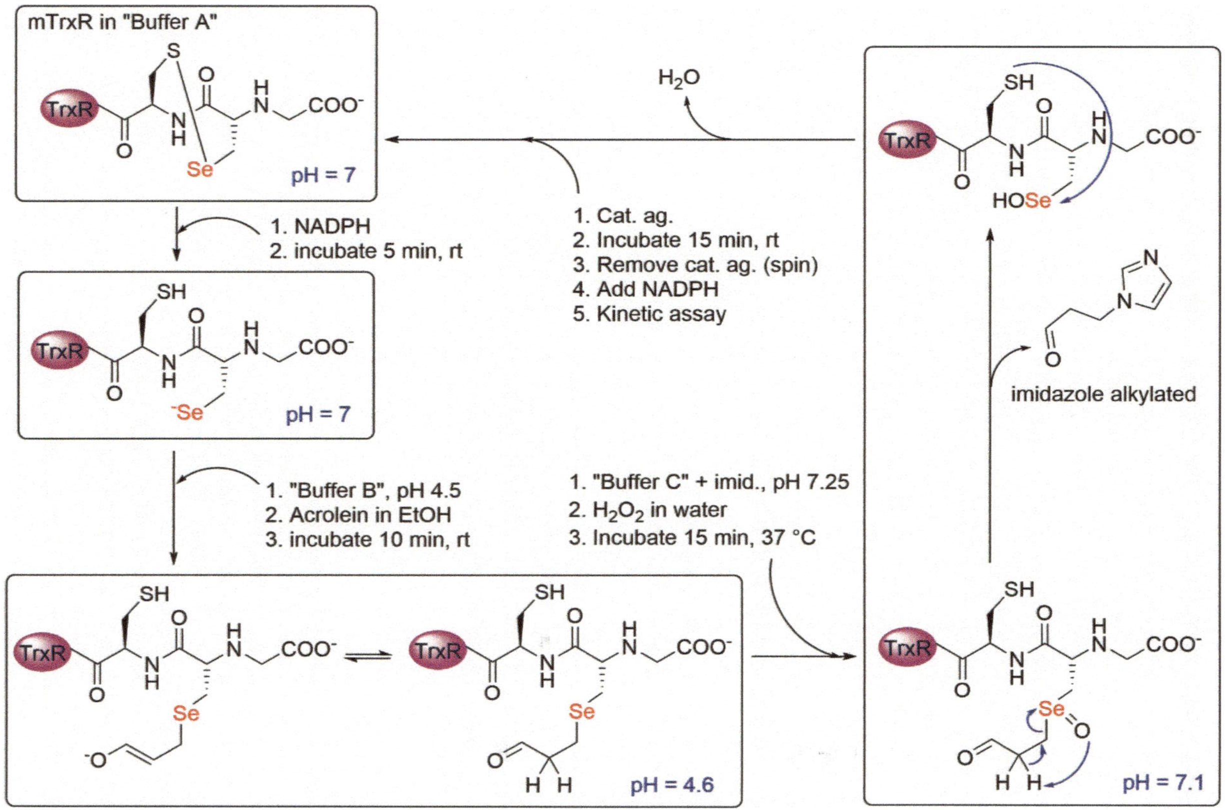

Scheme 4 See figure legend on next page.

and reagents to achieve the desired pH. Timing was also very important for this experiment; it was crucial that acrolein incubation was exactly 10 min for each trial to prevent variations in the inhibition step. It was also crucial that fresh acrolein stocks were used for every experiment, and that the acrolein used was recently purchased (i.e. not expired) as we noticed rapid decomposition and decreased potency of acrolein if it was old. It's also very important to note that any steps requiring acrolein were performed in a fume hood due to its potent toxicity, which further complicated this procedure. We also found that it was important to remove extra H_2O_2 from the assay before monitoring activity as H_2O_2 is also a substrate for mTrxR and makes activity of the enzyme appear artificially high if not removed. For this, we used catalase agarose (commercially available), which could be removed from the assay using a filtered spin column.

3.3.5 (αMe)Sec in a protein: Conclusions

Using (αMe)Sec as a chemical tool, we have extended the original concept that selenoenzymes resist inactivation by ROS (Maroney & Hondal, 2018) to a broader idea that selenoenzymes can resist electrophilic stress by RES as well (Ste Marie et al., 2020). We also used (αMe)Sec as a chemical tool to provide mechanistic evidence that suggests a selenoenzyme can undergo fast oxidative mediated repair via a β-syn selenoxide elimination to remove a RES adduct from Sec residues (Ste Marie et al., 2020). This is a significant gain of function that selenium adds to an enzyme relative to sulfur.

Scheme 4 Procedure for alkylation of mTrxR-CUG with acrolein, and subsequent reversal. Each box represents the Eppendorf tube, and the suspected state of the enzyme in the tube is shown along with the resulting pH for each reaction step (shown in blue text in the box). Starting at the top left corner, with mTrxR-CUG in Buffer A (pH 7), the enzyme is reduced with NADPH then treated with low pH buffer (Buffer B, pH 4.5), before addition of acrolein. Acrolein is allowed to react with the enzyme for 10 min at rt. Next, the reaction is treated with a higher pH buffer (Buffer C, pH 7.25) that contains imidazole (imid.), followed immediately by addition of H_2O_2, then incubated for 15 min at 37 °C. Next, catalase agarose (Cat. Ag.) is added and allowed to remove excess H_2O_2 for 15 min at rt. Cat. Ag. is then removed via centrifugation in filter units. Another dose of NADPH was added to the flow through, and the flow through (enzyme solution) was immediately assayed for Trx reductase activity using spectrophotometry. An almost identical procedure was used for mTrxR-C(αMe)UG mutant enzyme. Further experimental details are provided in materials and methods.

3.4 Other potential applications of (αMe)Sec

Beyond its role as a GPX mimic, we envision there will be many other applications for (αMe)Sec due to its unique stability. Thioester formation is a major limitation in commonly used native chemical ligation (NCL) reactions. One way to overcome N→S acyl shift limitations for thioester formation is to modify the Cys residue by targeting the amide bond to be alkylated to disrupt conjugation and increase the electrophilicity of the carbonyl amide (Bondalapati, Jbara, & Brik, 2016; Hou, Zhang, Li, & Liu, 2010; Kumar et al., 2011; Sato et al., 2013). Offer and coworkers used such an approach to develop an Fmoc compatible method to generate thioesters that would permit direct use in NCL (Burlina, Papageorgiou, Morris, White, & Offer, 2014). Offer and coworkers found that trityl side-chain protected (αMe)Cys was stable throughout peptide synthesis, and after deprotection, N→S acyl transfer and ligation could occur efficiently resulting in a peptide thioester (Burlina et al., 2014). As discussed previously, Macmillian and cowokers report that Se is a superior acyl transfer reagent over sulfur (Adams & Macmillan, 2013). Thus, a combination of the work done by Offer and Macmillian to generate C-terminal thioesters via utilization of (αMe)Sec and an N→Se acyl shift should be highly effective; (αMe)Sec should be a superior acyl-transfer reagent due to the increased nucleophilicity of Se to S for carbonyl attack and its unique steric properties that will increase electrophilicity of the carbonyl amide. These are only a few of the biotechnological applications we envision for (αMe)Sec, and there is still much work to be done in exploring the chemical abilities of this unique AA.

4. Conclusions

In this chapter, we have provided detailed procedures for the incorporation of (αMe)Sec into peptides and proteins; notably, we describe a novel method for coupling an incoming AA onto the sterically hindered N-terminal amine of (αMe)Sec during SPPS. We also provided detailed procedures for the alkylation of an (αMe)Sec-containing enzyme by acrolein, and subsequent reversal of the acrolein adduct using H_2O_2 and imidazole. We then provided two examples of how (αMe)Sec has been used as a chemical tool to study selenium redox biochemistry: (i) Incorporation of (αMe)Sec into a peptide that acted as a GPX mimic (Wehrle et al., 2019) and, (ii) Incorporation of (αMe)Sec into the active site of mTrxR as a replacement for Sec (Ste Marie et al., 2020). Together, these studies have provided important mechanistic information about the ability of selenoenzymes to resist ROS and RES through β-syn elimination reactions. Finally, we offered future applications of (αMe)Sec where we envision this chemical tool will find utility. We hope it will someday become commercially available as it is a powerful tool in peptide and protein chemistry.

References

Adams, A. L., & Macmillan, D. (2013). Investigation of peptide thioester formation via N- -Se acyl transfer. *Journal of Peptide Science*, *19*, 65–73.

Albericio, F., Cases, M., Alsina, J., Triolo, S. A., Carpino, L. A., & Kates, S. A. (1997). On the use of PyAOP, a phosphonium salt derived from HOAt, in solid-phase peptide synthesis. *Tetrahedron Letters*, *38*, 4853–4856.

Anestål, K., Prast-Nielsen, S., Cenas, N., & Arnér, E. S. (2008). Cell death by SecTRAPs: Thioredoxin reductase as a prooxidant killer of cells. *PLoS One*, *3*.

Atherton, E., Cameron, L. R., & Sheppard, R. C. (1988). Peptide synthesis: Part 10. Use of pentafluorophenyl esters of fluorenylmethoxycarbonylamino acids in solid phase peptide synthesis. *Tetrahedron*, *44*, 843–857.

Barber, D. R., & Hondal, R. J. (2019). Gain of function conferred by selenocysteine: catalytic enhancement of one-electron transfer reactions by thioredoxin reductase. *Protein Science*, *28*, 79–89.

Bondalapati, S., Jbara, M., & Brik, A. (2016). Expanding the chemical toolbox for the synthesis of large and uniquely modified proteins. *Nature Chemistry*, *8*, 407–418.

Burlina, F., Papageorgiou, G., Morris, C., White, P. D., & Offer, J. (2014). In situ thioester formation for protein ligation using α-methylcysteine. *Chemical Science*, *5*, 766–770.

Cai, W., Zhang, L., Song, Y., Wang, B., Zhang, B., Cui, X., et al. (2012). Small molecule inhibitors of mammalian thioredoxin reductase. *Free Radical Biology and Medicine*, *52*, 257–265.

Carpino, L. A., Abdel-Maksoud, A. A., Mansour, E., & Zewail, M. A. (2007). Segment coupling to a highly hindered N-terminal, alamethicin-related α-aminoisobutyric acid (Aib) residue. *Tetrahedron Letters*, *48*, 7404–7407.

Cho, C.-S., Lee, S., Lee, G. T., Woo, H. A., Choi, E.-J., & Rhee, S. G. (2010). Irreversible inactivation of glutathione peroxidase 1 and reversible inactivation of peroxiredoxin II by H_2O_2 in red blood cells. *Antioxidants & Redox Signaling*, *12*, 1235–1246.

Eckenroth, B., Harris, K., Turanov, A. A., Gladyshev, V. N., Raines, R. T., & Hondal, R. J. (2006). Semisynthesis and characterization of mammalian thioredoxin reductase. *Biochemistry*, *45*, 5158–5170.

Eckenroth, B. E., Lacey, B. M., Lothrop, A. P., Harris, K. M., & Hondal, R. J. (2007). Investigation of the C-terminal redox center of high M_r thioredoxin reductase by protein engineering and semisynthesis. *Biochemistry*, *46*, 9472–9483.

Flohe, L., Günzler, W., & Schock, H. (1973). Glutathione peroxidase: A selenoenzyme. *FEBS Letters*, *32*, 132–134.

Frérot, E., Coste, J., Pantaloni, A., Dufour, M.-N., & Jouin, P. (1991). PyBOP® and PyBroP: Two reagents for the difficult coupling of the α, α-dialkyl amino acid, Aib. *Tetrahedron*, *47*, 259–270.

Fu, Y. (2002). *Artificial Peptides Containing Ca, a-Disubstituted Amino Acids: Synthesis, Conformational Studies, and Application as β-Strand Mimics*. LSU Doctoral Dissertations. 2666.

Fu, Y., Hammarström, L. G., Miller, T. J., Fronczek, F. R., McLaughlin, M. L., & Hammer, R. P. (2001). Sterically hindered $C^{\alpha,\alpha}$-disubstituted α-amino acids: Synthesis from α-nitroacetate and incorporation into peptides. *The Journal of Organic Chemistry*, *66*, 7118–7124.

Fu, Y., & Hammer, R. P. (2002). Efficient acylation of the N-terminus of highly hindered $C^{\alpha,\alpha}$-disubstituted amino acids via amino acid symmetrical anhydrides. *Organic Letters*, *4*, 237–240.

Gan, F.-F., Kaminska, K. K., Yang, H., Liew, C.-Y., Leow, P.-C., So, C.-L., et al. (2013). Identification of Michael acceptor-centric pharmacophores with substituents that yield strong thioredoxin reductase inhibitory character correlated to antiproliferative activity. *Antioxidants & Redox Signaling*, *19*, 1149–1165.

Harris, K. M., Flemer, S., & Hondal, R. J. (2007). Studies on deprotection of cysteine and selenocysteine side-chain protecting groups. *Journal of Peptide Science*, *13*, 81–93.

Hondal, R. J., & Raines, R. T. (2002). Semisynthesis of proteins containing selenocysteine. *Methods in Enzymology*, *347*, 70.

Hondal, R. J., & Ruggles, E. L. (2011). Differing views of the role of selenium in thioredoxin reductase. *Amino Acids*, *41*, 73–89.

Hou, W., Zhang, X., Li, F., & Liu, C.-F. (2010). Peptidyl N,N-bis (2-mercaptoethyl)-amides as thioester precursors for native chemical ligation. *Organic Letters*, *13*, 386–389.

Humphrey, J. M., & Chamberlin, A. R. (1997). Chemical synthesis of natural product peptides: Coupling methods for the incorporation of noncoded amino acids into peptides. *Chemical Reviews*, *97*, 2243–2266.

Huse, M., Holford, M. N., Kuriyan, J., & Muir, T. W. (2000). Semisynthesis of hyperphosphorylated type I TGFβ receptor: Addressing the mechanism of kinase activation. *Journal of the American Chemical Society*, *122*, 8337–8338.

Iwaoka, M., Ito, S., Miyazaki, I., & Michibata, M. (2016). Synthesis of l-selenocysteine and α-methyl-l-selenocysteine derivatives using woollins' reagent and their application as chiral selenium catalysts. *Proceedings of the National Academy of Sciences, India Section A: Physical Sciences*, *86*, 499–509.

Kaiser, E., Colescott, R., Bossinger, C., & Cook, P. (1970). Color test for detection of free terminal amino groups in the solid-phase synthesis of peptides. *Analytical Biochemistry*, *34*, 595–598.

Katritzky, A. R., Todadze, E., Angrish, P., & Draghici, B. (2007). Efficient peptide coupling involving sterically hindered amino acids. *The Journal of Organic Chemistry*, *72*, 5794–5801.

Kehrer, J. P., & Biswal, S. S. (2000). The molecular effects of acrolein. *Toxicological Sciences*, *57*, 6–15.

Kil, J., Pierce, C., Tran, H., Gu, R., & Lynch, E. D. (2007). Ebselen treatment reduces noise induced hearing loss via the mimicry and induction of glutathione peroxidase. *Hearing Research*, *226*, 44–51.

Kumar, K., Bavikar, S. N., Spasser, L., Moyal, T., Ohayon, S., & Brik, A. (2011). Total chemical synthesis of a 304 amino acid K48-linked tetraubiquitin protein. *Angewandte Chemie International Edition*, *50*, 6137–6141.

Lacey, B. M., Eckenroth, B. E., Flemer, S., Jr., & Hondal, R. J. (2008). Selenium in thioredoxin reductase: A mechanistic perspective. *Biochemistry*, *47*, 12810–12821.

Lothrop, A. P., Ruggles, E. L., & Hondal, R. J. (2009). No selenium required: Reactions catalyzed by mammalian thioredoxin reductase that are independent of a selenocysteine residue. *Biochemistry*, *48*, 6213–6223.

Lothrop, A. P., Snider, G. W., Flemer, S., Jr., Ruggles, E. L., Davidson, R. S., Lamb, A. L., et al. (2014). Compensating for the absence of selenocysteine in high-molecular weight thioredoxin reductases: The electrophilic activation hypothesis. *Biochemistry*, *53*, 664–674.

Lynch, E., & Kil, J. (2009). Development of ebselen, a glutathione peroxidase mimic, for the prevention and treatment of noise-induced hearing loss. In *Seminars in Hearing* (pp. 047–055). ©Thieme Medical Publishers.

Ma, S., Caprioli, R. M., Hill, K. E., & Burk, R. F. (2003). Loss of selenium from selenoproteins: Conversion of selenocysteine to dehydroalanine in vitro. *Journal of the American Society for Mass Spectrometry*, *14*, 593–600.

Maiorino, M., Roveri, A., Coassin, M., & Ursini, F. (1988). Kinetic mechanism and substrate specificity of glutathione peroxidase activity of ebselen (PZ51). *Biochemical Pharmacology*, *37*, 2267–2271.

Makiyama, A., Komatsu, I., Iwaoka, M., & Yatagai, M. (2010). One-pot conversion of serine and α-methylserine derivatives to the corresponding cysteines and selenocystines by using chalcogenophosphate reagents. *Phosphorus, Sulfur and Silicon and the Related Elements.*, *186*, 125–133.

Maroney, M. J., & Hondal, R. J. (2018). Selenium versus sulfur: Reversibility of chemical reactions and resistance to permanent oxidation in proteins and nucleic acids. *Free Radical Biology and Medicine*, *127*, 228–237.

Masterson, D. S., Kedrowski, B. L., & Blair, A. (2010). An improved method for the preparation of protected (R)-2-methylcysteine: Solution-phase synthesis of a glutathione analogue. *Synlett*, *2010*, 2941–2943.

Maulik, N., Yoshida, T., & Das, D. K. (1999). Regulation of cardiomyocyte apoptosis in ischemic reperfused mouse heart by glutathione peroxidase. In *Stress Adaptation, Prophylaxis and Treatment* (pp. 13–21). Springer.

Reich, H. J., & Wollowitz, S. (1993). Preparation of α, β-Unsaturated Carbonyl Compounds and Nitriles by Selenoxide Elimination. *Organic Reactions.*

Rhee, S. G., & Cho, C.-S. (2010). Blot-based detection of dehydroalanine-containing glutathione peroxidase with the use of biotin-conjugated cysteamine. *Methods in Enzymology*, *474*, 23–34.

Rocher, C., Lalanne, J. L., & Chauidere, J. (1992). Purification and properties of a recombinant sulfur analog of murine selenium-glutathione peroxidase. *The FEBS Journal*, *205*, 955–960.

Saccoccia, F., Angelucci, F., Boumis, G., Carotti, D., Desiato, G., EMiele, A., et al. (2014). Thioredoxin reductase and its inhibitors. *Current Protein and Peptide Science*, *15*, 621–646.

Sato, K., Shigenaga, A., Kitakaze, K., Sakamoto, K., Tsuji, D., Itoh, K., et al. (2013). Chemical synthesis of biologically active monoglycosylated GM2-activator protein analogue using N-sulfanylethylanilide peptide. *Angewandte Chemie International Edition*, *52*, 7855–7859.

Snider, G. W., Ruggles, E., Khan, N., & Hondal, R. J. (2013). Selenocysteine confers resistance to inactivation by oxidation in thioredoxin reductase: Comparison of selenium and sulfur enzymes. *Biochemistry*, *52*, 5472–5481.

Ste Marie, E. J., & Hondal, R. J. (2018). Reduction of cysteine-S-protecting groups by triisopropylsilane. *Journal of Peptide Science*, *24*, e3130.

Ste Marie, E. J., Wehrle, R. J., Haupt, D. J., Wood, N. B., Van Der Vliet, A., Previs, M. J., et al. (2020). Can selenoenzymes resist electrophilic modification? Evidence from thioredoxin reductase and a mutant containing α-methylselenocysteine. *Biochemistry*, *59*, 3300–3315.

Stevens, J. F., & Maier, C. S. (2008). Acrolein: sources, metabolism, and biomolecular interactions relevant to human health and disease. *Molecular Nutrition & Food Research*, *52*, 7–25.

Thompson, R. E., & Muir, T. W. (2020). Chemoenzymatic semisynthesis of proteins. *Chemical Reviews*, *120*, 3051–3126.

Toppo, S., Vanin, S., Bosello, V., & Tosatto, S. C. (2008). Evolutionary and structural insights into the multifaceted glutathione peroxidase (Gpx) superfamily. *Antioxidants & Redox Signaling*, *10*, 1501–1514.

Wehrle, R. J., Ste Marie, E. J., Hondal, R. J., & Masterson, D. S. (2019). Synthesis of alpha-methyl selenocysteine and its utilization as a glutathione peroxidase mimic. *Journal of Peptide Science*, *25*, e3173.

Wilson, S. R., Zucker, P. A., Huang, R. R. C., & Spector, A. (1989). Development of synthetic compounds with glutathione peroxidase activity. *Journal of the American Chemical Society*, *111*, 5936–5939.

Zhang, J., Zhang, B., Li, X., Han, X., Liu, R., & Fang, J. (2019). Small molecule inhibitors of mammalian thioredoxin reductase as potential anticancer agents: An update. *Medicinal Research Reviews*, *39*, 5–39.

Zhong, L., & Holmgren, A. (2000). Essential role of selenium in the catalytic activities of mammalian thioredoxin reductase revealed by characterization of recombinant enzymes with selenocysteine mutations. *Journal of Biological Chemistry*, *275*, 18121–18128.

CHAPTER FOURTEEN

Modeling of selenocysteine-derived reactive intermediates utilizing a nano-sized molecular cavity as a protective cradle

Ryosuke Masuda and Kei Goto*
Department of Chemistry, School of Science, Tokyo Institute of Technology, Tokyo, Japan
*Corresponding author: e-mail address: goto@chem.titech.ac.jp

Contents

Abstract

In the biological functions of selenoproteins, various highly reactive species formed by oxidative modification of selenocysteine residues have been postulated to play crucial roles. Representative examples of such species are selenocysteine selenenic acids (Sec–SeOHs) and selenocysteine selenenyl iodides (Sec–SeIs), which have been widely recognized as important intermediates in the catalytic cycle of glutathione peroxidase (GPx) and iodothyronine deiodinase, respectively. However, examples of even spectroscopic observation of Sec–SeOHs and Sec–SeIs in either protein or small-molecule model systems remain elusive so far, most likely due to their notorious instability. For the synthesis of small-molecule model compounds of these reactive species, it is essential to suppress their very facile bimolecular decomposition such as self-condensation and disproportionation. Here we outline a novel method for the synthesis of stable small-molecule model compounds of the selenocysteine-derived reactive species, in which a nano-sized molecular cavity is used as a protective cradle to accommodate the reactive selenocysteine unit. Stabilization by the molecular cradle led to the

Methods in Enzymology, Volume 662
ISSN 0076-6879
https://doi.org/10.1016/bs.mie.2021.10.018

successful synthesis of Sec–SeOHs, which are stable in solution at low temperatures, and a Sec–SeI, which can be isolated as crystals. The catalytic cycle of GPx was investigated using the NMR-observable Sec–SeOH models, and all the chemical processes proposed for the catalytic cycle of GPx, including the bypass process from Sec–SeOH to the corresponding cyclic selenenyl amide, were experimentally confirmed. Detailed protocols for the syntheses of selenopeptide derivatives bearing the molecular cradle and for the spectroscopic monitoring of their reactions are provided.

1. Introduction

In the active site of selenoenzymes, various highly reactive intermediates have been postulated to be formed by oxidative modification of selenocysteine (Sec) residues and play crucial roles in their enzymatic reactions (Brigelius-Flohé & Maiorino, 2013; Flohé, 2020; Flohé, Günzler, & Schock, 1973; Kraus, Foster, & Ganther, 1983; Kryukov et al., 2003; Lubos, Loscalzo, & Handy, 2011). Glutathione peroxidase (GPx) is one of the most important selenoenzymes (Mills, 1957) due to its role in a diverse range of biological functions including the detoxification of H_2O_2 and hydroperoxides (Maiorino et al., 1995; Rotruck et al., 1973) to redox signaling and differentiation. Selenocysteine selenenic acids (Sec–SeOHs) (Brigelius-Flohé & Maiorino, 2013) have been recognized as key intermediates in the catalytic cycle of GPx and most discussions of the mechanism of GPx (Mauri et al., 2003) have been based on the formation of Sec–SeOH intermediates (Scheme 1A). Selenocysteine selenenyl iodides (Sec–SeIs) have also been attracting increasing attention as important intermediates of thyroid hormone deiodination by iodothyronine deiodinases (Dio)

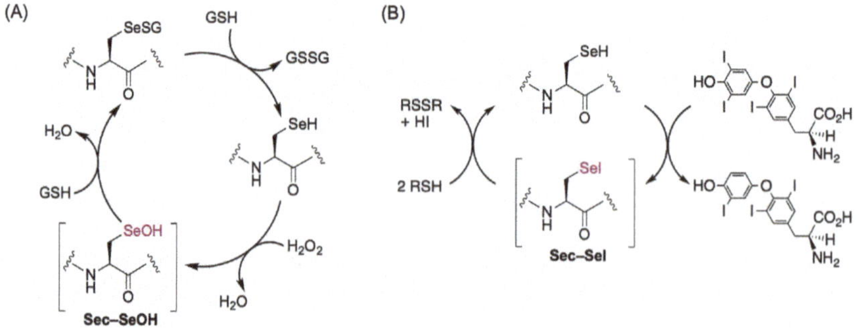

Scheme 1 Proposed mechanism for the catalytic cycle of selenoenzymes. (A) Glutathione peroxidase (GPx). (B) Type-1 Iodothyronine deiodinase (Dio-1).

(Berry, Banu, & Larsen, 1991; Köhrle, 1999) that catalyze regioselective deiodination of iodothyronines (Scheme 1B). However, examples of the direct observation of Sec–SeOHs and Sec–SeIs in proteins remain elusive. Even the trapping of Sec–SeOHs in a protein by chemical probes has been limited to very few examples (Liu & Rozovsky, 2013; Liu, Zhang, & Rozovsky, 2014; Payne, Barber, Ruggles, & Hondal, 2018; Scinto et al., 2019), and no experimental trapping of Sec–SeIs has been reported. To elucidate the mechanistic aspects of the catalytic functions of these selenoenzymes, small-molecule model compounds for the Sec-derived reactive intermediates with sufficient stability to study their chemical properties and reactivities are strongly desirable. Despite such demands, however, there have been no examples of even spectroscopic observations of small-molecule Sec–SeOHs and Sec–SeIs. This is largely due to the intrinsic instability of the –SeOH and –SeI species; in small-molecule systems, selenenic acids undergo extremely facile self-condensation to selenoseleninates (Reich, Hoger, & Wills, 1985), and selenenyl iodides readily disproportionate to diselenides and iodine (du Mont, Martens, Pohl, & Saak, 1990; Martens-V Salzen, Meyer, & du Mont, 1992) (Scheme 2). As a chemical tool to stabilize such reactive species, we have developed a nano-sized molecular cradle that can accommodate a reactive amino acid residue within its cavity and sterically suppress the bimolecular decomposition processes (Sano, Masuda, Sase, & Goto, 2021). This strategy provides a new, powerful tool for modeling the chemical behavior of reactive species derived from selenoproteins. In this chapter, we first discuss the molecular design of the protective cradle for stabilization of Sec-derived reactive species. We also discuss the application of this molecular cradle to model studies of the catalytic cycle of GPx. The latter part of this chapter provides detailed protocols for the syntheses of the selenopeptide derivatives bearing the molecular cradle and for the spectroscopic monitoring of their representative reactions (Masuda, Kimura, Karasaki, Sase, & Goto, 2021; Masuda, Kuwano, & Goto, 2021).

$$2\ \text{R-SeOH} \longrightarrow \text{R-Se-Se(O)-R} + H_2O$$

$$2\ \text{R-SI} \longrightarrow \text{R-Se-Se-R} + I_2$$

Scheme 2 Self-condensation of selenenic acids and disproportionation of selenenyl iodides.

2. Molecular design for stabilization of selenocysteine-derived reactive species

To suppress the bimolecular decomposition processes of reactive species such as self-condensation and disproportionation, kinetic stabilization utilizing bulky groups has been effectively employed (Goto, 2014). However, selenium-containing reactive species such as selenenic acids and selenenyl iodides are notoriously difficult to stabilize by steric protection. For example, the bimolecular decomposition of a selenenic acid cannot be prevented even with a 2,4,6-tri-*tert*-butyl group, which is known as a very effective steric protection group; selenenic acid **1** bearing this substituent disproportionated to the corresponding diselenide and seleninic acid via self-condensation (Scheme 3) (Reich & Jasperse, 1988).

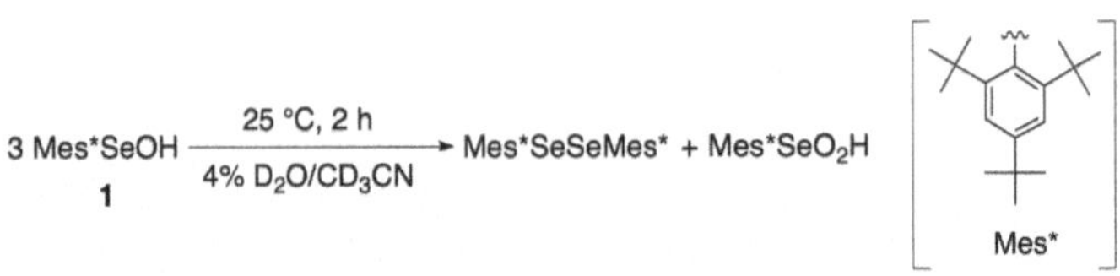

Scheme 3 Bimolecular decomposition of selenenic acid **1**.

Our and other groups have reported on the isolation of non-Sec-derivative selenenic acids as stable compounds by utilizing bulkier substituents (Goto et al., 2001; Ishii, Matsubayashi, Takahashi, & Nakayama, 1999; Saiki, Goto, & Okazaki, 1997; Sase, Kakimoto, & Goto, 2015; Sase, Kakimoto, Kimura, & Goto, 2015; Sase, Kimura, Masuda, & Goto, 2019; Zielinski, Presseau, Amorati, Valgimigli, & Pratt, 2014). Several examples of isolable selenenic acids, which we previously synthesized by utilizing cavity-shaped substituents, are shown in Fig. 1 (Goto et al., 2001; Saiki et al., 1997; Sase, Kakimoto, Kimura, & Goto, 2015). Compound **2** with a bridged calixarene framework is the first selenenic acid that was crystallographically analyzed (Saiki et al., 1997). Selenenic acid **3** was synthesized by direct oxidation of the corresponding selenol **5** with an equimolar

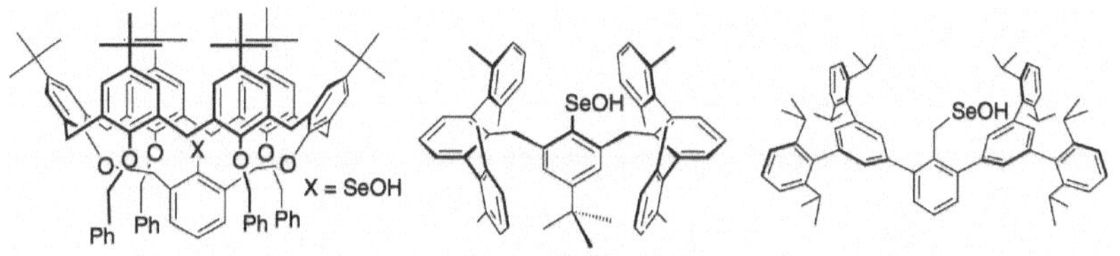

Fig. 1 Isolable non-Sec-derivative selenenic acids bearing cavity-shaped substituents.

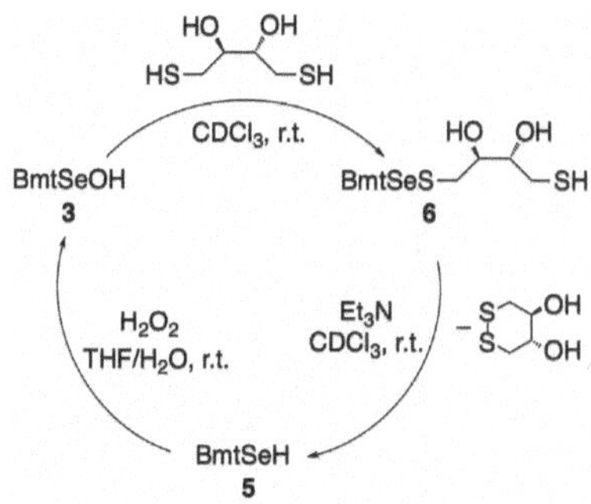

Scheme 4 Modeling of the chemical processes proposed for the canonical catalytic cycle of GPx.

amount of H_2O_2 and isolated in pure form (Scheme 4) (Goto et al., 2001). Treatment of **3** with dithiothreitol (DTT) reduced **3** to selenol **5** via selenenyl sulfide **6**, thus confirming the three chemical transformations included in the canonical catalytic cycle of GPx shown in Scheme 1A.

Non-Sec-derivative selenenyl iodides have also been synthesized as isolable compounds by kinetic or thermodynamic stabilization (du Mont et al., 1990; du Mont & Wagner, 1988; Goto, Sonoda, Shimada, Sase, & Kawashima, 2010; Mugesh, Panda, Singh, & Butcher, 1999; Sase, Kakimoto, Kimura, & Goto, 2015). We previously reported that the reaction of selenol **7** with a thyronine derivative in the presence of triethylamine afforded selenenyl iodide **8**, which was reduced to **7** by treatment with DTT (Scheme 5) (Goto et al., 2010). These reactions are consistent with the proposed chemical processes for Dio-1 shown in Scheme 1B.

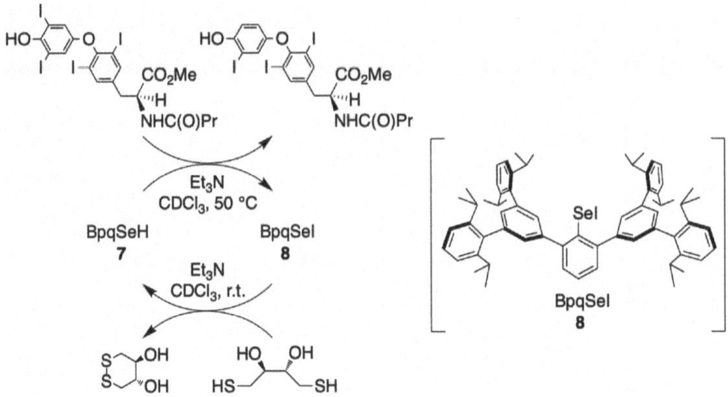

Scheme 5 Modeling of the chemical processes proposed for the canonical catalytic cycle of Dio-1.

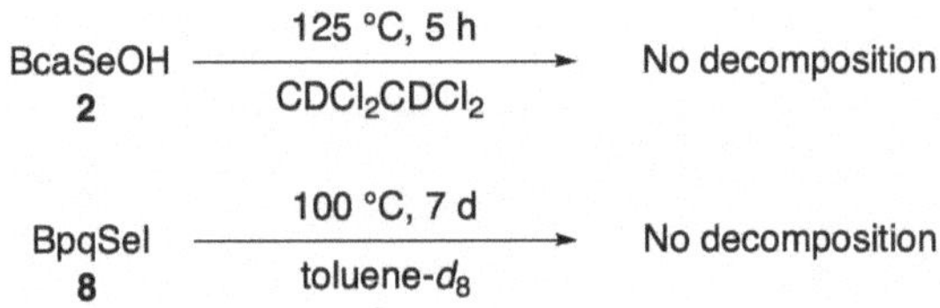

Scheme 6 Thermal stability of selenenic acid **2** and selenenyl iodide **8**.

It was found that selenenic acids and selenenyl iodides stabilized by the cavity-shaped substituents exhibited high thermal stability; heating selenenic acid **2** in tetrachloroethane-d_2 at 120 °C for 5 h and heating selenenyl iodide **8** in toluene-d_8 at 100 °C for 1 week caused no decomposition of these compounds (Scheme 6) (Goto et al., 2010; Saiki et al., 1997). These results indicate that –SeOH and –SeI functionalities themselves are thermally stable when isolated in inert environments. These non-Sec derivatives are useful for studying the basic chemical information of –SeOH and –SeI functionalities, but their structural and electronic properties are different from those of Sec-derived species of biological relevance.

For the kinetic stabilization of Sec-derived reactive species, it is necessary to sterically protect the reactive functionality located at the distal end of the Sec unit. The bulky groups that have been utilized so far have only a local protective effect limited to the directly attached reactive functionality and are therefore too ineffective in stabilizing a whole Sec unit with a –SeOH or –SeI functionality. We recently developed a molecular cavity based on an *m*-phenylene dendrimer framework of ca. 2-nm scale and applied it as a molecular cradle that can accommodate a reactive amino acid residue within its cavity (Fig. 2) (Sano et al., 2021).

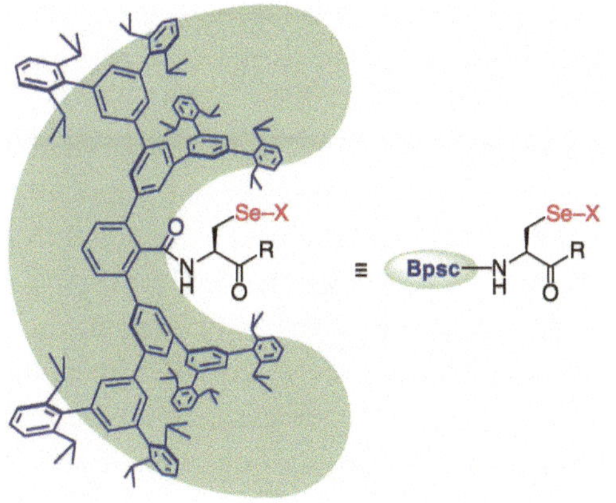

Fig. 2 Cradled selenocysteine model for stabilization of selenocysteine-derived reactive species.

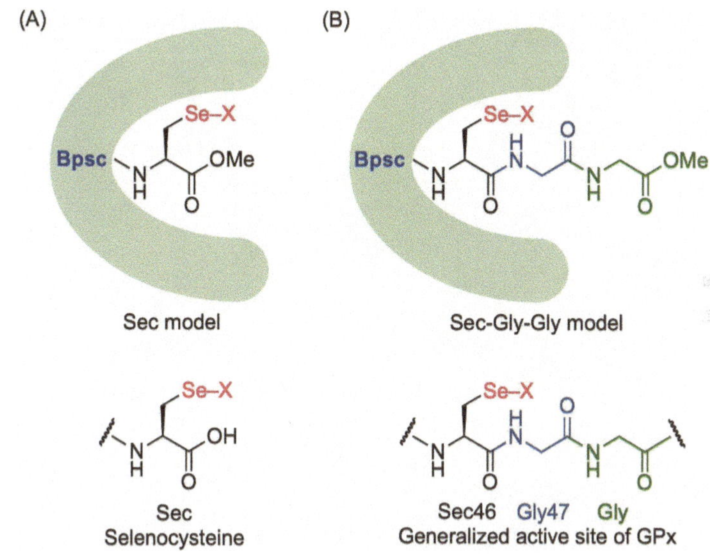

Fig. 3 Cradled selenocysteine (A) and selenopeptide (B) model compounds.

In this model, a cavity-shaped benzoyl group (henceforth denoted as Bpsc) is introduced to a Sec residue as an *N*-terminal protecting group. This novel Sec model, which is referred to as a cradled selenocysteine, is expected to be useful for the synthesis of reference compounds of highly reactive Sec derivatives such as Sec–SeOHs and Sec–SeIs.

We designed two model systems for different applications. One is the Sec model (Fig. 3A), in which a selenocysteine methyl ester is protected by a

Bpsc group, and the other is the Sec-Gly-Gly tripeptide model (Fig. 3B), which corresponds to the generalized form of the tripeptides in the catalytic site of GPx (Epp, Ladenstrin, & Wendel, 1983; Scheerer et al., 2007) and allows us to study the involvement of the amide groups near Sec in the reaction of the Sec-derived species. By utilizing these models, we succeeded in the direct observation of Sec–SeOHs by ^{1}H and ^{77}Se NMR spectroscopies as well as the isolation of a stable Sec–SeI and its crystallographic analysis. Sec–SeI **9** of the Sec-Gly-Gly model was isolated in the form of reddish-purple crystals, and its structure was determined by a single-crystal X-ray diffraction analysis (Fig. 4). The selenocysteine moiety of **9** is anchored in a large cavity produced by the Bpsc group (dimensions ca. 2.3 × 1.7 nm). Not only the Sec–SeI moiety but also the whole of Sec-Gly-Gly unit were found to be incorporated in the molecular cradle. Notably, Sec–SeI

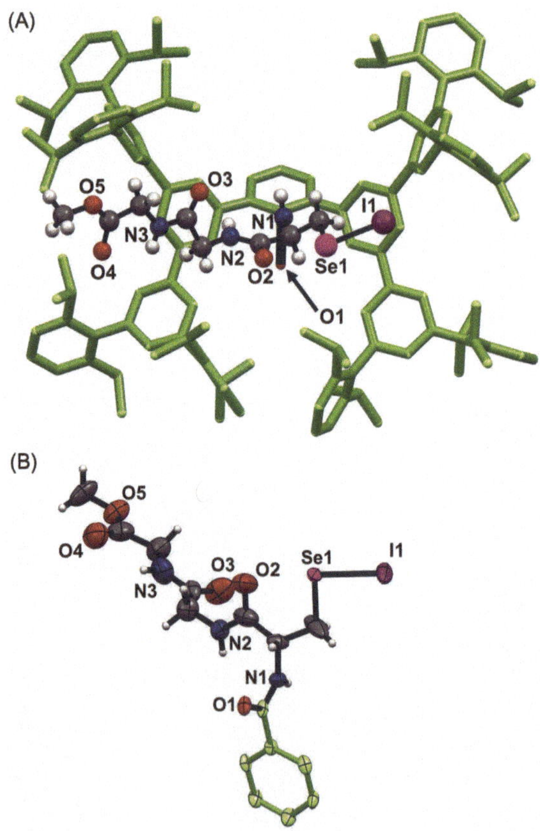

Fig. 4 Crystal structure of **9** (one of two independent molecules): (A) ball and stick representation (hydrogen atoms of the Bpsc group are omitted for clarity), (B) thermal ellipsoid representation of the selenopeptide unit (50% probability).

9 exhibited high thermal stability; in the solid state, it decomposed at a high temperature of 243 °C. Sec–SeI **9** is also stable in air and can be manipulated as a compound having "shelf-stability." Such stability of **9** indicates that Sec–SeIs are no longer unstable species as long as the disproportionation to diselenides is sterically inhibited. Thus, it is clearly demonstrated that the cradled selenocysteine model is very effective in stabilizing the Sec-derived reactive species by suppressing their bimolecular decomposition process.

3. Modeling the catalytic cycle of GPx by NMR spectroscopic analysis of Sec–SeOHs

By utilizing the cradled selenocysteine models, the chemical processes included in the catalytic cycle of GPx was investigated. As shown in the heating experiments of selenenic acid **2** (Scheme 6) (Saiki et al., 1997), the –SeOH group itself is a thermally stable functionality. However, a large body of experimental evidence suggests that Sec–SeOHs with a *β*-hydrogen atom readily undergo thermal deselenation (Ma, Caprioli, Hill, & Burk, 2003; Walter & Roy, 1971), which affords the corresponding dehydroalanines (Process III in Scheme 7) (Cho et al., 2010; Ma et al., 2003). As a mechanism by which GPx maintains its catalytic activity despite thermal

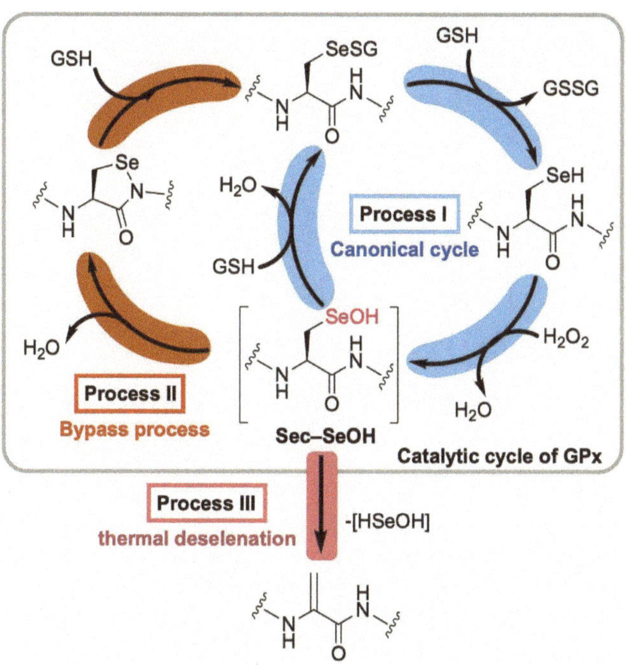

Scheme 7 Catalytic cycle of GPx including the bypass process proposed by Flohé et al.

instability of the Sec–SeOH intermediates, Flohé et al. postulated a bypass mechanism (Orian et al., 2015), in which Sec–SeOHs undergo intramolecular cyclization to the corresponding cyclic selenenyl amides (Bhowmick, Srivastava, D'Silva, & Mugesh, 2015; Fischer & Dereu, 1987; Reich & Jasperse, 1987; Sarma & Mugesh, 2005; Ungati, Govindaraj, Narayanan, & Mugesh, 2019) to prevent the thermal deselenation to dehydroalanines under glutathione-deficient conditions (Process II in Scheme 7). Their proposal was based on LC-MS/MS analysis of the oxidized forms of bovine GPx1 and rat GPx4 and DFT calculations; no experimental evidence for the formation of Sec–SeOH as intermediates was available.

The chemical processes proposed for the catalytic cycle of GPx shown in Scheme 7, including those for the bypass mechanism, were investigated in a stepwise manner by utilizing the Sec model (Fig. 3A) and the Sec-Gly-Gly model (Fig. 3B). When Sec–SeH **10a** of the Sec model was treated with H_2O_2 in the presence of NaOH at $-65\,°C$, generation of Sec–SeOH **11a** was observed by ^1H and ^{77}Se NMR spectroscopies at $-20\,°C$ (92% NMR yield) (Scheme 8A). Similarly, oxidation of Sec–SeH **10b** of the Sec-Gly-Gly model afforded Sec–SeOH **11b** (90% NMR yield), which was found to be stable at $-30\,°C$ for at least 24 h (Scheme 8B). Sec–SeOHs **11a** and **11b** were found to readily react with a cysteine thiol

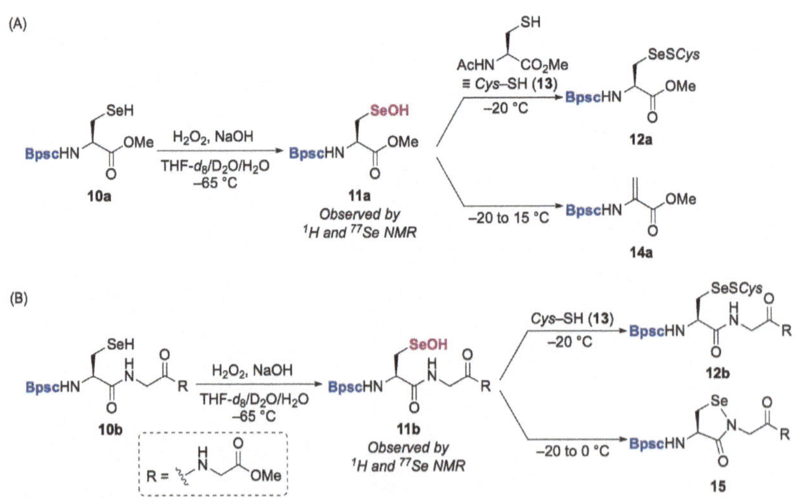

Scheme 8 Generation of Sec–SeOHs and their reactivities. (A) Generation of Sec–SeOH **11a** of the Sec model, its reaction with cysteine thiol **13** and thermal deselenation to dehydroalanine **14a.** (B) Generation of **11b** of the Sec-Gly-Gly model, its reaction with cysteine thiol **13** and intramolecular cyclization to cyclic selenenyl amide **15.**

even at the low temperature; treatment of **11a** and **11b** with *N*-acetylcysteine methyl ester (**13**) at −20 °C converted them to the corresponding selenenyl sulfides **12a** and **12b**, respectively. In the absence of a thiol, Sec–SeOH **11b** of the Sec-Gly-Gly model behaved differently from Sec–SeOH **11a**, which does not have an adjacent amide NH group. When a solution containing **11a** was warmed from −20 °C to 15 °C, **11a** underwent thermal deselenation to the corresponding dehydroalanine **14a** almost quantitatively (Scheme 8A). In sharp contrast, almost no such deselenation was observed in a similar experiment with **11b**. Instead, intramolecular cyclization of **11b** to cyclic selenenyl amide **15** was observed when a solution of **11b** was warmed from −20 to 0 °C (Scheme 8B). These results indicate that among the two thermal reaction processes of a Sec–SeOH in the absence of a thiol, the formation of a cyclic selenenyl amide proceeds faster than the deselenation to a dehydroalanine, which are consistent with the bypass mechanism as proposed by Flohé et al. By utilizing the cradled selenopeptide model, all the chemical processes that have been proposed for the catalytic cycle of GPx, including the bypass mechanism (Orian et al., 2015), were documented and confirmed (Scheme 9).

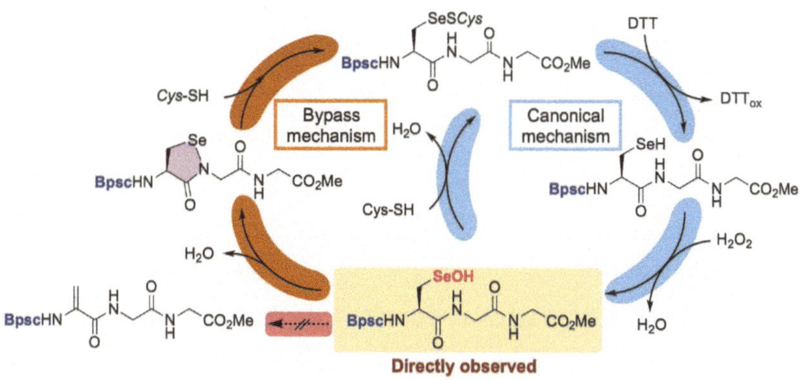

Scheme 9 Modeling of the chemical processes proposed for the catalytic cycle of GPx by the cradled selenopeptide.

In contrast with cysteine sulfenic acids (Cys–SOHs), sulfur analogues of Sec–SeOHs, for which detection methods using various chemical probes have been extensively studied, there have been very few reports on the trapping of Sec–SeOH in proteins (Liu et al., 2014; Liu & Rozovsky, 2013; Payne et al., 2018; Scinto et al., 2019). For all types of selenenic acids, there has been only one report of the reaction of chemical probes with a selenenic acid of observable stability; we previously reported that a non-Sec-derivative selenenic acid bearing a cavity-shaped substituent reacts rapidly with 1,3-diketones such as dimedone (**16**) and acetylacetone (**17**). The reactivity of Sec–SeOH towards 1,3-diketones were examined using the NMR-observable Sec–SeOHs **11a** and **11b**. After Sec–SeOH **11a** is generated at −65 °C, 1,3-diketones **16** or **17** was added at −20 °C to produce the corresponding selenides in good conversion yields (Scheme 10). Similarly, Sec–SeOH **11b** of the Sec-Gly-Gly model reacted with **17** to give the corresponding selenide at −20 °C. These results provide experimental evidence that Sec–SeOH reacts readily with 1,3-diketones **16** and **17** even at low temperature of −20 °C to produce the corresponding selenide, indicating the utility of the 1,3-diketones as a chemical probe for Sec–SeOH.

In the following sections, the experimental details of the syntheses of the cradled selenopeptides of the Sec-Gly-Gly model and the NMR spectroscopic monitoring of their representative reactions are described.

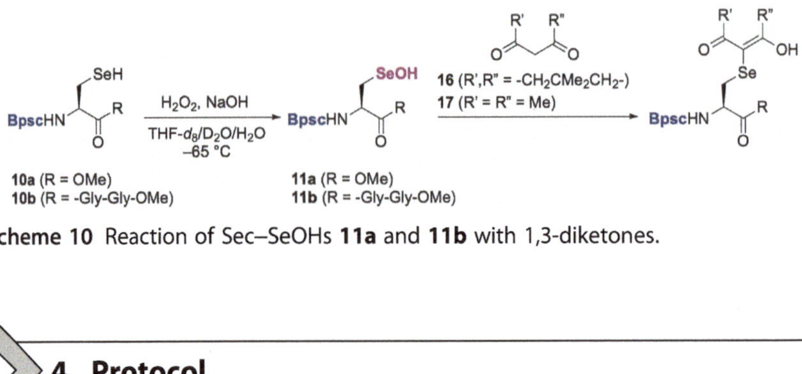

10a (R = OMe)
10b (R = -Gly-Gly-OMe)

11a (R = OMe)
11b (R = -Gly-Gly-OMe)

Scheme 10 Reaction of Sec–SeOHs **11a** and **11b** with 1,3-diketones.

4. Protocol

4.1 Synthetic pathway for the cradle-type protective unit

As a reagent for the introduction of the cradle-type protective unit, we synthesized a cavity-shaped benzoyl chloride, Bpsc–Cl (**23**) (Scheme 11). The *m*-phenylene dendrimer framework of the Bpsc unit was constructed via Hart reaction (Du, Hart, & Ng, 1986) in our first report (Sano et al., 2021), but we recently developed an efficient synthetic route of Bpsc–OH (**22**) via a Suzuki-Miyaura cross coupling (Miyaura & Suzuki, 1995; Miyaura, Yamada, & Suzuki, 1979) followed by removal of a protecting group of the carboxylic acid moiety (Masuda, Kuwano, & Goto, 2021) (Scheme 11). After conversion of bromoquinquephenyl **18** to boronic ester **19** by a lithiation-borylation sequence, the Suzuki-Miyaura cross coupling

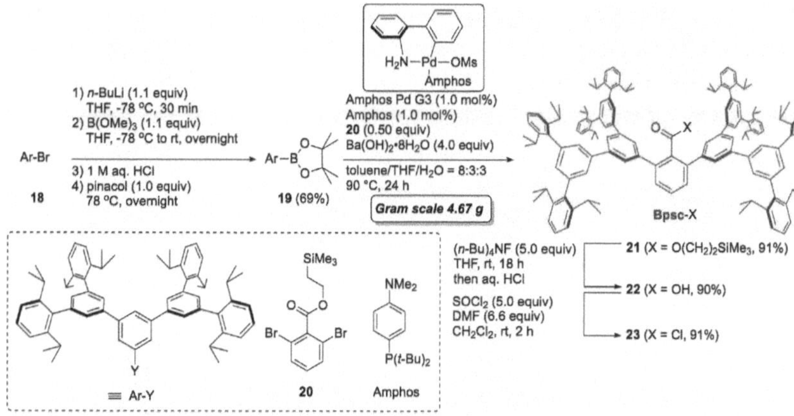

Scheme 11 Synthesis of Bpsc–Cl (**23**) via Suzuki-Miyaura cross coupling.

with dibromide **20**, which have a 2-(trimethylsilyl)ethyl ester moiety, afforded ester **21** bearing the Bpsc framework in excellent yield. The use of Buchwald's G3 precatalyst (Bruno, Tudge, & Buchwald, 2013) of Amphos (Guram et al., 2006, 2007), i.e., Amphos Pd G3, was crucial for the rapid formation of the carbon-carbon bonds before decomposition of the base-sensitive dibromide **20**. Deprotection of a 2-(trimethylsilyl)ethyl group of **21** by treatment with tetrabutylammonium fluoride yielded the desired Bpsc–OH (**22**) in 90% yield. Bpsc–Cl (**23**) was obtained by chlorination of **22** with $SOCl_2$ in 91% yield. Isolation of **23** as colorless crystals was important for the following *N*-benzoylation of selenocysteine derivatives with **23**. The protocol for the synthesis of **23** is described in our previous report (Masuda, Kuwano, & Goto, 2021; Sano et al., 2021).

4.2 Synthetic route for selenocysteine selenols bearing the molecular cradle

For the synthesis of selenocysteine selenols (Sec–SeHs) bearing a Bpsc group, we investigated several synthetic routes and found that Sec–SeHs can be obtained in pure form by the route via the corresponding selenenyl iodides (Sec–SeIs) (Masuda, Kimura, et al., 2021). As noted in Section 2, we revealed that, with the protective cradle, Sec–SeIs are no longer unstable species and can be isolated in pure form by simple protocols. We also reported that Sec–SeIs can be readily reduced to Sec–SeHs.

Recently, Iwaoka et al. reported the synthesis of cyclic diselenides of selenopeptides via oxidation of the corresponding *Se*-diphenylmethyl (henceforth denoted as Dpm) derivatives (Shimodaira & Iwaoka, 2017)

(Scheme 12). We assumed that these diselenides were formed via disproportionation of Sec–SeIs generated in situ by cleavage of the C(Dpm)–Se bond by iodine. Based on these findings, we devised the synthetic route for Sec–SeH **10b** of the Sec-Gly-Gly model, which involves the introduction of a Bpsc group into *Se*-Dpm-protected selenopeptide **32**, followed by conversion to Sec–SeI **9** and subsequent reduction to Sec–SeH **10b** (Scheme 13).

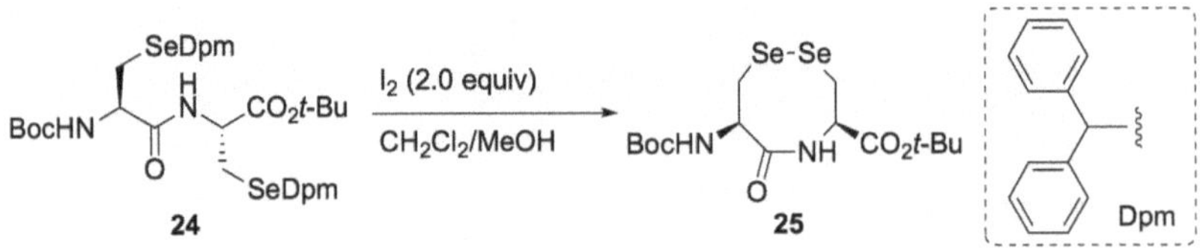

Scheme 12 Synthesis of a cyclic diselenide via oxidation by iodine reported by Iwaoka et al.

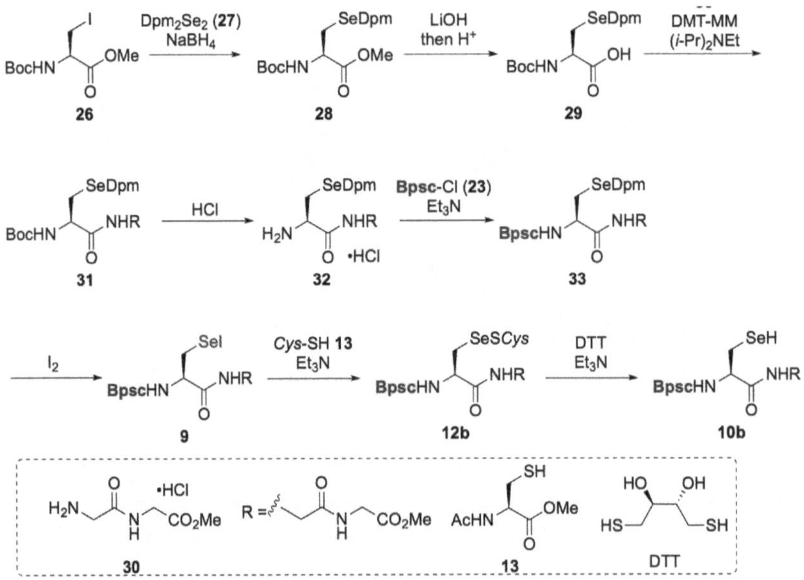

Scheme 13 Synthetic route for Sec–SeH **10b** of the Sec-Gly-Gly model.

4.3 Protocols for the synthesis and derivatization of selenopeptides

4.3.1 Protocol for the synthesis of a Dpm-protected selenocysteine

As a selenocysteine unit, *N*-Boc-L-Sec(Dpm)-OMe (**28**) was synthesized by a S_N2 reaction of *N*-Boc-L-iodoalanine methyl ester (**26**) (Stempel, Kaml,

Table 1 Synthesis of *Se*-Dpm selenocysteine derivative **28**.

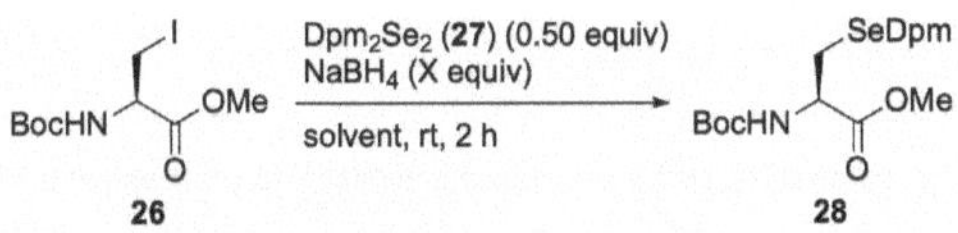

Entry	X (equiv)	Solvent	Yield (%)
1	3.0	Et_2O/MeOH = 3:2	45[a]
2	1.2	Et_2O/MeOH = 3:2	14
3	2.4	THF/EtOH = 3:2	71
4	2.4	DMF/EtOH = 3:2	67
5	1.2	THF/EtOH = 3:2	66[b]

[a] Starting material **26** and dehydroalanine **34** were obtained in 37% and 17%, respectively. [b] 3.93 g (8.77 mmol) of **28** was obtained.

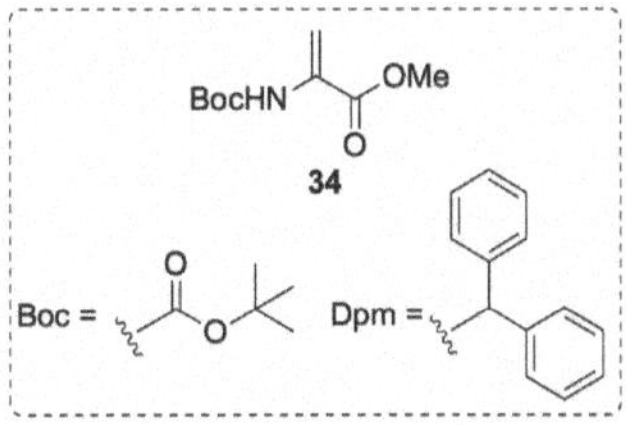

Budisa, & Kalesse, 2018) with selenolate anion (Dpm-Se^-), which was in situ generated by reduction of bis(diphenylmethyl) diselenide (**27**) (Kim, Lee, & Park, 2015) with $NaBH_4$ (Table 1). It was reported that using an iodoalanine as a substrate rather than an *O*-tosylserine was suitable for preparation of selenocysteine derivatives in terms of both yields and enantiopurity of the products (Shimodaira & Iwaoka, 2017). We optimized the reaction conditions for the preparation of **28** (Table 1). Treatment of **26** with **27** (0.50 equiv) in the presence of $NaBH_4$ (3.0 equiv) in Et_2O/MeOH (v/v = 3:2) under argon atmosphere afforded the desired product **28** but its yield was low (45%; entry 1). In this reaction, the starting material **26** and dehydroalanine **34** were also obtained in 37% and 17% yields, respectively. Reducing the amount of $NaBH_4$ to 1.2 equiv decreased the yield of **28** (14%; entry 2). Optimization of the solvents was then carried out under the reaction conditions using 2.4 equiv of $NaBH_4$ because the solubility of **27** and/or its selenolate form was considered to be crucial for this type of reaction. With THF (tetrahydrofuran)/EtOH (v/v = 3:2) or DMF (*N,N*-dimethylformamide)/EtOH (v/v = 3:2), **28** was obtained in 71% and 67% yields, respectively (entries 3 and 4). Finally, treatment of **26** with **27** (0.50 equiv) in the presence of $NaBH_4$ (1.2 equiv) in THF/EtOH (v/v = 3:2) was found to be optimal for this reaction in terms of avoiding the use of a large excess amount of $NaBH_4$. Under these conditions, 3.93 g (66% yield) of **28** was obtained after purification by silica-gel chromatography (eluent: hexane/$CHCl_3$ = 1:6). The obtained oil after chromatography was crystalized in a refrigerator and **28** was isolated as colorless crystals. To avoid decomposition of **28**, storage in a refrigerator was required. It is strongly recommended that this reaction be carried out using degassed EtOH and THF. In addition, the crude product should be purified as soon as possible.

4.3.2 Protocol for the synthesis of Sec-Gly-Gly tripeptides

For the derivatization of selenocysteine methyl ester **28** to Sec-Gly-Gly tripeptides, hydrolysis of **28** to the carboxylic acid, *N*-Boc-L-Sec(Dpm)-OH (**29**), was then examined. Methyl esters of selenopeptides are known to be labile in the presence of bases such as LiOH, NaOH, and KOH due to their facile deselenation. For example, it was reported that, under standard hydrolysis conditions using LiOH, *N*-Alloc-Ala-Sec(Ph)-OMe undergoes decomposition (Nicolaou, Estrada, Zak, Lee, & Safina, 2005). Although Me_3SnOH was reported to be effective for hydrolysis of selenocysteine methyl esters (Nicolaou et al., 2005), simpler reaction conditions are desirable considering the toxicity and price of Me_3SnOH. We found that the hydrolyzed product **29** could be obtained in 90% NMR yield by treatment of **28** with 1.0 equiv of LiOH in degassed THF/H_2O (v/v=2:1) for 1 h under argon atmosphere (Scheme 14A). To prevent the base-promoted decomposition of the selenocysteine unit to the corresponding dehydroalanine derivative, not only the use of just one equivalent of LiOH·H_2O but also the short reaction time were critical factors in this hydrolysis reaction. After the reaction was quenched by 10% aq. $KHSO_4$, a rapid work-up operation was required to avoid undesired side

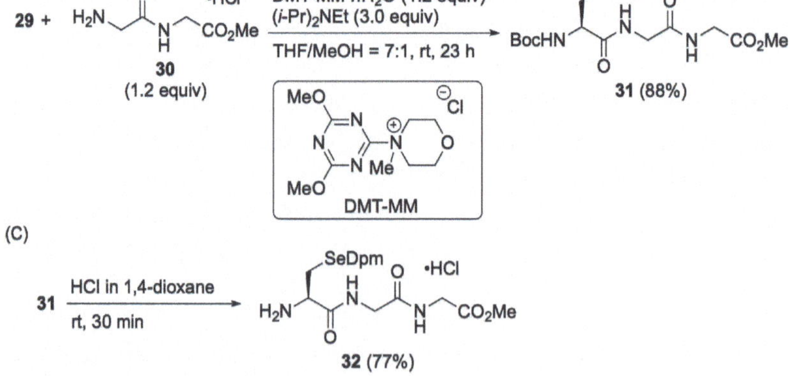

Scheme 14 (A) Synthesis of **29** by hydrolysis of **28.** (B) Synthesis of **31** by condensation of **29** with **30.** (C) Synthesis of **32.**

reactions such as deprotection of the Boc group and deselenation. The obtained oily crude product was recrystallized from Et_2O-hexane to afford **29** as colorless crystals in 61% yield. This recrystallization was important to prevent its decomposition during storage as well as to improve the yield and enantiopurity of the next condensation process. As the second process, the condensation of **29** with H-Gly-Gly-OMe•HCl (**30**) was examined (Scheme 14B). We found that DMT-MM (Kunishima, Kawachi, Hioki, Terao, & Tani, 2001; Kunishima, Kawachi, Iwasaki, Terao, & Tani, 1999) was effective for this type of condensation because the reaction can be carried out in the presence of MeOH, which is a good solvent to dissolve an amine salt such as **30**, and also because this reagent has a significant inhibitory effect on racemization/epimerization. Treatment of **29** with **30** (1.2 equiv) in the presence of DMT-MM (1.2 equiv) and $(i\text{-Pr})_2NEt$ (3.0 equiv) in THF/MeOH (v/v=7:1) afforded the desired *N*-Boc-Sec(Dpm)-Gly-Gly-OMe (**31**) in 88% yield after purification by silica-gel chromatography (CH_2Cl_2/MeOH=50:1). Purification of the substrate **29** by recrystallization before use was crucial for this condensation step. Removal of a Boc group from selenopeptide **31** by treatment of hydrogen chloride in 1,4-dioxane for 30 min yielded amine salt **32** in 77% (Scheme 14C), which was used for the next reaction without further purification.

4.3.3 Protocol for the synthesis of N-Bpsc Selenopeptides

The introduction of a Bpsc group to Sec-Gly-Gly selenopeptide **32** was then examined. The reaction of Bpsc–Cl (**23**) with selenopeptide **32** (1.2 equiv) was optimized in the presence of Et_3N (10 equiv) in several solvents at room temperature (Table 2). When CH_2Cl_2, a good solvent for dissolving **23**, was used, the yield of the *N*-Bpsc product **33** was only 43%, and the remaining

Table 2 Optimization of the synthesis of *N*-Bpsc selenopeptide **33**.

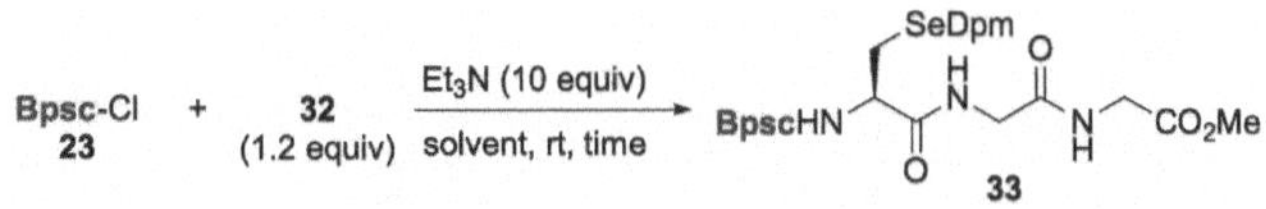

Entry	Solvent	Time (h)	Yield (%)
1	CH_2Cl_2	14	43
2	1,4-Dioxane	14	69
3	1,4-Dioxane	48	93

56% was the starting material **23** (entry 1). In contrast, when 1,4-dioxane was used as a solvent instead of CH_2Cl_2, the yield of **33** increased to 69% although the system became a pale-yellow suspension due to the low solubility of **23** (entry 2). Finally, the reaction in 1,4-dioxane for 2 days resulted in a high conversion; **33** was isolated in 93% yield after purification by silica-gel chromatography (hexane/CH_2Cl_2 = 1:4; ca. 400 mg scale; entry 3). This reaction was efficient and reproducible, but the yield decreased significantly without any of the following procedures: (1) isolation of Bpsc–Cl (**23**), (2) use of distilled Et_3N and 1,4-dioxane, and (3) drying of amine salt **32** prior to the reaction.

4.3.4 Protocol for the synthesis of Sec–SeI

Sec–SeI **9** of the Sec-Gly-Gly model was synthesized by iodine oxidation of the Dpm-protected peptide **33** bearing a Bpsc group with cleavage of the C(Dpm)–Se bond (Table 3). To a CH_2Cl_2/MeOH (1.0 mL/0.1 mL) solution of **33** (47.5 mg, 20.6 μmol) in a Schlenk tube, elemental iodine (5.8 mg, 21 μmol) was added in one portion under argon atmosphere, and the tube was carefully sealed by a glass cap with silicon grease. The color of the solution turned to reddish purple. After the mixture was stirred for 40 h in dark, CH_3CN was added under argon atmosphere and then 90% of the solvent was slowly removed in vacuo without heating. This slow evaporation was very important for the isolation of Sec–SeI **9**, because complete removal of the solvent often caused its decomposition such as deselenation to a dehydroalanine and formation of a symmetric diselenide. After addition of CH_3CN and cooling, the suspension was filtered through a Kiriyama funnel in air, and the resulting solid was washed with cold CH_3CN to give **9** as

Table 3 Optimization of the synthesis of Sec–SeI **9**.

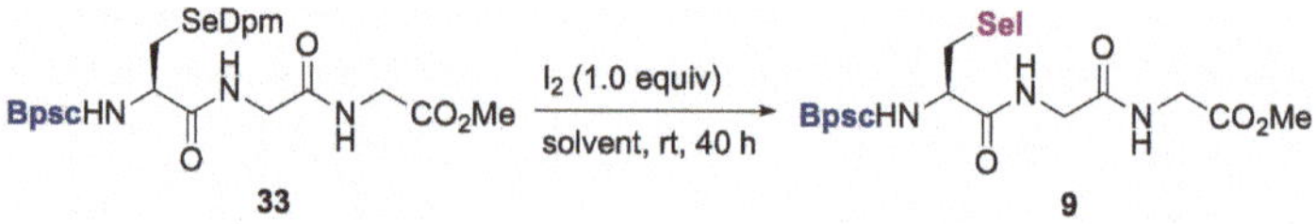

Entry	33 (mmol)	Solvent	Conv. (%)	Isolated yield (%)
1	0.021	CH_2Cl_2/MeOH = 10:1	99	83
2	0.185	CH_2Cl_2/MeOH = 10:1	70	–
3	0.243	CH_2Cl_2	85	–
4	0.350	CH_2Cl_2	55	–

reddish-purple crystals. Yield 38.5 mg (17.0 μmol, 83%). At this scale, this experiment was reproducible. When scaled up to 0.185 mmol, however, the starting material **33** was found to remain (70% conversion). In a single solvent system of CH_2Cl_2, a similar decrease in conversion rate was observed; 85% conversion at 0.243 mmol scale and 55% conversion at 0.350 mmol scale. As described in Section 2, the structure of the Sec–SeI **9** was determined by a single-crystal X-ray diffraction analysis (Fig. 4).

4.3.5 Protocol for the synthesis of Sec–SeH

Sec–SeH **10b** of the Sec-Gly-Gly model was prepared from Sec–SeI **9** (Scheme 15). Although we initially attempted the direct conversion of **9** to **10b** by DTT under basic conditions via the selenenyl sulfide intermediate **35**, only 38% yield of **10b** was obtained and reproducibility was poor. After some optimizations, the reduction of the corresponding cyclic selenenyl amide **15**, which was synthesized by cyclization of **9** in the presence of NaOH in 86% yield, was found to give the desired product **10b** in 68% yield. The reduction of selenenyl sulfide **12b**, which was prepared by the reaction of **9** with *N*-acetyl cysteine methyl ester (*Cys*–SH; **13**) in 81% yield, was also an effective route for the preparation of Sec–SeH **10b**. To a THF (3.0 mL) solution of selenenyl sulfide **12b** (109 mg, 46.9 μmol) was added Et_3N (23 μL, 0.16 mmol) followed by DTT (11 mg, 70 μmol). After carefully sealing the tube with a glass cap with silicon grease, the

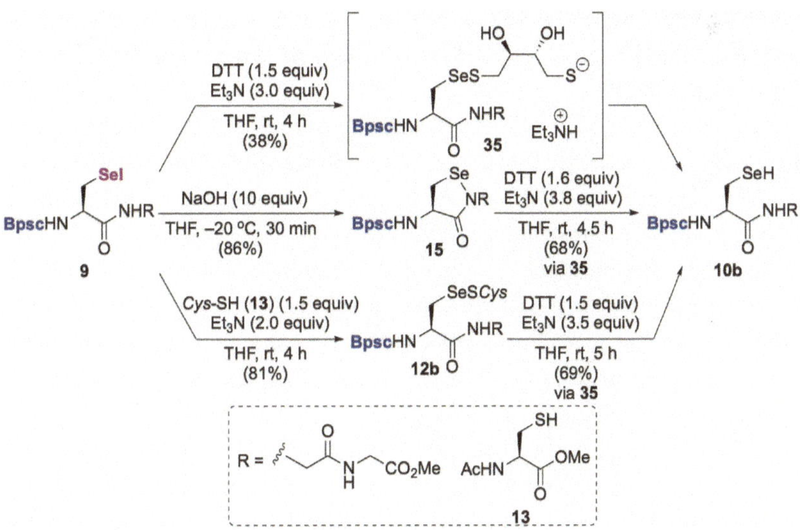

Scheme 15 Synthesis of Sec–SeH **10b**.

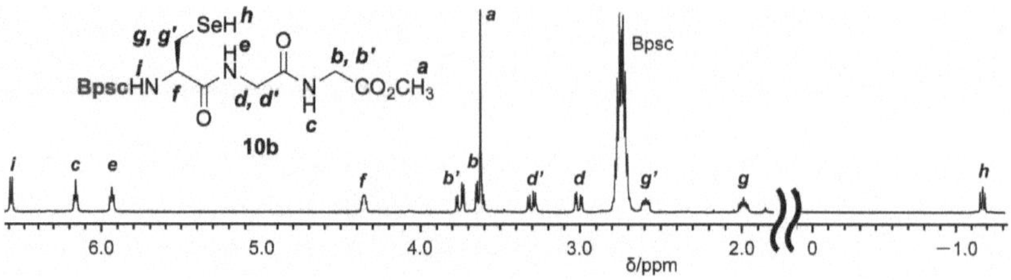

Fig. 5 ^{1}H NMR spectrum of **10b** (500 MHz, $CDCl_3$).

resulting solution was stirred at room temperature for 5 h. The following workup was performed under argon atmosphere. After dilution with degassed Et_2O, degassed brine was added to the solution. The two layers were separated and the aqueous layer was extracted with Et_2O (3 × 4 mL) using Schlenk tubes and Teflon tubes. The combined organic layer was dried over Na_2SO_4 and filtered using a glass filter with a Celite layer. The filtrate in a Schlenk tube was concentrated in vacuo; the solvent was removed by a vacuum pump without exposure to air. The crude product was immediately purified by argon-flash column chromatography on silica gel (hexane/CH_2Cl_2 = 8:1) to give 68.9 mg of **10b** as colorless crystals (69% yield). In the ^{1}H NMR spectrum of **10b**, a Se<u>H</u> proton was observed at −1.07 ppm (Fig. 5).

4.4 Protocols for the monitoring of the generation and reactions of Sec–SeOHs

4.4.1 Protocol for the ^{1}H NMR monitoring of the generation of Sec–SeOH

The protocol for the ^{1}H NMR monitoring of the generation of Sec–SeOH **11a** of the Sec model by H_2O_2 oxidation of Sec–SeH **10a** was investigated (Masuda, Kimura, et al., 2021). After several examinations, we found that two processes that must be prevented to achieve the direct observation of the Sec–SeOH; one is the bimolecular reaction between the formed **11a** and the unreacted material **10a** to form the corresponding symmetric diselenide **36a** (Kice & Chiou, 1986), and the other is thermal deselenation of **11a** to dehydroalanine **14a** (Scheme 16). To prevent these two undesirable processes, oxidation at low temperatures was found to be effective in suppressing both diselenide formation and thermal deselenation. Furthermore, it was found that, to prevent the formation of diselenide **36a** by promoting the conversion of Sec–SeH **10a** to Sec–SeOH **11a**, a solution of **10a** should be added dropwise to a solution containing an excess

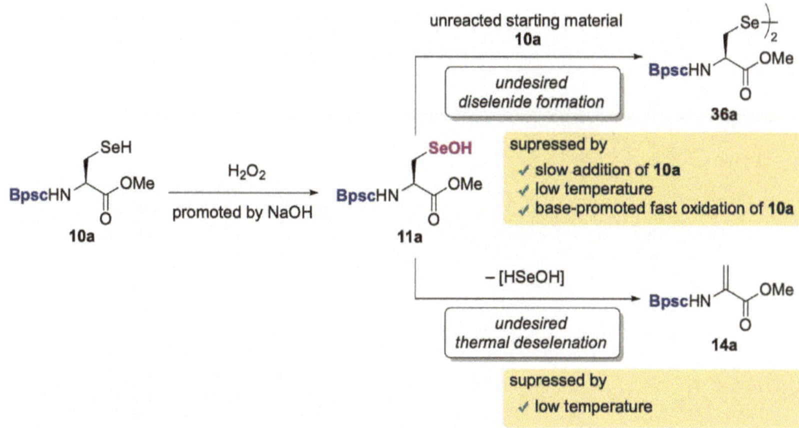

Scheme 16 Undesirable reaction processes involving Sec–SeOH **11a** and their preventive measures.

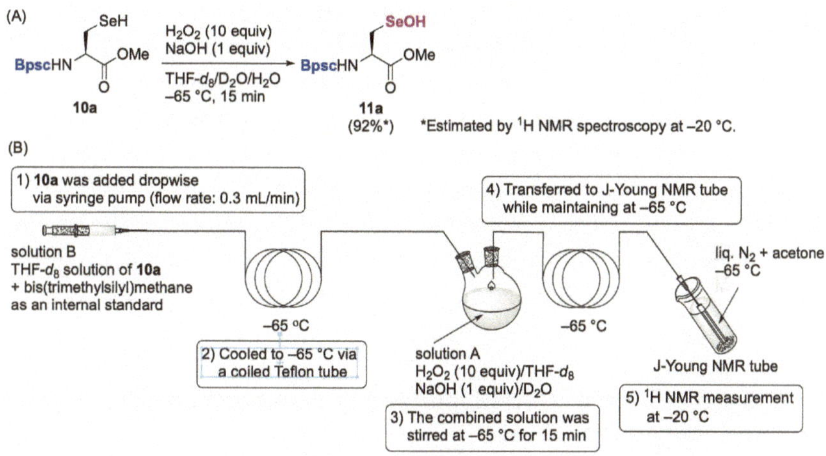

Fig. 6 (A) Generation of Sec–SeOH **11a** by H_2O_2 oxidation of Sec–SeH **10a**. (B) Protocol for the ^1H NMR monitoring of **11a**.

amount of H_2O_2 and NaOH. The established protocol is illustrated in Fig. 6. A solution of H_2O_2 (30%, 80 μL, 0.78 mmol) in THF-d_8 (degassed by argon bubbling, 1.0 mL) and a solution of NaOH (2.38 mmol) in D_2O (1.00 mL) were prepared under argon atmosphere, respectively, prior to the reaction. After evacuated and backfilled with argon, THF-d_8 (0.50 mL) was placed in a Schlenk flask, and the solution of NaOH (2.38 M, 2.6 μL, 6.2 μmol) followed by the solution of H_2O_2 (0.78 M, 71 μL, 55 μmol) was

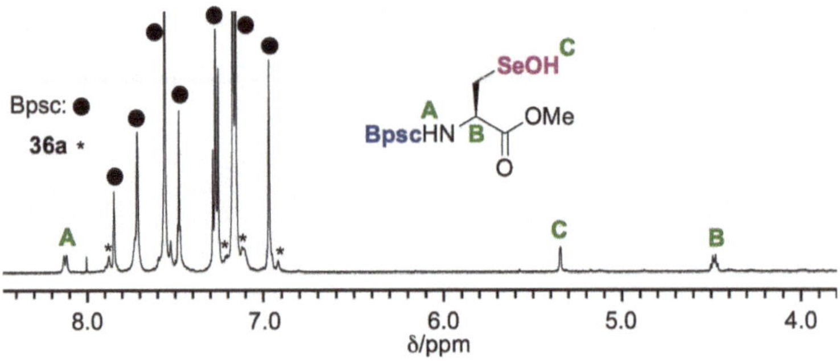

Fig. 7 Part of the ^{1}H NMR spectrum of Sec–SeOH **11a** (500 MHz, THF-d_8, −20 °C).

added. The obtained mixture was cooled to −65 °C (solution A). A solution of **10a** (5.53 μmol) in THF-d_8 (0.40 mL) containing bis(trimethylsilyl)methane as an internal standard (solution B), which was prepared in another *J*-Young Schlenk tube, was filled in a gas tight syringe and then connected to a Teflon tube. Solution B was added dropwise to solution A at −65 °C through a Teflon tube cooled to the same temperature by a syringe pump (flow rate: 0.3 mL/min). The resulting solution was stirred at −65 °C for 15 min and transferred to a *J*-Young NMR tube through a Teflon tube cooled to the same temperature as soon as possible, and the ^{1}H NMR spectrum was recorded at −20 °C. From preparation of the NMR sample to the spectroscopic measurement, the sample was cooled in a Dewar bottle at −65 °C. The formation of Sec–SeOH **11a** in 92% NMR yield was observed in the ^{1}H NMR spectrum (Fig. 7).

4.4.2 Protocol for the ^{1}H NMR monitoring of the reaction of Sec–SeOH with dimedone

We examined the ^{1}H NMR monitoring of the reaction of Sec–SeOH **11a** with dimedone (**16**) (Scheme 17) (Masuda, Kimura, et al., 2021). After generation of Sec–SeOH **11a** by oxidation of Sec–SeH **10a** (14.6 mg, 7.21 μmol) according to the protocol described in Section 4.4.1, a ^{1}H NMR spectrum of the resulting mixture was recorded at −20 °C (Sec–SeOH **11a** and diselenide **36a** were observed in 65% and 11% NMR yields, respectively) and then the *J*-Young NMR sample tube containing the mixture was cooled to −65 °C in a Dewar bottle. After 10 min, dimedone (**16**) (9.2 mg, 66 μmol, 9.1 equiv) was added to the mixture at −65 °C.

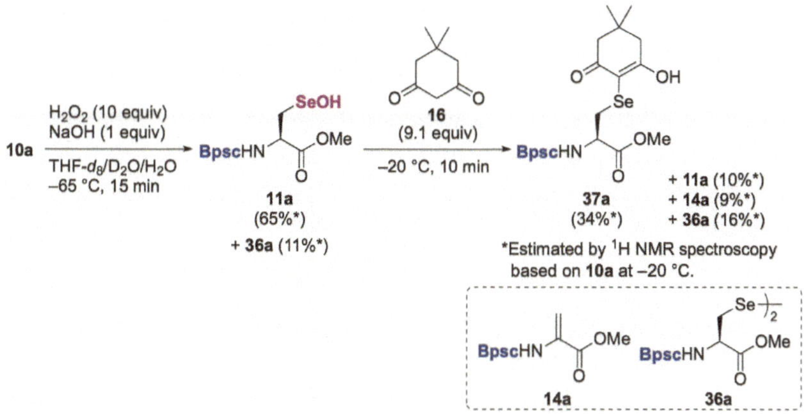

Scheme 17 Reaction of Sec–SeOH **11a** with dimedone (**16**).

After 10 min, a ^{1}H NMR spectrum was recorded again at −20 °C. The formation of selenide **37a**, dehydroalanine **14a**, diselenide **36a**, and Sec–SeOH **11a** was observed in 34%, 9%, 16%, and 10% NMR yields, respectively.

4.4.3 Protocol for the ^{1}H NMR monitoring of the reaction of Sec–SeOH with acetylacetone

We also investigated the ^{1}H NMR monitoring of the reaction of Sec–SeOH **11a** with acetylacetone (**17**) (Scheme 18) (Masuda, Kimura, et al., 2021). After generation of Sec–SeOH **11a** by oxidation of Sec–SeH **10a** (12.3 mg, 6.12 μmol) according to the protocol described in Section 4.4.1, a ^{1}H NMR spectrum of the resulting mixture was recorded at −20 °C (Sec–SeOH **11a** and diselenide **36a** were observed in 69% and 16% NMR yields,

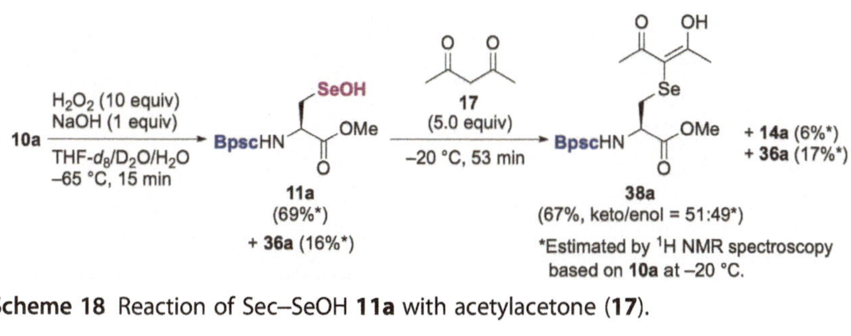

Scheme 18 Reaction of Sec–SeOH **11a** with acetylacetone (**17**).

respectively) and then the *J*-Young NMR sample tube containing the mixture was cooled to −65 °C in a Dewar bottle. After 10 min, acetylacetone (**17**) (3.0 μL, 29 μmol, 5.0 equiv) was added to the mixture at −65 °C. After 53 min, a ^{1}H NMR spectrum was recorded again at −20 °C. The formation of selenide **38a** (keto/enol = 51:49), dehydroalanine **14a**, diselenide **36a**, and Sec–SeOH **11a** were observed in 67%, 6%, 17%, and 0% NMR yields, respectively.

4.4.4 Protocol for the ^{1}H NMR monitoring of the reaction of Sec–SeOH with a cysteine thiol

As discussed in Section 3, the reaction of Sec–SeOH with GSH has been recognized as a part of the canonical catalytic cycle of GPx under conditions of sufficient concentration of GSH (Orian et al., 2015). We examined the reaction of Sec–SeOH **11a** with *N*-acetylcysteine methyl ester (**13**) (Scheme 19) (Masuda, Kimura, et al., 2021). After generation of Sec–SeOH **11a** by oxidation of Sec–SeH **10a** (11.2 mg, 5.53 μmol) according to the protocol described in Section 4.4.1, a ^{1}H NMR spectrum of the resulting mixture was recorded at −20 °C (Sec–SeOH **11a** and diselenide **36a** were observed in 92% and 8% NMR yields, respectively) and then the *J*-Young NMR sample tube containing the mixture was cooled to −65 °C in a Dewar bottle. A THF-d_8 solution of cysteine thiol **13** (0.10 M, 111 μL, 11 μmol, 2.0 equiv) was added to the mixture at −65 °C. After 8 min, a ^{1}H NMR spectrum was recorded again at −20 °C. The formation of selenenyl sulfide **12a** and diselenide **36a** were observed in 84% and 9% NMR yields, respectively. Notably, the formation of dehydroalanine **14a** was not detected in this reaction. These results demonstrate that the reaction of a Sec–SeOH with a thiol to produce a selenenyl sulfide is rapid even at low temperature where the thermal deselenation to a dehydroalanine does not proceed.

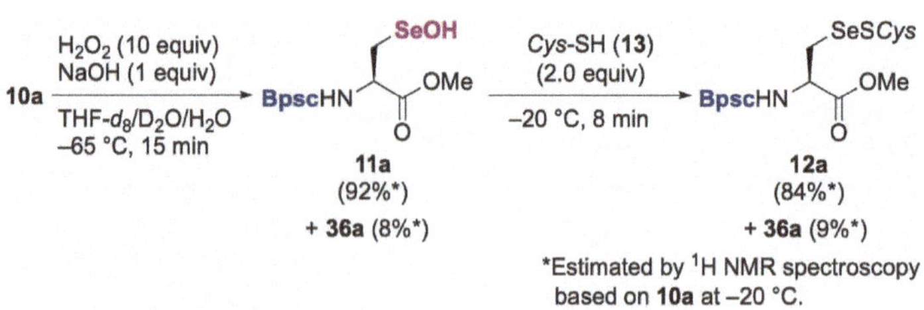

Scheme 19 Reaction of Sec–SeOH **11a** with cysteine thiol **33**.

4.4.5 Protocol for the ^{1}H NMR monitoring of the intramolecular cyclization of Sec–SeOH of the Sec-Gly-Gly model to the cyclic selenenyl amide

As discussed in Section 3, the intramolecular cyclization of Sec–SeOH to the corresponding cyclic selenenyl amide was proposed by Flohé to be a bypass process in the catalytic cycle of GPx (Orian et al., 2015). Based on the protocol established using the Sec model, we performed the model study of the bypass process using Sec–SeOH **11b** of the Sec-Gly-Gly model (Masuda, Kimura, et al., 2021). Sec–SeOH **11b** was found to be successfully generated by oxidation of Sec–SeH **10b** according to the same protocol as Sec–SeOH **11a** of the Sec model. After generation of Sec–SeOH **11b** by oxidation of Sec–SeH **10b** (13.9 mg, 6.50 μmol) according to the protocol described in Section 4.4.1, a ^{1}H NMR spectrum of the resulting mixture was recorded at −20 °C. The formation of Sec–SeOH **11b** and symmetric diselenide **36b** was observed in 87% and 11% NMR yields, respectively (Fig. 8A).

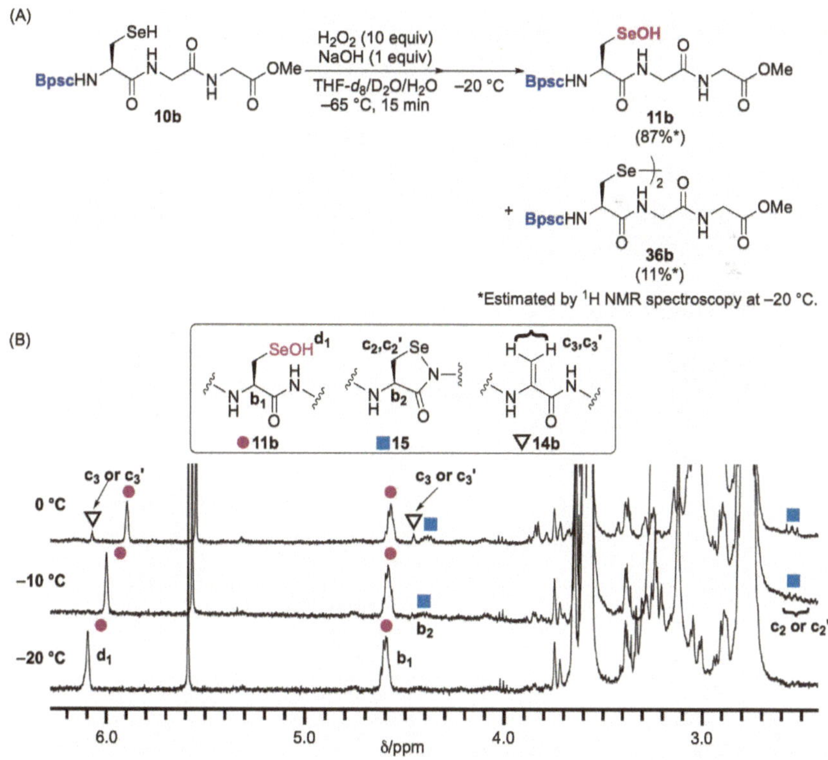

Fig. 8 (A) Generation of Sec–SeOH **11b**. (B) VT ^{1}H NMR spectra (400 MHz, −20 up to 0 °C) of a solution containing **11b** in THF-d_8/D_2O.

To investigate the thermal behavior of Sec–SeOH **11b**, the mixture was warmed from −20 °C to 0 °C and ^{1}H NMR spectra at selected temperatures were recorded (Fig. 8B). At −10 °C, the formation of cyclic selenenyl amide **15** with a five-membered ring in 6% yield was observed while the ratio of **11b** decreased from 87% to 79% (Table 4). Notably, dehydroalanine **14b** was not detected at −10 °C. At 0 °C, the ratio of **15** increased to 12%, while that of **11b** decreased to 70%, and only 2% of **14b** was detected.

4.4.6 Protocol for the ^{77}Se NMR analysis of Sec–SeOHs

The ^{77}Se NMR spectroscopy has been utilized as a reliable method for characterization of selenenic acids. We succeeded in the measurement of the ^{77}Se NMR spectra of Sec–SeOHs for the first time. To obtain clear ^{77}Se NMR spectra, the formation of Sec–SeOHs at high concentration is required. After generation of Sec–SeOH **11a** by oxidation of Sec–SeH **10a** (83.9 mg, 41.4 μmol) in THF-d_8/D_2O according to the protocol described in Section 4.4.1, a ^{77}Se NMR spectrum of the resulting mixture was recorded at −20 °C. A signal of **11a** was observed at 1298 ppm. In the case of Sec–SeOH **11b**, a signal of **11b** was observed at 1082 ppm when Sec–SeH **10b** (87.3 mg, 40.8 μmol) was used as a starting material. The ^{77}Se NMR chemical shifts of Sec–SeHs **10a** and **10b**, Sec–SeOHs **11a** and **11b**, selenenyl sulfides **12a** and **12b** and cyclic selenenyl amide **15** as well as non-Sec-derivative selenenic acid **4** with a primary alkyl substituent (Sase, Kakimoto, Kimura, & Goto, 2015) are shown in Table 5.

Table 4 Intramolecular cyclization of **11b** to **15**.

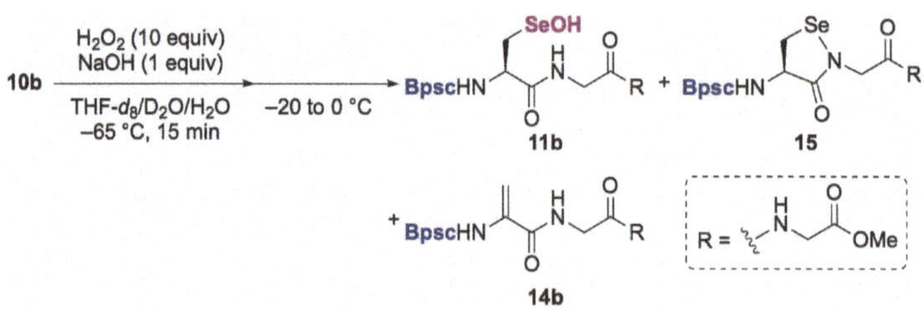

	NMR yield (%)		
Temp. (°C)	**11b**	**15**	**14b**
−20	87	n.d.	n.d.
−10	79	6	n.d.
0	70	12	2

Table 5 ^{77}Se NMR chemical shifts of several selenocysteine derivatives including Sec–SeOHs and a non-Sec-derivative selenenic acid.

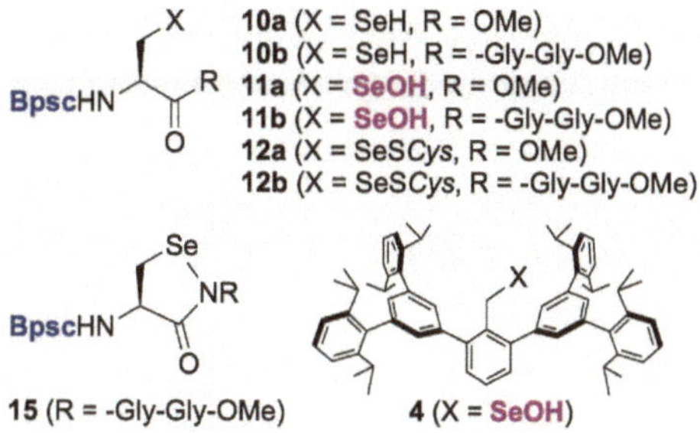

Compound	Solvent	Chemical shift (ppm)
10a	$CDCl_3$	−83
10b	$CDCl_3$	−79
11a	THF-d_8/D_2O	1298[a]
11b	THF-d_8/D_2O	1082[b]
12a	$CDCl_3$	384
12b	$CDCl_3$	382
15	$CDCl_3$	876
4	$CDCl_3$	1261

[a]At −20 °C.
[b]At −30 °C.

5. Conclusion

In this chapter, we have outlined how to stabilize Sec-derived reactive species in small-molecule model systems. The molecular cradle proved to be very effective in preventing the bimolecular decomposition processes of the Sec-derived reactive species as demonstrated by the isolation of Sec–SeI of the Sec-Gly-Gly model as crystals with "shelf-stability." The bimolecular decomposition of Sec–SeOHs was also prevented by the cradle, allowing us to experimentally examine the various reactivities of Sec–SeOHs by NMR monitoring for both reactions with other reagents and spontaneous thermal reactions. All the chemical processes proposed for the catalytic cycle of GPx, i.e., not only those of the canonical catalytic cycle, but also those involved in the bypass mechanism, including the intramolecular cyclization of Sec–SeOH to a selenenyl amide, were confirmed in a stepwise manner. Although these experiments were carried out in organic solvents or mixtures of an organic solvent with water, the availability of basic information on the intrinsic reactivity of the reactive species would be significant in mechanistic studies of the selenoenzymes. The strategy of the cradled selenopeptide is expected to provide a powerful tool for modeling a variety of other reactive species that occur in selenoproteins.

Acknowledgments

This work was partly supported by Grants-in-Aid for Scientific Research (B) (No. 15H03776 and 19H02698) from the Ministry of Education, Culture, Sports, Science, and Technology (Japan).

References

Berry, M. J., Banu, L., & Larsen, P. R. (1991). Type I iodothyronine deiodinase is a selenocysteine-containing enzyme. *Nature, 349*, 438–440.

Bhowmick, D., Srivastava, S., D'Silva, P., & Mugesh, G. (2015). Highly efficient glutathione peroxidase and peroxiredoxin mimetics protect mammalian cells against oxidative damage. *Angewandte Chemie International Edition, 54*, 8449–8453.

Brigelius-Flohé, R., & Maiorino, M. (2013). Glutathione peroxidases. *Biochimica et Biophysica Acta, General Subjects, 1830*, 3289–3303.

Bruno, N. C., Tudge, M. T., & Buchwald, S. L. (2013). Design and preparation of new palladium precatalysts for C-C and C-N cross-coupling reactions. *Chemical Science, 4*, 916–920.

Cho, S.-C., Lee, S., Lee, G. T., Woo, A. H., Choi, J.-E., & Rhee, G. S. (2010). Irreversible inactivation of glutathione peroxidase 1 and reversible inactivation of peroxiredoxin II by H_2O_2 in red blood cells. *Antioxidants & Redox Signaling, 12*, 1235–1246.

Du, C. J. F., Hart, H., & Ng, K. K. D. (1986). A one-pot synthesis of *m*-terphenyls via a two-aryne sequence. *The Journal of Organic Chemistry, 51*, 3162–3165.

du Mont, W. W., Martens, A., Pohl, S., & Saak, W. (1990). Reversible dismutation and coordination of bis(2,4,6-triisopropylphenyl) diselenide with iodine—A model study that relates to iodine intercalation between selenium chains. *Inorganic Chemistry, 29*, 4847–4848.

du Mont, W. W., & Wagner, I. (1988). Iodo[tris(trimethylsilyl)methyl]selane: Synthesis of the first alkylselenenyl iodide. *Chemische Berichte, 121*, 2109–2110.

Epp, O., Ladenstrin, R., & Wendel, A. (1983). The refined structure of the selenoenzyme glutathione peroxidase at 0.2-nm resolution. *European Journal of Biochemistry, 133*, 51–69.

Fischer, H., & Dereu, N. (1987). Mechanism of the catalytic reduction of hydroperoxides by Ebselen: A selenium—^{77}Se NMR study. *Bulletin des Sociétés Chimiques Belges, 96*, 757–768.

Flohé, L. (2020). Looking back at the early stages of redox biology. *Antioxidants, 9*, 1254.

Flohé, L., Günzler, W. A., & Schock, H. H. (1973). Glutathione peroxidase: A selenoenzyme. *FEBS Letters, 32*, 132–134.

Goto, K. (2014). Synergy of reactivity and stability in nanoscale molecular architectures. In T. Nabeshima (Ed.), *Synergy in supramolecular chemistry* (pp. 191–218). London: CRC Press.

Goto, K., Nagahama, M., Mizushima, T., Shimada, K., Kawashima, T., & Okazaki, R. (2001). The first direct oxidative conversion of a selenol to a stable selenenic acid: Experimental demonstration of three processes included in the catalytic cycle of glutathione peroxidase. *Organic Letters, 3*, 3569–3572.

Goto, K., Sonoda, D., Shimada, K., Sase, S., & Kawashima, T. (2010). Modeling of the 5'-deiodination of thyroxine by iodothyronine deiodinase: Chemical corroboration of a selenenyl iodide intermediate. *Angewandte Chemie International Edition, 49*, 545–547.

Guram, A. S., King, A. O., Allen, J. G., Wang, X., Schenkel, L. B., Chan, J., et al. (2006). New air-stable catalysts for general and efficient Suzuki-Miyaura cross-coupling reactions of heteroaryl chlorides. *Organic Letters, 8*, 1787–1789.

Guram, A. S., Wang, X., Bunel, E. E., Faul, M. M., Larsen, R. D., & Martinelli, M. J. (2007). New catalysts for Suzuki-Miyaura coupling reactions of heteroatom-substituted heteroaryl chlorides. *The Journal of Organic Chemistry, 72*, 5104–5112.

Ishii, A., Matsubayashi, S., Takahashi, T., & Nakayama, J. (1999). Preparation of a selenenic acid and isolation of selenoseleninates. *The Journal of Organic Chemistry, 64*, 1084–1085.

Kice, J. L., & Chiou, S. (1986). Rates of oxidation of *o*-nitrobenzeneselenenyl compounds by *m*-chloroperoxybenzoic acid and the rate of reaction of *o*-nitrobenzeneselenol with *o*-nitrobenzeneselenenic acid. *The Journal of Organic Chemistry, 51*, 290–294.

Kim, C., Lee, J., & Park, M. S. (2015). Synthesis of new diorganodiselenides from organic halides: Their antiproliferative effects against human breast cancer MCF-7 cells. *Archives of Pharmacal Research, 38*, 659–665.

Köhrle, J. (1999). Local activation and inactivation of thyroid hormones: The deiodinase family. *Molecular and Cellular Endocrinology, 151*, 103–119.

Kraus, R. J., Foster, S. J., & Ganther, H. E. (1983). Identification of selenocysteine in glutathione peroxidase by mass spectroscopy. *Biochemistry, 22*, 5853–5858.

Kryukov, G. V., Castellano, S., Novoselov, S. V., Lobanov, A. V., Zehtab, O., Guigó, R., et al. (2003). Characterization of mammalian selenoproteomes. *Science, 300*, 1439–1443.

Kunishima, M., Kawachi, C., Hioki, K., Terao, K., & Tani, S. (2001). Formation of carboxamides by direct condensation of carboxylic acids and amines in alcohols using a new alcohol- and water-soluble condensing agent: DMT-MM. *Tetrahedron, 57*, 1551–1558.

Kunishima, M., Kawachi, C., Iwasaki, F., Terao, K., & Tani, S. (1999). Synthesis and characterization of 4-(4,6-dimethoxy-1,3,5-triazin-2-yl)-4-methylmorpholinium chloride. *Tetrahedron Letters, 40*, 5327–5330.

Liu, J., & Rozovsky, S. (2013). Contribution of selenocysteine to the peroxidase activity of selenoprotein S. *Biochemistry, 52*, 5514–5516.

Liu, J., Zhang, Z., & Rozovsky, S. (2014). Selenoprotein K form an intermolecular diselenide bond with unusually high redox potential. *FEBS Letters, 588*, 3311–3321.

Lubos, E., Loscalzo, J., & Handy, D. E. (2011). Glutathione peroxidase-1 in health and disease: From molecular mechanisms to therapeutic opportunities. *Antioxidants & Redox Signaling, 15*, 1957–1997.

Ma, S., Caprioli, R. M., Hill, K. E., & Burk, R. F. (2003). Loss of selenium from selenoproteins: Conversion of selenocysteine to dehydroalanine in vitro. *Journal of the American Society for Mass Spectrometry, 14*, 593–600.

Maiorino, M. F., Brigelius-Flohé, R., Aumann, K. D., Roveri, A., Schomburg, D., & Flohé, L. (1995). Diversity of glutathione peroxidases. *Methods in Enzymology, 252*, 38–53.

Martens-V Salzen, A., Meyer, H.-U., & du Mont, W.-W. (1992). Diselenides and iodine: Influence of solution equilibria between covalent compounds and charge-transfer complexes. *Phosphorus, Sulfur and Silicon and the Related Elements, 67*, 67–71.

Masuda, R., Kimura, R., Karasaki, T., Sase, S., & Goto, K. (2021). Modeling the catalytic cycle of glutathione peroxidase by nuclear magnetic resonance spectroscopic analysis of selenocysteine selenenic acids. *Journal of the American Chemical Society, 143*, 6345–6350.

Masuda, R., Kuwano, S., & Goto, K. (2021). Late-stage functionalization of the periphery of oligophenylene dendrimers with various arene units via fourfold C − H borylation. *The Journal of Organic Chemistry, 86*, 14433–14443.

Mauri, P., Benazzi, L., Flohé, L., Maiorino, M., Pietta, G. P., Pilawa, S., et al. (2003). Versatility of selenium catalysis in PHGPx unraveled by LC/ESI-MS/MS. *Biological Chemistry, 384*, 575–588.

Mills, G. C. (1957). Hemoglobin catabolism I. Glutathione peroxidase, an erythrocyte enzyme which protects hemoglobin from oxidative breakdown. *Journal of Biological Chemistry*, *229*, 189–197.

Miyaura, N., & Suzuki, A. (1995). Palladium-catalyzed cross-coupling reactions of organoboron compounds. *Chemical Reviews*, *95*, 2457–2483.

Miyaura, N., Yamada, K., & Suzuki, A. (1979). A new stereospecific cross-coupling by the palladium-catalyzed reaction of 1-alkenylboranes with 1-alkenyl or 1-alkynyl halides. *Tetrahedron Letters*, *20*, 3437–3440.

Mugesh, G., Panda, A., Singh, H. B., & Butcher, R. J. (1999). Intramolecular Se⋯N non-bonding interactions in low-valent organoselenium derivatives: A detailed study by ^{1}H and ^{77}Se NMR spectroscopy and X-ray crystallography. *Chemistry—A European Journal*, *5*, 1411–1421.

Nicolaou, K. C., Estrada, A. A., Zak, M., Lee, S. H., & Safina, B. S. (2005). A mild and selective method for the hydrolysis of esters with trimethyltin hydroxide. *Angewandte Chemie International Edition in English*, *44*, 1378–1382.

Orian, L., Mauri, P., Roveri, A., Toppo, S., Benazzi, L., Bosello-Travain, V., et al. (2015). Selenocysteine oxidation in glutathione peroxidase catalysis: An MS-supported quantum mechanics study. *Free Radical Biology & Medicine*, *87*, 1–14.

Payne, N. C., Barber, D. R., Ruggles, E. L., & Hondal, R. J. (2018). Can dimedone be used to study selenoproteins? An investigation into the reactivity of dimedone toward oxidized forms of selenocysteine. *Protein Science*, *28*, 41–55.

Reich, H. J., Hoger, C. A., & Wills, W. W. (1985). Organoselenium chemistry: A study of intermediates in the fragmentation of aliphatic ketoselenoxides. Characterization of selenoxides, selenenamides and selenolseleninates by ^{1}H-,^{13}C-and ^{77}Se-NMR. *Tetrahedron*, *41*, 4771–4779.

Reich, H. J., & Jasperse, C. P. (1987). Organoselenium chemistry. Redox chemistry of selenocysteine model systems. *Journal of the American Chemical Society*, *109*, 5549–5551.

Reich, H. J., & Jasperse, C. P. (1988). Organoselenium chemistry. Preparation and reactions of 2,4,6-tri-*Tert*-butylbenzeneselenenic acid. *The Journal of Organic Chemistry*, *53*, 2389–2390.

Rotruck, T. J., Pope, L. A., Ganther, E. H., Swanson, B. A., Hafeman, G. D., & Hoekstra, G. W. (1973). Selenium: Biochemical role as a component of glutathione peroxidase. *Science*, *179*, 588–590.

Saiki, T., Goto, K., & Okazaki, R. (1997). Isolation and X-ray crystallographic analysis of a stable selenenic acid. *Angewandte Chemie International Edition in English*, *36*, 2223–2224.

Sano, T., Masuda, R., Sase, S., & Goto, K. (2021). Isolable small-molecule cysteine sulfenic acid. *Chemical Communications*, *57*, 2479–2482.

Sarma, B. K., & Mugesh, G. (2005). Glutathione peroxidase (GPx)-like antioxidant activity of the organoselenium drug Ebselen: Unexpected complications with thiol exchange reactions. *Journal of the American Chemical Society*, *127*, 11477–11485.

Sase, S., Kakimoto, R., & Goto, K. (2015). Synthesis of a stable selenoaldehyde by self-catalyzed thermal dehydration of a primary-alkyl-substituted selenenic acid. *Angewandte Chemie International Edition*, *54*, 901–904.

Sase, S., Kakimoto, R., Kimura, R., & Goto, K. (2015). Synthesis of a stable primary-alkyl-substituted selenenyl iodide and its hydrolytic conversion to the corresponding selenenic acid. *Molecules*, *20*, 21415–21420.

Sase, S., Kimura, R., Masuda, R., & Goto, K. (2019). Model study on trapping of protein selenenic acids by utilizing a stable synthetic congener. *New Journal of Chemistry*, *43*, 6830–6833.

Scheerer, P., Borchert, A., Krauss, N., Wessner, H., Gerth, C., Höhne, W., et al. (2007). Structural basis for catalytic activity and enzyme polymerization of phospholipid hydroperoxide glutathione Peroxidase-4 (GPx4). *Biochemistry*, *46*, 9041–9049.

Scinto, S. L., Ekanayake, O., Seneviratne, U., Pigga, J. E., Boyd, S. J., Taylor, M. T., et al. (2019). Dual-reactivity *trans*-cyclooctenol probes for sulfenylation in live cells enable temporal control via bioorthogonal quenching. *Journal of the American Chemical Society, 141*, 10932–10937.

Shimodaira, S., & Iwaoka, M. (2017). Improved synthetic routes to the selenocysteine derivatives useful for Boc-based peptide synthesis with benzylic protection on the selenium atom. *Arkivoc*, 260–271.

Stempel, E., Kaml, R. F. X., Budisa, N., & Kalesse, M. (2018). Painting Argyrins blue: Negishi cross-coupling for synthesis of deep-blue tryptophan analogue β-(1-Azulenyl)-L alanine and its incorporation into argyrin C. *Bioorganic & Medicinal Chemistry, 26*, 5259–5269.

Ungati, H., Govindaraj, V., Narayanan, M., & Mugesh, G. (2019). Probing the formation of a seleninic acid in living cells by the fluorescence switching of a glutathione peroxidase mimetic. *Angewandte Chemie International Edition, 58*, 8156–8160.

Walter, R., & Roy, J. (1971). Selenomethionine, a potential catalytic antioxidant in biological systems. *The Journal of Organic Chemistry, 36*, 2561–2563.

Zielinski, Z., Presseau, N., Amorati, R., Valgimigli, L., & Pratt, D. (2014). Redox chemistry of selenenic acids and the insight it brings on transition state geometry in the reactions of peroxyl radicals. *Journal of the American Chemical Society, 136*, 1570–1578.

CHAPTER FIFTEEN

Diselenide-selenoester ligation in the chemical synthesis of proteins

Max J. Bedding[a,b,†], **Sameer S. Kulkarni**[a,b,†], **and Richard J. Payne**[a,b,*]
[a]School of Chemistry, The University of Sydney, Camperdown, NSW, Australia
[b]Australian Research Council Centre of Excellence for Innovations in Peptide and Protein Science, The University of Sydney, Camperdown, NSW, Australia
*Corresponding author: e-mail address: richard.payne@sydney.edu.au

Contents

Abstract

Peptides and proteins represent an important class of biomolecules responsible for a plethora of structural and functional roles *in vivo*. Following their translation on the ribosome, the majority of eukaryotic proteins are post-translationally modified, leading to a proteome that is much larger than the number of genes present in a given organism.

† These authors contributed equally.

Methods in Enzymology, Volume 662
ISSN 0076-6879
https://doi.org/10.1016/bs.mie.2021.10.017

In order to understand the functional role of a given protein modification, it is necessary to access peptides and proteins bearing homogeneous and site-specific modifications. Accordingly, there has been significant research effort centered on the development of peptide ligation methodologies for the chemical synthesis of modified proteins. In this chapter we outline the discovery and development of a contemporary methodology called the diselenide-selenoester ligation (DSL) that enables the rapid and efficient fusion of peptide fragments to generate synthetic proteins. The practical aspects of using DSL for the preparation of chemically modified peptides and proteins in the laboratory is described. In addition, recent advances in the application of the methodology are outlined, exemplified by the synthesis and biological evaluation of a number of complex protein targets.

1. Introduction

Peptides and proteins are responsible for a host of functional and structural roles in living systems. The properties of a protein are determined by the sequence of 21 proteinogenic amino acids that make up their primary structure. The information required for the sequence of proteins in the "proteome" is stored within the genetic code of an organism, with the flow of this information from DNA to RNA to proteins being the foundation of the central dogma of molecular biology (Crick, 1970). Decoding the information stored within DNA has been the focus of extensive research in the 20th and 21st centuries and it is now thought that almost 20,000 genes in the human genome encode for over a million unique protein products (Nurk et al., 2021).

A commonality of all life is the ability to adapt to changes in the environment. In response to environmental factors, the constituent proteins within the proteome can change in an effort to maintain cellular homeostasis. This can be achieved through the translation of new proteins, degradation of existing proteins, or chemical installation of post-translational modifications (PTMs) to modulate protein function. PTMs can involve the addition or removal of any chemical functionality from a protein following its translation by the ribosome. They are diverse in both their structure and their effect on protein function. For example, cysteine (Cys) residues can be prenylated to increase the hydrophobicity of a protein, facilitating immobilization into cell membranes (Chen, Sun, Niu, Jarugumilli, & Wu, 2018; Wang & Casey, 2016), asparagine (Asn) can be modified with complex glycans which can induce the display of a protein onto the cell surface for signaling to other cells (Bieberich, 2014), and lysine (Lys) can be ligated to ubiquitin, which results in protein degradation by the proteasome

(Swatek & Komander, 2016) (outlined in Fig. 1). Since the modification of proteins requires significant expenditure of energy, it has become widely accepted that PTMs must play an essential role in maintaining cellular homeostasis. The extraordinary complexity of potential modifications has opened new avenues of investigation, both from a chemical and biological standpoint. While remarkable progress has been made in identifying and characterizing a plethora of PTMs, for the vast majority the precise function of a given modification across all organisms remains unknown.

Due to their dynamic enzymatic transfer and removal, the study of PTMs in native biological systems can be difficult due to the presence of heterogeneous proteoforms in the cellular milieu. The ability to access homogenous proteins bearing a specific post-translational modification can simplify investigations by allowing the comparison of a modified and unmodified sample when evaluating a hypothesized physiological function. The burgeoning field of chemical biology involves the application of synthetic chemistry, semi-synthetic and chemoenzymatic methods as tools to interrogate and manipulate biological systems. This approach to studying biochemical pathways has resulted in a high demand for proteins bearing non-natural modifications for thorough biological investigation. These may include proteins modified with fluorophores to monitor the movement of proteins within cells, affinity handles such as biotin so that proteins may be separated from complex mixtures for target validation purposes, or PEG

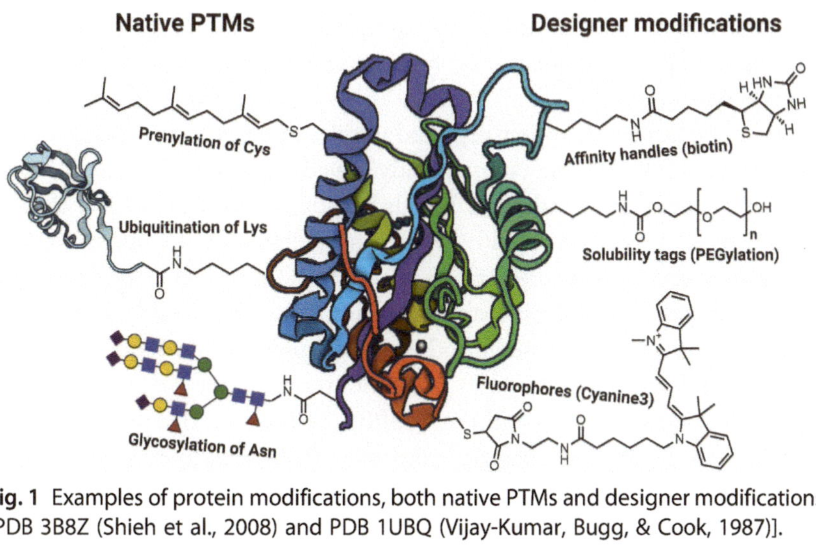

Fig. 1 Examples of protein modifications, both native PTMs and designer modifications [PDB 3B8Z (Shieh et al., 2008) and PDB 1UBQ (Vijay-Kumar, Bugg, & Cook, 1987)].

groups to modulate solubility or immunogenicity (see Fig. 1). Similar to native PTMs, these tailored modifications need to be precisely installed into a target protein for meaningful biological studies to be conducted. As such, development of methodologies to reliably prepare homogenously modified protein samples has been an important area of research for both chemists and biologists.

Large polypeptides and proteins can be readily accessed through recombinant expression in both prokaryotic and eukaryotic systems. Although expression in mammalian cell lines can be used to generate proteins with PTMs which resemble those found natively in humans, they are often produced as heterogeneous mixtures of isoforms (*vide supra*), making interrogation of a specific PTM difficult. Alternatively, innovative genetic code expansion methods can be utilized to incorporate non-proteinogenic amino acids into large, expressed proteins through reprogramming of an unassigned codon (often the UAG amber codon), enabling incorporation of the desired amino acid by ribosomal translation using orthogonal aminoacyl-tRNA synthetase-tRNA pairs (Chin, 2017; Liu & Schultz, 2010). This technique has been used to efficiently incorporate native PTMs (Italia et al., 2020; Neumann, Peak-Chew, & Chin, 2008) and non-natural residues into proteins, however, there is a limit to the functionality that can be installed through this ribosome-based methodology (Antonczak, Simova, & Tippmann, 2009). Chemical synthesis has emerged as a powerful tool for accessing homogeneously modified proteins. Importantly, whether it be a native PTM or a non-natural designer modification, chemical synthesis offers precise control over the nature and position of the installed modification, usually through the incorporation of specifically functionalized amino acid cassettes into peptide synthesis.

In 1963, Bruce Merrifield reported the development of solid-phase peptide synthesis (SPPS) which subsequently served as a pioneering methodology for the synthesis of peptides (for which he was awarded the Nobel Prize in Chemistry in 1984) (Merrifield, 1963). This technology involves immobilization of a suitably protected amino acid *via* its C-terminal carboxylate onto a solid support followed by chain elongation through iterative deprotection and coupling cycles to assemble the desired polypeptide sequence. Based on the protecting group used to mask the α-amine, the method is divided into two categories, Boc- and Fmoc-SPPS. In the case of Fmoc-SPPS, the α-amine is protected using base-labile 9-fluorenylmethyloxycarbonyl (Fmoc) group, while the side chain functional groups are orthogonally protected using protecting groups that are labile to acid treatment. The fully extended peptide is then subjected to acidic cleavage followed by purification using reversed-phase high-performance liquid chromatography (HPLC), affording the peptide of interest in pure form. This approach offers precise control over the chemical functionalities that can be incorporated at any point in a sequence, allowing access to peptides and small proteins bearing a wide array of both native and non-native modifications, which would be otherwise inaccessible by recombinant methods. However, the nature of the SPPS methods means that it is typically limited to the preparation of relatively small polypeptides (generally <50 amino acid residues) (Kent et al., 2012). For larger systems, aggregation of the peptide chains can lead to incomplete deprotection or coupling steps, along with deleterious side reactions or epimerization, making synthesis and purification of the final product nearly impossible.

As a result, efforts have been made in the preparation of polypeptides and proteins by fusing smaller peptide fragments through efficient amide bond forming "ligation" reactions (Agouridas et al., 2019; Bondalapati, Jbara, & Brik, 2016; Kulkarni, Sayers, Premdjee, & Payne, 2018; Tan, Wu, Wei, & Li, 2020). Initial work invoked condensation reactions between two side-chain protected fragments, which has been utilized in the preparation of a number of polypeptide targets, including the large-scale synthesis of therapeutic peptide enfuvirtide (Bray, 2003). However, inefficiencies surrounding poor solubility of protected fragments and propensity of C-terminal amino acid epimerization limits the utility of this strategy. The most significant advance in the field of peptide ligation came with the advent of native chemical ligation (NCL). Specifically, in a seminal report, Dawson, Muir, Clark-Lewis, and Kent (1994) disclosed that a peptide bearing a C-terminal thioester could be reacted chemoselectively with another peptide fragment featuring an N-terminal Cys residue *via* a reversible transthioesterification step, facilitated by the nucleophilic Cys side chain, followed by a facile $S \rightarrow N$ acyl shift to generate the native amide bond (outlined in Fig. 2A). Importantly, this reaction can be performed on unprotected peptide substrates in mild aqueous conditions making it amenable to a variety of protein targets (Kent, 2009; Muir & Kent, 1993). While the original method relied on the use of Cys residues, one of the least abundant amino acids in native proteins (estimated to be 1.8%), the scope of this powerful synthetic platform has been greatly expanded through the development of radical desulfurization chemistry (Wan & Danishefsky, 2007; Yan & Dawson, 2001). For Cys ligations this chemistry enables conversion to alanine following the ligation, but with the introduction of synthetic thiolated variants of amino acids NCL-desulfurization chemistry can now be

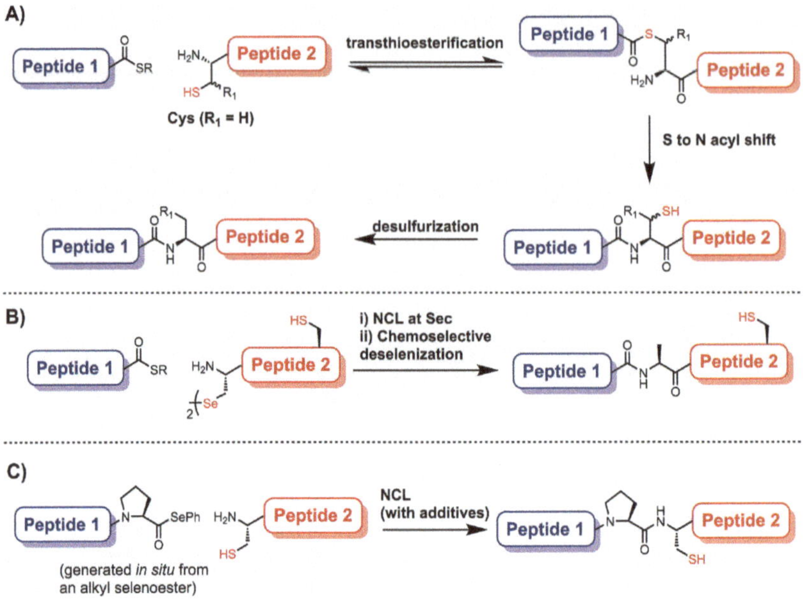

Fig. 2 Expansion of the NCL manifold through the incorporation of Sec or selenoesters. (A) NCL proceeds *via* a transthioesterification followed by an S → N acyl shift. The ligated product can be readily desulfurized to afford native residues. (B) NCL at Sec facilitates rapid ligation and chemoselective deselenization can be performed in the presence of native cysteines. (C) Peptide selenoesters are highly reactive acyl donors, highlighted by NCL at the challenging Pro-Cys junction.

performed at most proteinogenic amino acids (Cergol et al., 2014; Chen et al., 2008; Crich & Banerjee, 2007; Haase et al., 2008; Harpaz, Siman, Kumar, & Brik, 2010; Malins, Giltrap, Dowman, & Payne, 2015; Malins & Payne, 2014, 2015; Thompson, Chan, Radom, Jolliffe, & Payne, 2013; Wan & Danishefsky, 2007; Yan & Dawson, 2001; Yang, Pasunooti, Li, Liu, & Liu, 2009). It is important to note that a semi-synthetic variant of NCL also exists which is commonly called expressed protein ligation (EPL). In this approach, first reported by Muir et al. (1998), thioesters of recombinantly expressed large proteins are generated using intein technology which can be ligated with site-specifically modified synthetic peptides under standard NCL conditions. Taken together, these ligation technologies have enabled the synthesis and semi-synthesis of a number of complex protein targets to date. Some examples include chemically-ubiquitinated histone H2B (McGinty, Kim, Chatterjee, Roeder, & Muir, 2008), glycosylated interferon-β (Sakamoto et al., 2012), tetraubiquitin (Kumar et al., 2011), antifreeze glycoproteins (Wilkinson et al., 2012), and D-protein enantiomers of DNA polymerase IV (Jiang et al., 2017) and *Pfu* DNA polymerase (Fan, Deng, & Zhu, 2021), capable of amplifying sequences of L-DNA for bioorthogonal information storage. The complexity of these targets highlights the power of chemical ligation in preparing homogenous proteins, which would be otherwise inaccessible by other methods.

While the utility of NCL has been exemplified by the successful synthesis of a number of modified proteins since the first report of the methodology, a number of limitations have hindered its applicability to certain protein targets and classes. Firstly, although NCL reactions proceed efficiently at mM concentrations, at higher dilution (μM), ligation reactions either proceed sluggishly or not at all. This limits its utility to soluble peptide fragments and makes poorly water-soluble targets such as lipidated proteins difficult to generate using this methodology. Secondly, while the development of thiolated amino acids has expanded the scope of NCL beyond Cys and Ala, ligations at hindered junctions, such as those containing β-branched amino acids or Pro, are often plagued by poor reaction kinetics resulting in significant thioester hydrolysis and low yields. Finally, while desulfurization allows conversion of Cys or a thiolated amino acid to the native amino acid post ligation, this transformation is not chemoselective and results in undesired desulfurization of native Cys residues elsewhere in the sequence. While protecting groups, such as the acetamidomethyl (Acm) (Veber, Milkowski, Denkewalter, & Hirschmann, 1968; Veber, Milkowski, Varga, Denkewalter, & Hirschmann, 1972) can be used to preserve native Cys residues, this strategy requires additional synthetic manipulations and cannot be implemented for systems involving expressed proteins. Therefore, protein chemists have sought to develop alternative ligation methodologies that complement NCL for the preparation of complex and highly functionalized proteins.

In parallel to the advances made in NCL, researchers have also explored the use of selenocysteine (Sec, U), the 21st amino acid (Bock et al., 1991). Sec is the chalcogenic analog of Cys and possesses a number of unique chemical properties. Firstly, its lower reduction potential (−381 mV) compared to Cys (−180 mV), means that Sec readily oxidizes in air and exists predominantly as a diselenide dimer (selenocystine) (Besse, Siedler, Diercks, Kessler, & Moroder, 1997; Guenther, 1967). Secondly, owing to the lower pK_a of Sec (5.2–5.6) compared to Cys ($pK_a = 8.2$), Sec retains its nucleophilicity over a much broader pH range, albeit only in its monomeric form (*i.e.*, selenol/selenoate). This superior nucleophilicity has piqued significant interest in Sec-mediated ligations as a means to improve the efficiency and versatility of NCL.

The first explorations into the use of selenocysteine in peptide ligation chemistry were reported independently by three research groups, Hondal et al. (2001), Quaderer and Hilvert (2002), Quaderer et al. (2001), and Gieselman et al. (2001). In these studies the authors demonstrated that peptide thioesters can undergo ligation with another peptide bearing an N-terminal Sec, proceeding *via* an NCL-like mechanism (as outlined in Fig. 2B). Notably, since selenocystine (the oxidized form of selenocysteine) is non-nucleophilic, addition of a reductant, such as tris(2-carboxyethyl) phosphine (TCEP) or thiophenol, was required to generate the active selenoate. Pleasingly, in all instances, the reaction proceeded far more rapidly and across a wider pH range than NCL. Hondal et al. (2001) impressively applied this ligation manifold to access the C110U mutant of RNase A through ligation between an expressed thioester fragment and a synthetic peptide diselenide, demonstrating the compatibility of Sec with expressed protein ligation. Additionally, Quaderer and Hilvert (2002) showed that the ligated product could be deselenized using Raney Ni and TCEP under an atmosphere of H_2 to afford an Ala residue at the ligation junction. Metanis, Keinan, and Dawson (2010) subsequently reported that the deselenization transformation can be performed under milder conditions (without any transition metal catalyst), using TCEP and dithiothreitol (DTT) as a hydrogen atom donor. Notably, the absence of a radical initiator in this transformation enabled chemoselective deselenization of Sec in the presence of native unprotected Cys residues, providing a unique advantage over the NCL-desulfurization approaches described above.

While investigating the utility of selenoesters as acyl donors, Durek and Alewood (2011) also found that peptidyl phenyl selenoesters, generated *in situ* from preformed alkyl selenoesters, readily react with peptides bearing an N-terminal Cys residue. As anticipated, based on the enhanced electrophilicity of selenoesters compared to thioesters, this transformation displayed superior kinetics compared to NCL. This rate enhancement was particularly significant when ligation was performed at Pro selenoesters, which are typically unreactive with thioester acyl donors due to electronic ($n \rightarrow \pi^*$) disarming effects. Indeed, the authors showed that ligation reactions proceed 350 times faster at these selenoesters compared to the corresponding thioester (as outlined in Fig. 2C).

Recently, Mitchell et al. (2015) showed that by combining the enhanced electrophilicity of selenoesters and nucleophilicity of Sec ligation reactions proceed at unprecedented rates. This reaction dubbed the diselenide-selenoester ligation (DSL) can be combined with *in situ* deselenization chemistry for the rapid and efficient synthesis of proteins (with or without modifications). Throughout this chapter, we will outline the practical aspects of DSL along with recent advances in the scope of this methodology, highlighted by the synthesis of complex modified proteins to reveal the functional consequences of a range of PTMs.

2. Advent of DSL

Based on the improved reactivity offered by Sec-mediated NCL and selenoesters as the acyl donor, Mitchell et al. (2015) hypothesized that the rate of ligation reactions could be substantially enhanced through the use of Sec on the N-terminus of one peptide segment and an aryl selenoester (instead of an alkyl selenoester) on the other reacting peptide segment. To validate this hypothesis, a series of control experiments were designed, wherein a model peptide with C-terminal phenyl selenoester was reacted with another peptide diselenide bearing N-terminal Sec under aqueous conditions. To avoid undesired phosphine-mediated deselenization, the authors explored the possibility of using electrochemical reduction of Sec to initiate the ligation through generation of the nucleophilic selenoate that was deemed necessary to initiate the ligation reaction. However, surprisingly it was discovered that the diselenide-selenoester ligation could be performed by simply mixing the substrates in a denaturing buffer without any exogenous reductant. Remarkably, this "additive-free" DSL reaction proceeded to completion on extraordinary fast timescales (1–10 min), even at sterically hindered junctions (as outlined in Fig. 3A). This methodology is also reported to possess additional unique and advantageous features (Kulkarni et al., 2019). For example, DSL reactions can be carried out over a broad pH range (3–7), due to the lower pK_a of Sec compared to Cys. It is worth mentioning that the ligation product can be observed as a mixture of three species: (1) a symmetrical diselenide dimer product (**I**), (2) an asymmetrical diselenide product (**II**) and (3) a product selenoester (**III**), wherein the selenol sidechain of Sec is acylated with the N-terminal peptide fragment. However, all these intermediates converge to the final product after deselenization, affording Ala at the ligation junction. Importantly, as highlighted above, the ability to perform the deselenization step chemoselectively in the presence

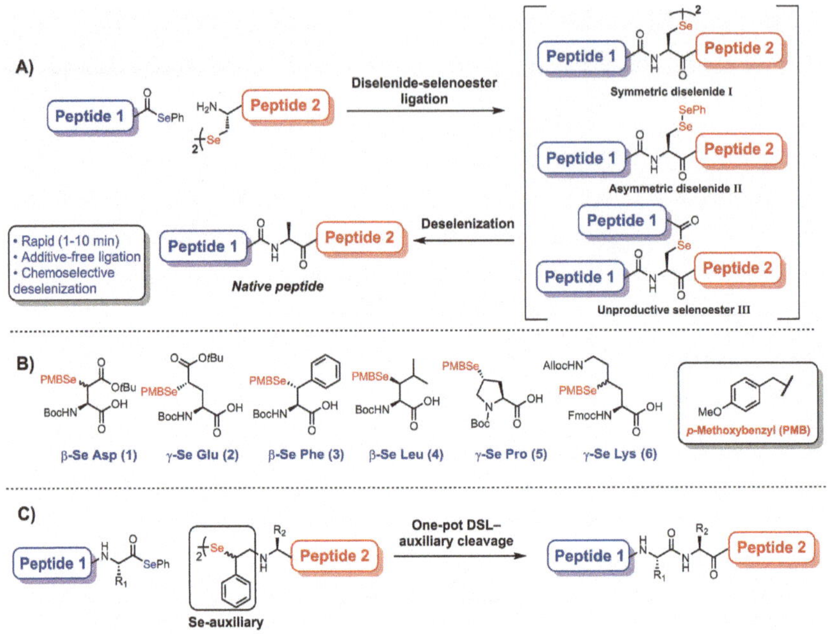

Fig. 3 (A) Diselenide-selenoester ligation (DSL) affords a mixture of symmetric diselenide **I**, asymmetric diselenide **II**, and unproductive selenoester **III** intermediates which converge to the native peptide product following chemoselective deselenization. (B) Selenylated amino acids to facilitate DSL beyond Ala/Ser. (C) An auxiliary-mediated DSL approach.

of unprotected Cys residues in a one-pot fashion provides a key advantage over traditional NCL approaches. As an alternative to this reductive deselenization of Sec to Ala, the Payne and Metanis research groups have independently demonstrated that Sec can also be converted to serine after ligation *via* an oxidative deselenization pathway (Dery et al., 2015; Malins, Mitchell, McGowan, & Payne, 2015; Mitchell et al., 2017).

2.1 Expanding the scope of DSL

Since its inception in 2015, two avenues have been predominantly explored to expand the scope of this methodology beyond Ala/Ser residues at the ligation junction. Firstly, after the original discovery, a number of synthetic selenylated amino acid building blocks were developed that can be employed as Sec surrogates in DSL reactions. These building blocks, wherein the selenium moiety is protected with a *p*-methoxybenzyl (PMB) group, are compatible with Fmoc-SPPS and have been successfully demonstrated to be competent in DSL-deselenization reaction sequences upon PMB deprotection (Kulkarni et al., 2018). To date, six selenylated amino acids, β-seleno Asp (**1**), γ-seleno Glu (**2**), β-seleno Phe (**3**), β-seleno Leu (**4**), γ-seleno Pro (**5**), and γ-seleno Lys (**6**) have been reported (as outlined in Fig. 3B). This has enabled a larger range of disconnection sites beyond Ala/Ser residues and, as such, the availability of these building blocks has greatly expanded the scope of the DSL methodology (Clayton et al., 2020; Dardashti et al., 2020; Malins & Payne, 2012; Mitchell et al., 2017; Sayers et al., 2018; Townsend et al., 2012; Wang, Ashhurst, et al., 2020; Wang, Corcilius, et al., 2020; Wang, Sanchez, Stone, & Payne, 2017). In addition to enhancing retrosynthetic flexibility, these selenylated amino acids offer some exciting features, which have been exemplified through the synthesis of proteins that would have been difficult to access using other methodologies. Examples will be discussed in detail in the latter half of this chapter.

More recently, Yin, Lu, Wang, and Wang (2019) reported an auxiliary-mediated approach to facilitate DSL (as shown in Fig. 3C). For this purpose, 2-seleno-2-phenyl acetaldehyde was first prepared in one step from commercially available phenyl acetaldehyde, wherein the selenium moiety was incorporated as the PMB selenoether. With this aldehyde in hand, the auxiliary could be installed at the desired ligation junction *via* reductive amination at the N-terminus of the resin-bound peptide, facilitated by sodium cyanoborohydride. An acidolytic cleavage of the peptide from resin along with concomitant PMB deprotection (using a cocktail of trifluoroacetic acid (TFA)/2,2′-dithiobis(5-nitropyridine) (DTNP)/*i*-Pr_3SiH/H_2O) followed by HPLC purification afforded the dimeric diselenide fragment, which can take part in a DSL reaction. Subsequently, the authors illustrated the generality of this auxiliary-mediated DSL methodology using model peptides and various ligation junctions, including examples involving sterically hindered Ile and Val residues. Importantly, after ligation, the auxiliary could be cleaved without intermediary purification using tris(2-carboxyethyl)phosphine (TCEP) and morpholine. The utility of this protocol was also showcased through the synthesis of two small proteins, peptidyl-prolyl *cis–trans* isomerase protein (FKBP11, 47 residues) and B-cell linker protein isoform (BCL, 65 residues). While it offers greater retrosynthetic freedom, some of the drawbacks of this auxiliary-based method include reduced ligation efficiency at hindered junctions, due to a sluggish Se → N acyl shift step. In addition, the auxiliary cleavage step is slow, which could lead to undesired desulfurization in targets that possess native Cys residues.

2.2 Kinetically controlled ligations *via* tandem DSL and NCL

Another key benefit of DSL is that the enhanced reaction rate over NCL creates an avenue for one-pot, kinetically controlled ligation reactions where the two methods are used sequentially for protein synthesis (Agten et al., 2021; Mitchell et al., 2015; Premdjee, Andersen, Larance, Conde-Frieboes, & Payne, 2021). While this orthogonality between DSL and NCL was first demonstrated in the seminal DSL report through the synthesis of the *Mycobacterium tuberculosis* early secretory antigen ESAT-6 (Mitchell et al., 2015), recent development of various selenylated amino acids has broadened the scope of this one-pot, multi-segment assembly approach, enabling access to numerous post-translationally modified proteins. For example, Agten et al. (2021) highlighted the utility of the DSL-NCL manifold through efficient synthesis of highly potent, site-specifically sulfated trivalent thrombin inhibitors. Thrombin is regarded as the vital enzyme in the blood coagulation cascade, and small molecules or peptides exhibiting inhibitory activity against thrombin offer tremendous potential as anticoagulants for the treatment of stroke and other thrombotic disorders (Crawley et al., 2007). Interestingly, several hematophagous organisms (*e.g.*, leeches, ticks, and flies) are known to produce a variety of peptides and small proteins that are modified at one or more site with tyrosine sulfate PTMs. These molecules possess potent antithrombotic activity to ensure uninterrupted supply of a blood meal from the host (Brahma, Blanchet, Kaur, Kini, & Doley, 2017; Figueiredo et al., 2012; Watson et al., 2018). Based on several naturally occurring thrombin inhibitors, which are known to interact with one of two binding sites on thrombin (exosite I or exosite II) in addition to the catalytic site, a series of hybrid inhibitors that simultaneously targeted all three sites of the protease were designed. Specifically, a bivalent tsetse fly-derived thrombin inhibitor (TTI) (Calisto et al., 2021), which targets exosite II and the catalytic site was selected along with exosite I binders, variegin (Var) and a truncated analog of anophelin (Ano). For the TTI-Var hybrids **7–9**, a three-component ligation strategy based on G13–D14 (TTI) and T14–A15 (Var) disconnections was devised, wherein sulfotyrosine residues—Tyr9, Tyr12 (TTI) and Tyr27 (Var)—were located within the N- and C-terminal fragments. The bifunctional fragments **10–12** possessed a C-terminal alkyl thioester and a β-seleno Asp residue at the N-terminus, contained residues from both TTI and variegin, joined *via* three different linker units, L-alanine, *N*-methylglycine (sarcosine), and 2-aminoisobutyric acid (Aib). Of these, sarcosine and Aib were incorporated to generate hybrids resistant to proteolytic cleavage by thrombin. These fragments **10–12** were first subjected to rapid, additive-free DSL reactions with the N-terminal fragment **13** bearing a phenyl selenoester, followed by deselenization. Notably, since the C—Se bond is significantly weakened in this case, owing to the radical stabilization through an adjacent carboxylate moiety, complete deselenization was achieved in just 10 min. Without purification, the intermediates **14–16** bearing C-terminal thioesters were subsequently treated separately with fragment **17** bearing an N-terminal Cys to facilitate NCL. This reaction employed trifluoroethanethiol (TFET) (Thompson et al., 2014) as a thiol additive and, following desulfurization, afforded the TTI-Var hybrids **7–9** in 22–31% yields over four steps after a single HPLC purification step (outlined in Fig. 4). A slightly different

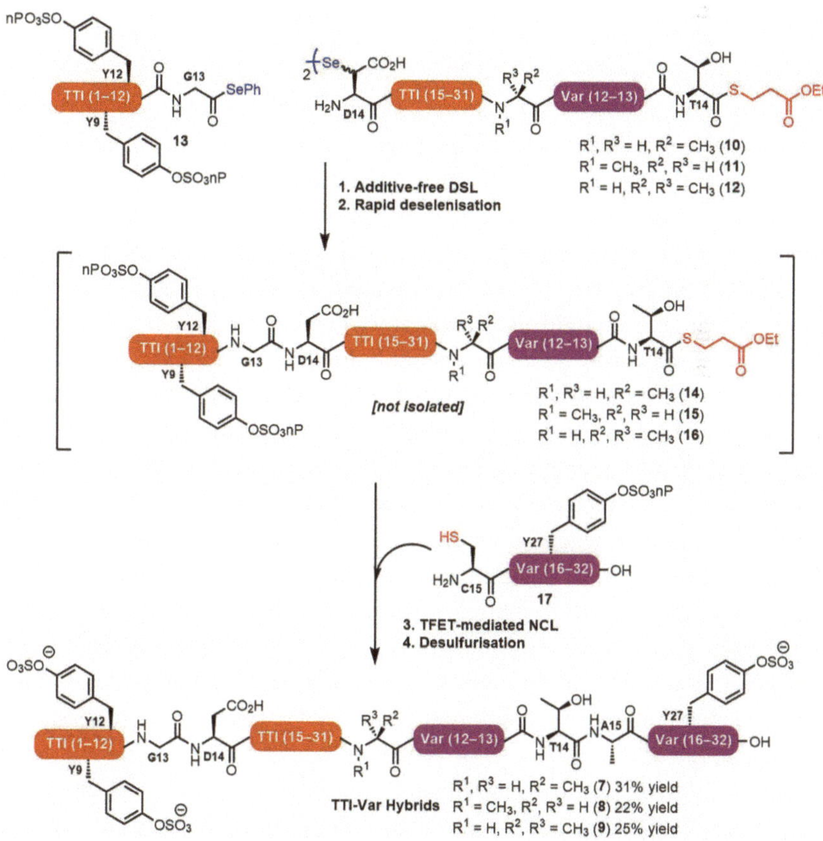

Fig. 4 Synthesis of sulfated TTI-Var hybrid thrombin inhibitors **7–9** featuring kinetically controlled ligations employing tandem DSL and NCL.

strategy was employed to access the TTI-Ano hybrids, wherein the bifunctional fragment featured a thiazolidine (as a masked Cys surrogate) at the C-terminus and β-seleno Asp residue at the N-terminus. The reaction sequence in this case involved DSL, deselenization, followed by thiazolidine deprotection to enable NCL and desulfurization, furnishing TTI-Ano hybrids in 12–17% yield over four steps. Importantly, this structure-guided protein engineering approach led to the discovery of femtomolar inhibitors of thrombin that possessed potent anticoagulant activity *in vitro* and *in vivo*.

2.3 DSL at high dilution

While DSL chemistry boasts superior kinetics at millimolar segment concentration, until recently, the reaction manifold was reportedly inefficient at lower (μM and nM) concentrations. It was subsequently disclosed that this was likely attributed to a distinct initiation step in the mechanistic cycle of the reaction (Chisholm et al., 2020). Specifically, following mechanistic studies it was proposed that a small amount of phenylselenoate, generated through hydrolysis/aminolysis of the phenyl selenoester, is involved in redox exchange with the diselenide fragment, affording the monomeric peptide selenoate *in situ*. After this initiation, the ligation proceeds *via* trans-selenoesterification and Se → N acyl shift, following an NCL-like mechanism (as outlined in Fig. 5A). Since the bimolecular phenylselenoate-mediated redox exchange is substantially slower at low concentrations, it was proposed that the DSL reactions were not operable due to a failure to initiate at these concentrations. Based on this hypothesis, the authors arrived at a solution to this problem by introducing two additives to the reaction mixture, 30 mM TCEP and 20 mM diphenyldiselenide (DPDS). While TCEP was included to quantitatively reduce the peptide diselenide to an active selenoate, DPDS served as a radical scavenger to suppress premature deselenization of Sec during the ligation reaction. This additive-based approach, dubbed reductive DSL (rDSL) was capable of facilitating ligations at nM concentrations, creating opportunities to access aggregation-prone, hydrophobic protein targets, particularly lipoproteins. In this work, Chisholm et al. (2020) also unveiled that these ligations can be combined with a novel photodeselenization step, expediting the overall process. The power of this technology was exemplified through the synthesis of differentially palmitylated analogs of the transmembrane lipoprotein, phospholemman (FXYD1). FXYD1 is natively palmitylated at C40 and C42 residues and plays a critical role in regulating the Na^+/K^+ pump (Tulloch et al., 2011). Although palmitylation is required to

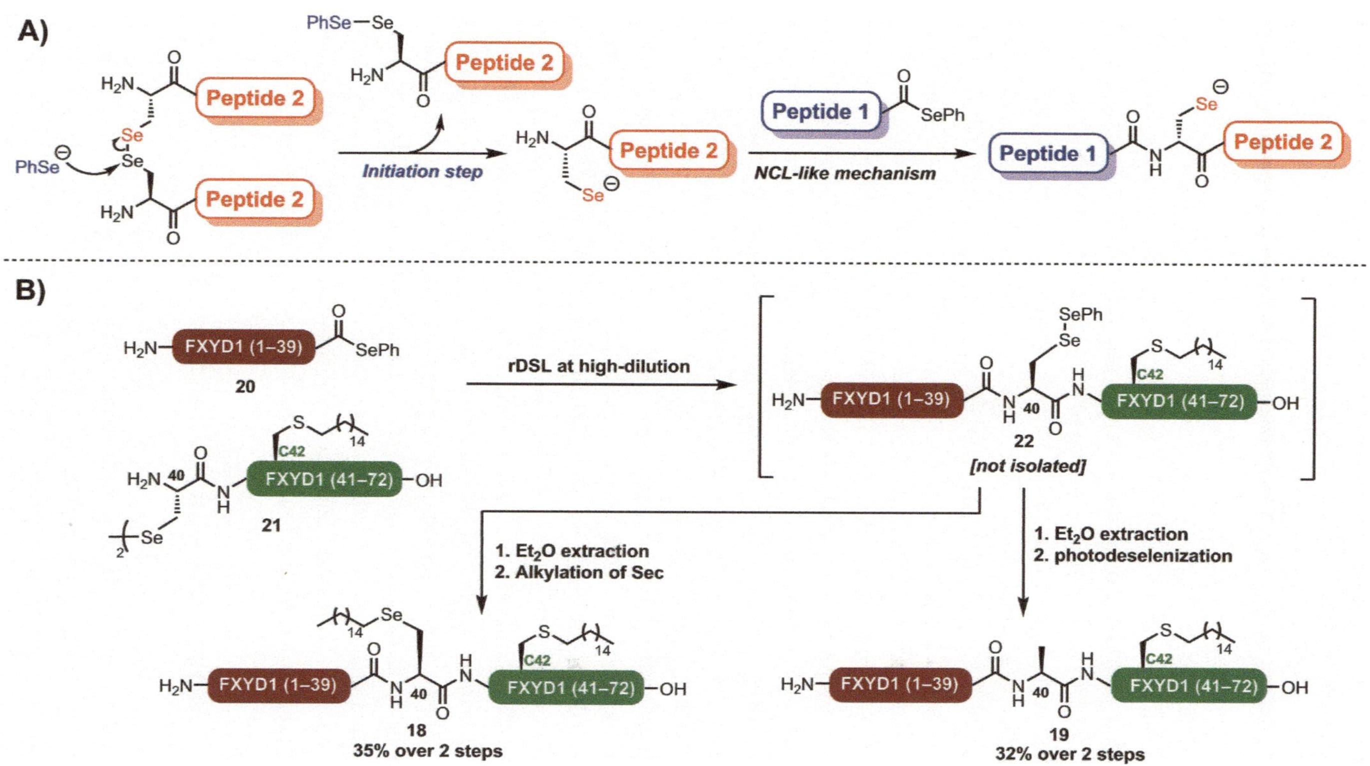

Fig. 5 (A) Mechanism of DSL. (B) Chemical synthesis of differentially lipidated FXYD1 using rDSL conditions followed by either alkylation of Sec to afford **18** or photodeselenization, affording **19**.

retain the inhibitory effect of FXYD1, accessing the protein bearing this reversible modification has been a long-standing synthetic challenge. It was postulated that the stable variants of FXYD1 (**18** and **19**) could be assembled by incorporating palmitylated C42 (*i.e.*, palmityl group attached to the Cys side chain) during Fmoc-SPPS, with C40 substituted with Sec to enable an rDSL reaction at the R39–U40 junction (as outlined in Fig. 5B). Following the ligation, the Sec residue could be subjected to late stage palmitylation at U40 rather than deselenization to Ala to generate the lipoprotein. The two peptide fragments bearing C-terminal selenoester (**20**) and N-terminal Sec (**21**), respectively, were first prepared using standard Fmoc SPPS techniques. The ligation was initiated by dissolving the fragments in ligation buffer with additives (6M Gdn•HCl, 0.2M BIS-TRIS, 30mM TCEP, 20mM DPDS, pH 5.1). Due to the highly hydrophobic nature of the fragments, the ligation was performed at 60°C under high dilution (120μM concentration with respect to the monomeric C-terminal fragment) and reached completion in 4h. From this point, two lipidated analogs of FXYD1 could be conveniently prepared from ligation product **22** by implementing different set of conditions in a divergent manner. Specifically, after extraction of residual DPDS with Et_2O, the reaction mixture was treated with 1-bromohexadecane and DTT to afford doubly palmitylated FXYD1 **18** in 35% yield over two steps after HPLC purification. Alternatively, after Et_2O extraction, the reaction mixture was treated with an equal volume of deselenization solution (6M Gdn•HCl, 800mM TCEP, 80mM glutathione (GSH)) followed by photoirradiation ($\lambda = 254$ nm, 35W, 1h), furnishing the mono palmitylated FXYD1 **19** (Ala instead of Cys40) in 32% yield over two steps. The effect of palmitylation on the Na^+/K^+-ATPase activity of FXYD1 was also investigated. Interestingly, mono palmitylated FXYD1 (**19**) (palmitylation at Cys42) was found to possess greater inhibitory activity compared to the doubly palmitylated analog **18**. Based on this work, it can be envisioned that this innovative rDSL-photodeselenization platform will find widespread application for rapid access to previously intractable lipoprotein targets in the future.

3. General DSL and rDSL protocols

In this section, we provide detailed general protocols to set up and monitor DSL and rDSL reactions along with tips for troubleshooting and for cases that employ selenylated amino acids. The ligation-deselenization procedure is operationally simple, highly efficient and can be performed without any formal training or synthetic expertise. Please note that the methods to access selenylated amino acids and suitably functionalized substrates (*i.e.*, peptide selenoesters and peptide diselenide dimers) have been previously reported and will not be discussed in this chapter (Griffiths & Mitchell, 2021; Kulkarni et al., 2019; Watson, Malins, & Payne, 2021).

Chemicals

1. **Additive-free DSL**—Requisite peptide fragments (TFA or formate salt), Milli-Q® Type 1 water, Gdn•HCl, Na_2HPO_4, aqueous NaOH (1.0 M).
2. **Reductive DSL (rDSL)**—Requisite peptide fragments (TFA or formate salt), Milli-Q® Type 1 water, Gdn•HCl, BIS-TRIS, TCEP, DPDS, aqueous NaOH (1.0 M).
3. **Deselenization**—Milli-Q® Type 1 water, Et_2O, argon, Gdn•HCl, Na_2HPO_4, TCEP, DTT, aqueous NaOH (1.0 M).
4. **Photodeselenization**—Milli-Q® Type 1 water, Et_2O, argon, Gdn•HCl, Na_2HPO_4, TCEP, reduced glutathione (GSH), aqueous NaOH (1.0 M).
5. **Reversed-phase HPLC purification**—Milli-Q® Type 1 water with 0.1 vol% TFA ($\geq$99.0%) ***or*** 0.1 vol% formic acid (98–100%), MeCN (gradient grade $\geq$99.9%) with 0.1 vol% TFA ($\geq$99.0%) ***or*** 0.1 vol% formic acid (98–100%).

Equipment

1. Analytical balance: dual range; Mettler Toledo.
2. Eppendorf tubes: Protein LoBind® 1.5, 2 mL.
3. Set of mechanical pipettes: Eppendorf Research® plus; 0.1–2.5, 2–20, 20–200, 100–1000 μL.
4. Pipette tips: Gilson DIAMOND® Standard Tips; D10, D200, D1000.
5. pH meter: LAQUAtwin compact pH meter.
6. Sonicator: Elma ultrasonic cleaner.
7. Photoreactors: (1) Penn PhD Photoreactor M2 with 450 nm light source. (2) Rayonet RPR-100 with 35 W, 254 nm light source.
8. Lyophilizer: Labconco, FreeZone 4.5 Plus model.
9. Lyophilizer vessels: 900, 300, 150 mL.
10. Preparative reverse-phase HPLC system: Waters multisolvent delivery system with a Rheodyne injection valve (4 mL loading loop) and Waters pump with a Waters programmable wavelength detector operating at 214, 230, 254 or 280 nm.
11. HPLC columns: (1) Waters Sunfire; 19 × 150 mm, C18, 5 μm, 130 Å, (2) X-Bridge; 10 × 250 mm, C18, 5 μm, 300 Å, (3) X-Bridge; 30 × 150 mm, C18, 5 μm, 130 Å.

12. Analytical UPLC system: UPLC system equipped with photodiode array (PDA) eλ detector ($\lambda = 210$–400 nm), Sample Manager FAN and Quaternary Solvent Manager (H-class) modules.
13. UPLC-MS system: equipped with an LC-M20A pump, a Shimadzu 2020 ESI mass spectrometer operating in positive mode unless otherwise stated, and an SPD-M30A diode array detector.
14. MALDI-TOF system: Bruker autoflex speed LRF MALDI-TOF mass spectrometer.

3.1 General protocol for additive-free DSL followed by deselenization

Timing: 19 h + overnight lyophilization (for cases involving standard deselenization); 2.5 h + overnight lyophilization (for cases involving photodeselenization)

1. Weigh out lyophilized peptide selenoester (1.2 equiv.) and peptide diselenide dimer (1.0 equiv. with respect to the monomer) fragments into separate Eppendorf tubes.

   ***Tip*:** For peptide selenoesters with a hindered C-terminal residue (*e.g.*, Val, Ile, Thr, Pro), 1.5–2.0 equiv. of this fragment might be required for the ligation to reach completion.
2. Dissolve the peptide diselenide dimer fragment in ligation buffer (6 M Gdn•HCl, 0.1 M Na_2HPO_4, pH 7.2, 5 mM concentration with respect to the monomer) and add the resulting solution to the selenoester fragment. Please note that quick sonication (10–15 s) could be required to ensure complete dissolution and formation of a clear solution.
3. Immediately measure the pH of the ligation mixture and adjust it to ~5.8–6.5. While DSL reactions can be performed over a broad pH range, optimal ligation rates are achieved at pH ~ 6.0.

   ***Tip*:** The pH is normally adjusted with 1.0 M aqueous NaOH solution due to the acidity of the peptide segments following HPLC purification (that uses acidic eluents).

   Critical step: The pH adjustment must be performed very carefully as rapid selenoester hydrolysis occurs at a pH > 7.0, leading to incomplete ligation and low overall yield.
4. Incubate the reaction mixture at room temperature for 10 min and analyze by UPLC-MS to monitor product formation. Specifically, take a 2 μL aliquot, dilute it to 50 μL with H_2O:MeCN (1:1 v/v) and use 10 μL of this sample for UPLC-MS analysis.

Tip: The reaction mixture usually turns cloudy at this stage, due to the formation of DPDS, a water-insoluble by-product generated during ligation.

Tip: The ligation product is generally observed as a mixture, consisting of symmetric diselenide, asymmetric diselenide and unproductive selenoester (highlighted in Fig. 3A). Absence of the diselenide dimer fragment starting material indicates that the ligation is complete.

Tip: The additive-free DSL at Pro selenoesters usually requires several hours to reach completion (Sayers et al., 2018).

5. Extract the DPDS generated during ligation by treating the ligation mixture with Et_2O ($3 \times 0.5\,mL$).

   Critical step: It is important to note that DPDS acts as a radical quencher and needs to be removed from the ligation mixture before proceeding with the deselenization step.

6. Prepare the deselenization solution in ligation buffer (6 M Gdn•HCl, 0.1 M Na_2HPO_4, 250 mM TCEP, 25 mM DTT) and degas it thoroughly by sparging argon for 10 min.

   Critical step: Thorough degassing is critical during this step to avoid undesirable oxidative deselenization.

7. Add an equal volume of this freshly prepared deselenization solution to the ligation mixture [final concentrations: 2.5 mM ligation product (with respect to the monomer), 125 mM TCEP and 12.5 mM DTT], sparge the resulting solution for 5 min and incubate the reaction mixture at room temperature for 16 h to ensure complete deselenization.

   Tip: All the ligation products converge to a single peak after this step. Please note that the unproductive selenoester first undergoes thiolysis with GSH or DTT and the resulting selenol can then undergo deselenization.

   Tip: At this stage, for examples involving multiple fragments and more than one ligation step, the reaction mixture can be simply subjected to the next set of conditions to carry out NCL-desulfurization without an intermediate purification step.

   Tip: This transformation can also be carried out using recently reported photodeselenization conditions (with a final concentration of 200 mM TCEP and 40 mM reduced glutathione (GSH), pH 7.0, UV irradiation (254 nm, 35 W), ~5 min). It should be noted that photodeselenization is not chemoselective and should not be used in cases where there are unprotected Cys residues elsewhere in the sequence.

8. After UPLC-MS analysis, dilute the reaction mixture with H_2O or ligation buffer and purify using reversed-phase HPLC.
9. After careful analysis, pure fractions can be combined and lyophilized to obtain the pure protein.

3.2 General protocol for rDSL followed by deselenization

Timing: 21–22 h + overnight lyophilization (for cases involving standard deselenization); 4.5–5.5 h + overnight lyophilization (for cases involving photodeselenization)

1. Weigh out Gdn•HCl (6 M, 570 mg), BIS-TRIS (200 mM, 42 mg), TCEP (30 mM, 8.6 mg) and DPDS (20 mM, 6.2 mg) and dissolve components in Milli-Q water to make up 1 mL final volume.

   Tip: Note that the DPDS does not fully dissolve in aqueous solutions and requires prolonged sonication (15 min) to achieve saturation.
2. Adjust the pH of the resulting solution to 5.0–5.3 with aqueous NaOH (1.0 M), and thoroughly degas the buffer with simultaneous sonication for 20 min.
3. Dissolve the fragments using freshly prepared ligation buffer and adjust the pH of the reaction mixture to 5.0–5.3.

   Critical step: The pH adjustment must be performed very carefully as rapid selenoester hydrolysis occurs at a pH > 7.0, leading to incomplete ligation and low overall yield.
4. Allow the ligation to proceed at 37 °C followed by UPLC-MS analysis to ensure complete conversion.
5. Extract the reaction mixture with Et_2O (7×0.5 mL) to completely remove DPDS to avoid issues during deselenization.
6. At this stage, the ligation product can be subjected to either standard deselenization or photodeselenization depending on the presence or absence of Cys residues, respectively. For photodeselenization, add an equal volume of freshly prepared and degassed solution of 400 mM TCEP and 80 mM reduced GSH in ligation buffer (6 M Gdn•HCl, 0.2 M BIS-TRIS, pH 7.5), adjust the pH to 7.0 with aqueous NaOH (1.0 M) and irradiate the solution in a photoreactor ($\lambda = 254$ nm, 35 W) at room temperature to achieve complete deselenization. The reaction should be monitored by UPLC-MS and is typically complete in 5 min.

3.3 Safety considerations and standards

1. TFA and formic acid used for the preparation of HPLC buffers are highly corrosive and should be handled very carefully. Appropriate PPE must be used (gloves, safety glasses and lab coat) while handling these chemicals.

2. Diphenyl diselenide (DPDS) is highly toxic and should be handled carefully with appropriate PPE.
3. Phenylselenoate (generated *in situ* during the preparation of rDSL buffer) is highly malodorous and should only be handled in a fume hood.
4. Diethyl ether (Et_2O) is highly flammable and should only be handled in a fume hood.

4. Assembly of proteins through ligations at challenging junctions

As discussed earlier, NCL reactions typically suffer from sluggish reaction rates at sterically hindered junctions, resulting in significant thioester hydrolysis and poor yields. Given the markedly improved reaction rates offered by DSL, complex protein targets that necessitate disconnection at challenging junctions can be assembled using this methodology. This has been made possible through the recent development of a number of selenylated amino acid building blocks **1–6** that have expanded the number of disconnection sites that can be accessed through DSL-deselenization chemistry.

4.1 DSL at sterically hindered junctions

For example, the utility of β-seleno Leu (**4**) was illustrated by Wang et al. (2017) in a synthesis of the human cytomegalovirus (HCMV) protein UL22A (**23**), and its sulfated analogs **24–26** (as outlined in Fig. 6A). UL22A is an 83-residue chemokine binding protein (CKBP) and has been shown to interact with CC chemokine ligand 5 (CCL5, also known as RANTES), an inflammatory chemokine that signals through a variety of chemokine receptors, recruiting leukocytes to sites of inflammation (see Fig. 6B) (Wang, Bresnahan, & Shenk, 2004). Importantly, UL22A contains two tyrosine residues (Tyr65 and Tyr69) embedded within an acidic stretch of amino acids, indicative of post-translational sulfation (Monigatti, Gasteiger, Bairoch, & Jung, 2002; Stone & Payne, 2015). As tyrosine sulfation is effected *in vivo* by tyrosylprotein sulfotransferases (TPSTs) in the Golgi apparatus, expression of UL22A in mammalian cells usually results in a heterogeneous mixture of sulfoforms (Maxwell & Payne, 2020). Therefore, in order to precisely ascertain the effect of tyrosine sulfation on UL22A-RANTES interaction, homogenous sulfoforms of UL22A **23–26** were generated through chemical synthesis. Due to a lack of cysteines or suitably placed alanine residues, accessing this protein by traditional ligation methods was not feasible. As a result, a sterically encumbered Val-Leu junction was chosen to assemble

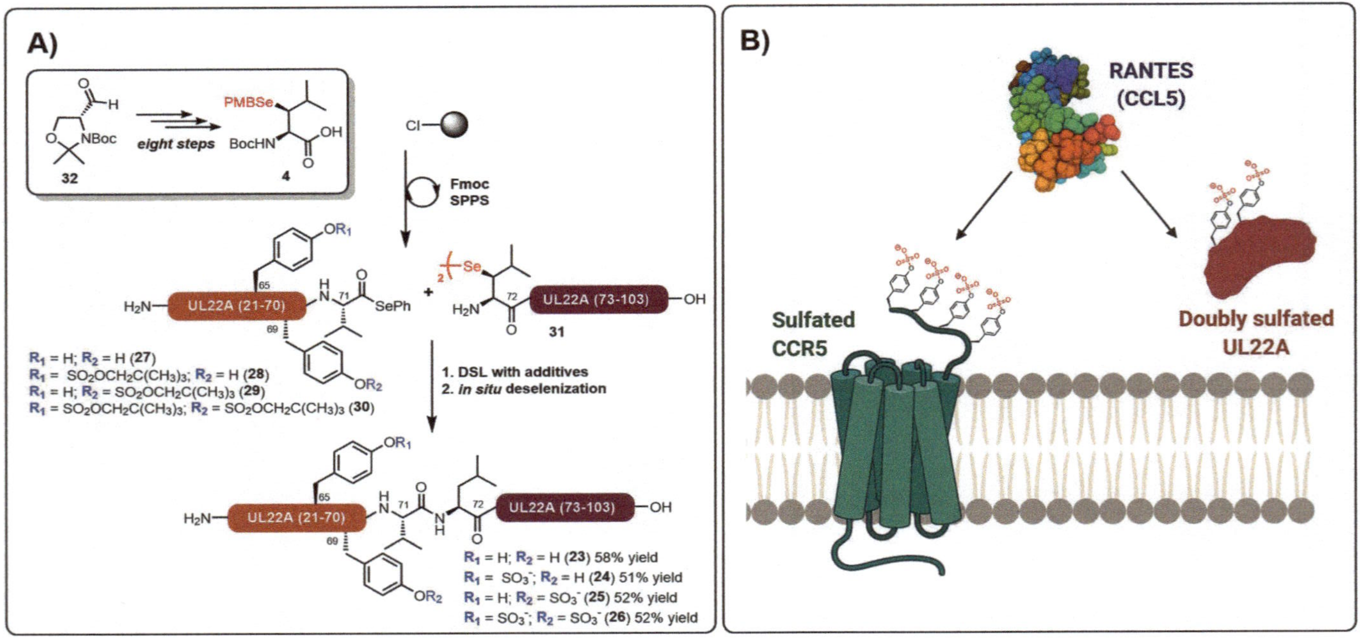

Fig. 6 (A) Chemical synthesis of four homogenous sulfoforms of UL22A **23–26** *via* DSL-deselenization chemistry at a sterically hindered Val-Leu junction. (B) Cartoon representation of sulfated UL22A binding RANTES (CCL5), mimicking its interaction with sulfated chemokine receptor (CCR5).

the full-length protein, taking advantage of the efficiency of DSL-deselenization chemistry even at this hindered junction. This strategy resulted in five requisite fragments: selenoester fragments **27–30** with a C-terminal Val and the two putative Tyr sulfation sites (Tyr65 and Tyr69), and a diselenide fragment **31** bearing an N-terminal β-seleno Leu residue. The suitably protected β-seleno Leu building block **4** could be accessed in eight steps from Garner's aldehyde (**32**) and was incorporated into the C-terminal fragment **31** by Fmoc-SPPS (Wang et al., 2020). The N-terminal fragments **27–30** (possessing all four possible sulfation patterns) were prepared by Fmoc-SPPS, cleaved from the solid support under mildly acidic conditions and subjected to solution-phase selenoesterification followed by acidolytic deprotection of the sidechain protecting groups and HPLC purification. Notably, the acid-labile sulfate was protected with a neopentyl group to prevent acidolysis of the highly acid labile modification during the resin cleavage and side chain deprotection step that employs a TFA-based cocktail. An initial, additive-free DSL proved to be unsuccessful likely owing to the exceptionally hindered Val-Leu junction. However, when performed under reductive conditions with tris(2-carboxyethyl)phosphine (TCEP) and diphenyl diselenide (DPDS) as additives, the ligations proceeded smoothly, reaching completion within 1 h. *In situ* deselenization resulted in the effective conversion of the β-seleno Leu at the ligation junction to the native Leu and concomitant removal of the neopentyl group from the Tyr sulfate residues, furnishing the homogenous sulfoforms of UL22A (**23–26**) in excellent yields following HPLC purification. The doubly sulfated UL22A (**26**) was shown to bind RANTES with almost 400 times the affinity of the unmodified protein **23**, validating the importance of tyrosine sulfation in UL22A-RANTES protein-protein interactions (outlined in Fig. 6B). Importantly, the speed and efficiency of the ligation reactions to assemble UL22A exemplifies the utility of DSL in the formation of sterically hindered peptide bonds during protein synthesis.

4.2 DSL at the sterically hindered and electronically disarmed Pro-Pro junction

Despite not possessing a sterically hindered side chain, ligations that involve proline thioesters fail to initiate under standard NCL conditions due to electronic deactivation (*vide supra*). Specifically, the constrained *trans* conformation of the proline amide bond results in an $n \rightarrow \pi^*$ interaction between adjacent carbonyl groups, drastically reducing the electrophilicity of a proline thioester (Newberry & Raines, 2017; Pollock & Kent, 2011).

To solve this issue, Gui et al. (2016) reported the preparation of a γ-thiolated prolyl thioester, wherein the electrophilicity of the proline acyl donor is greatly improved due to the formation of a bicyclic thiolactone intermediate. While this is a very promising advancement, the need to remove the thiol moiety through desulfurization makes the approach impractical for examples wherein native Cys residues need to be retained within a given peptide or protein sequence. Previously, Durek and Alewood (2011) reported that Pro-Cys junctions can be accessed by replacing the proline thioester with an alkyl selenoester counterpart, whereas Townsend et al. (2012) demonstrated that the peptides functionalized with an N-terminal *trans*-γ-seleno Pro can undergo NCL reactions with a range of thioesters, with the exception of proline; ligation reactions at a proline-proline junction had therefore remained elusive. In 2018, Sayers et al. were able to forge Pro-Pro junctions for the first by employing DSL, leveraging the significantly enhanced reactivities of both the prolyl aryl selenoesters and γ-seleno Pro diselenides (Sayers et al., 2018).

Synthesis of the *trans*-γ-seleno Pro building block **5** was carried out in four steps from γ-hydroxy-L-proline (**33**), wherein the selenol moiety was protected as a PMB selenoether. This building block could be conveniently incorporated into a model peptide by Fmoc-SPPS, followed PMB deprotection using 20 vol% dimethyl sulfoxide (DMSO) in TFA, affording the corresponding diselenide peptide dimer upon HPLC purification. During initial experiments based on the model systems, γ-seleno Pro was found to react rapidly with a variety of peptide selenoesters, including those containing a C-terminal proline, which ligated with similar efficiency under both additive and additive-free conditions and could be smoothly deselenized after ligation. Surprisingly, when the reaction was repeated with the *cis* isomer of γ-seleno Pro (**34**), the branched selenoester intermediate was observed instead of the ligation product. This outcome implies that, while the trans-selenoesterification step proceeded readily, the usually facile Se → N acyl shift—required to generate the native amide bond—could not take place (outlined in Fig. 7). Computational studies revealed that the *trans*-isomer could position the branched selenoester in a favorable conformation to facilitate acyl transfer to the α-NH, but such a conformation was not possible for the *cis*-configured γ-seleno Pro, corroborating the experimental results.

After validating and optimizing the ligation-deselenization strategy on a model system, the authors turned their attention to the synthesis of lumbricin-1 (**35**). This 62-residue peptide with a 15% proline content was first isolated from an earthworm (*Lumbricus rubellus*) and exhibits *in vitro* antimicrobial activity against a broad spectrum of Gram-positive

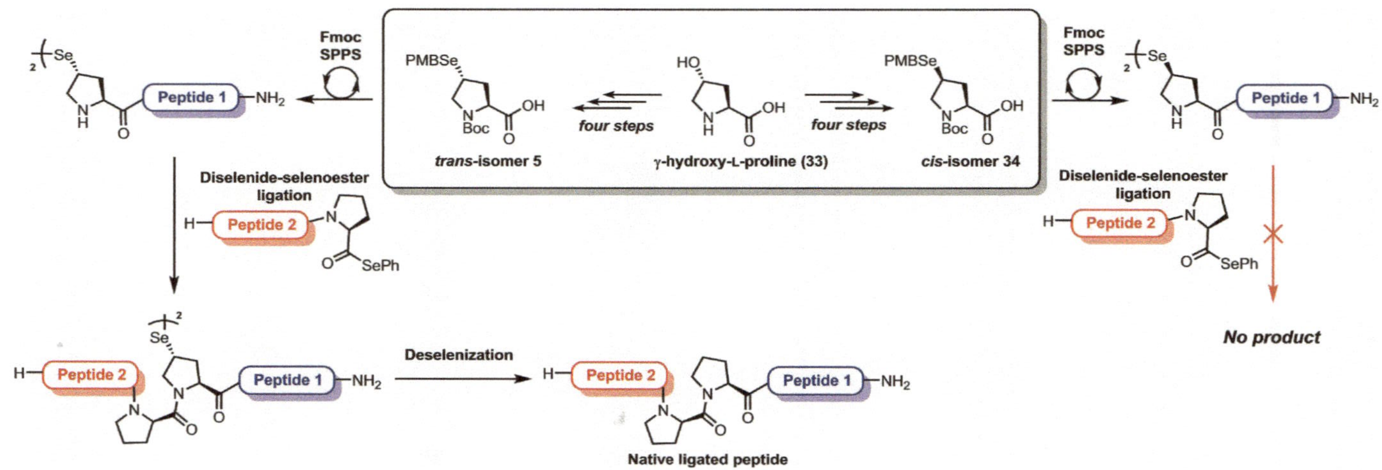

Fig. 7 Synthesis of the suitably protected *trans*- and *cis*-diastereomers of γ-seleno Pro **5** and **34**. Subsequent Pro-Pro DSL-deselenization reactions of the γ-seleno Pro derivatives on a model peptide system revealing that amide bond formation was only possible when the *trans* isomer was used.

and -negative bacteria (Cho et al., 1998). As such, the protein was disconnected at the P33–P34 junction, requiring the synthesis of two peptide fragments for ligation: a 33-amino-acid fragment **36** bearing a C-terminal Pro selenoester and a peptide dimer **37** containing an N-terminal *trans*-γ-seleno Pro, which were accessed by standard Fmoc-SPPS (as outlined in Fig. 8). These fragments underwent DSL under additive-free conditions furnishing the ligation intermediate within 16 h. However, the deselenization step was found to be inefficient, presumably because the unproductive selenoester of the ligation product (which does not undergo deselenization), could not be thiolyzed by DTT, leading to incomplete conversion. To avoid this problem, the ligation was repeated, and the ligation product was treated with 2 vol% hydrazine hydrate to ensure complete hydrazinolysis of the unproductive selenoester. This crude reaction mixture, containing ligation product **38** exclusively as a diselenide, was then subjected to deselenization, followed by HPLC purification, delivering lumbricin-1 (**35**) in 52% yield. This exemplifies the utility of the DSL-deselenization manifold as the first and only method to tackle electronically disarmed and sterically encumbered Pro-Pro ligation junctions.

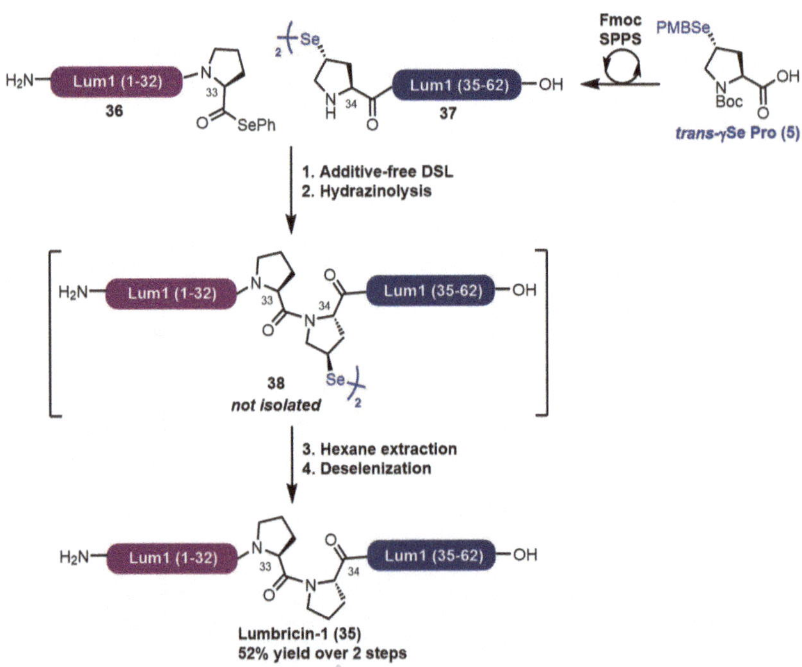

Fig. 8 Synthesis of the lumbricin-1 **35** *via* a Pro-Pro DSL-deselenization. The *trans* isomer of γ-seleno Pro (**5**) was required to facilitate Se→N acyl shift.

5. Rapid and selective deselenization

5.1 Rapid deselenization in protein library generation

As peptides and proteins are getting more widely approved as therapeutic agents, efficient methods to generate libraries of site-specifically modified proteins are in high demand for establishing structure-activity relationships (SAR), leading to the discovery of new drugs. Undoubtedly, addition of DSL to the ligation toolbox has served as a promising step in this regard, vastly improving the speed and efficiency with which the ligations can be performed. However, while DSL reactions are usually complete within minutes, the relatively sluggish deselenization of selenocysteine (usually 6–16 h) (Metanis, Keinan, & Dawson, 2010; Mitchell et al., 2015) abolishes the prospects of accelerating the overall process for rapid protein library assembly on unprecedented timescales. Interestingly, during the development of selenylated Asp and Glu amino acids to expand the scope of DSL, Mitchell et al. (2017) observed that the deselenization of these residues post ligation could be achieved in <1 min. Based on mechanistic and computational studies, it was proposed that the β-centered radical intermediate, generated during this transformation can be stabilized by the adjacent carboxylate moiety, making the C—Se bond extremely labile. These findings sparked two exciting opportunities for these building blocks in chemical protein synthesis. Firstly, capitalizing on the rapid deselenization step, ligations involving β-seleno Asp or γ-seleno Glu junctions can facilitate rapid assembly of proteins in a matter of minutes (Clayton et al., 2020; Mitchell et al., 2017) rather than several days that would be necessary with other synthetic methods. The speed in which proteins can be assembled through one-pot DSL-deselenization chemistry was recently exemplified through the synthesis of a library of 34 homogeneously sulfated proteins. In this work, Watson et al. (2019) applied the β-seleno Asp-mediated DSL-deselenization manifold to rapidly generate differentially sulfated analogs of tick-derived proteins—hyalomins, andersonins, and madanin-like proteins, to gauge the effect of sulfation on their thrombin inhibitory activity. Subsequent biological evaluation of this library revealed that Tyr sulfation led to significantly enhanced (up to two orders of magnitude) inhibitory activity of these proteins and led to the identification of key structure-activity relationships, akin to small molecule medicinal chemistry.

5.2 Selective deselenization at β-seleno Asp

An additional benefit of the superior deselenization kinetics at β-seleno Asp is the ability to perform chemoselective deselenization at Asp in the presence of an unprotected Sec, presenting opportunities to access native selenoproteins. There are 25 known genes which encode for the mammalian selenoproteome. These proteins facilitate a variety of physiological functions including cellular redox homeostasis (Kryukov et al., 2003; Lu & Holmgren, 2009). However, their isolation and characterization has been hampered, primarily due to the laborious and low-yielding recombinant methods relying on endogenous translational machinery (Liu et al., 2017; Peeler et al., 2020). To circumvent these issues, chemical synthesis has been explored as an attractive alternative to construct selenoproteins in order to probe their physiological function (Dery et al., 2017). Mitchell et al. (2017) showcased the utility of chemoselective deselenization of β-seleno Asp through the efficient synthesis of selenoprotein K (SelK). SelK (**39**) is a 93-residue membrane-bound protein which exhibits peroxidase activity and has been implicated in cellular calcium flux and endoplasmic reticulum-associated protein degradation (Shchedrina et al., 2011; Verma et al., 2011). Interestingly, the protein exists as a homodimer with the selenocysteine residue close to the C-terminus (Sec92) forming an intermolecular diselenide (Liu, Zhang, & Rozovsky, 2014). Given the retrosynthetically unfavorable position of Sec92, the protein was disconnected at the Y60–D61 junction and the resulting peptide fragments **40** and **41** were prepared using standard Fmoc-SPPS conditions (as outlined in Fig. 9).

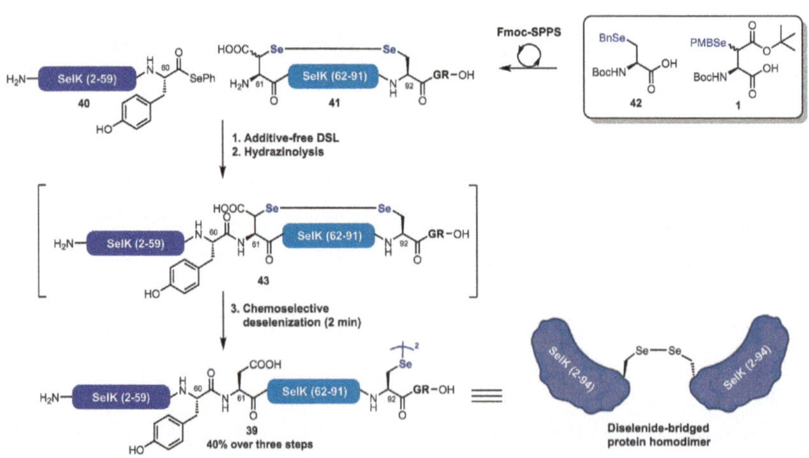

Fig. 9 Synthesis of selenoprotein K (SelK) **39** by DSL followed by chemoselective deselenization of β-seleno Asp in the presence of Sec. Intramolecular diselenide fragment **41** was synthesized from suitably protected Sec **42** and β-seleno Asp (**1**) building blocks.

For the C-terminal fragment, suitably protected β-seleno Asp (PMB)(**1**) and Sec (benzyl)(**42**) were incorporated during solid phase synthesis, and the resulting peptide **41** was isolated as the intramolecular diselenide following removal of these selenol-protecting groups. An additive-free DSL between fragments **40** and **41** reached completion in 15 min, affording the ligation intermediate **43** as an intramolecular diselenide. The reaction mixture was then treated with TCEP for 2 min to accomplish chemoselective deselenization of β-seleno Asp in the presence of the Sec residue and after immediate HPLC purification, SelK (**39**) was isolated as a homodimer in 40% yield. In summary, this work establishes β-seleno Asp or γ-seleno Glu-mediated DSL as a viable and exciting option for the synthesis of libraries of modified proteins of therapeutic interest, as well as native selenoproteins.

6. Protein assembly *via* one-pot iterative DSL reactions

While the development and expansion of DSL chemistry has provided innovative solutions to some of the long-standing problems associated with NCL, a distinct advantage of NCL is the ability to assemble proteins *via* iterative ligations in a one-pot fashion (Bang et al., 2004; Tang et al., 2017, 2015). Such a multi-component strategy relies on the availability of protecting groups to mask the N-terminal Cys residue within a bifunctional fragment bearing a thioester at the C-terminus, preventing undesired self-ligation or polymerization. In this way, a first ligation can be performed and the N-terminal Cys can be unmasked, before proceeding with the next NCL reaction involving another peptide thioester. Notable examples of these protecting groups include, thiazolidine (Thz) (Bang et al., 2004), acetamidomethyl (Acm) (Veber et al., 1972), and trifluoromethyl acetamidomethyl (TfAcm) (Tang et al., 2015). Importantly, since the reagents required to unmask these groups are compatible with the NCL conditions, purification steps between subsequent ligations can sometimes be omitted, making the process efficient and high yielding (Bang et al., 2004; Maity et al., 2016; Tang et al., 2015). In contrast, Sec-mediated ligations (*i.e.*, thioester-Sec NCL and DSL) rely on a smaller set of protecting groups, namely PMB and the selenazolidine (Sez) group; these either require harsh conditions for removal or require reagents that are incompatible with subsequent ligations. To circumvent this issue, Kambanis, Chisholm, Kulkarni, and Payne (2021) designed a photolabile protecting group, which can be rapidly and quantitatively removed *via* photoirradiation ($\lambda = 450$ nm, 36 W, 1–10 min) with a view to develop a strategy that would enable iterative DSL reactions without intermediary purification steps. This novel building block **44**, wherein the Sec side chain was masked with a 7-diethylamino-3-methyl coumarin (DEAMC) moiety, was synthesized from readily available starting materials in six steps (see Fig. 10A). To validate the utility of this protecting group, the authors selected an unmodified variant of the 20-residue variable number tandem repeat (VNTR) domain of the trans-membrane glycoprotein mucin 1 (MUC1) as a model system. Specifically, using suitably functionalized fragments, including a bifunctional fragment featuring C-terminal selenoester and N-terminal Sec(DEAMC), 60- and 80-residue MUC1 polypeptides were assembled *via* multiple DSL reactions on very short time scales and in impressive overall yields. For example, synthesis of 80-residue MUC1 was accomplished in 35% yield over six steps (three DSL reactions, two photodeprotection steps and a global deselenization step) and required only 3h of reaction and a single purification. The speed and efficiency of this streamlined approach was further exploited through the synthesis of a challenging protein target, apolipoprotein C-III (ApoCIII). ApoCIII is a lipoprotein, predominantly expressed by hepatocytes and enterocytes in humans. While ApoCIII exerts its function

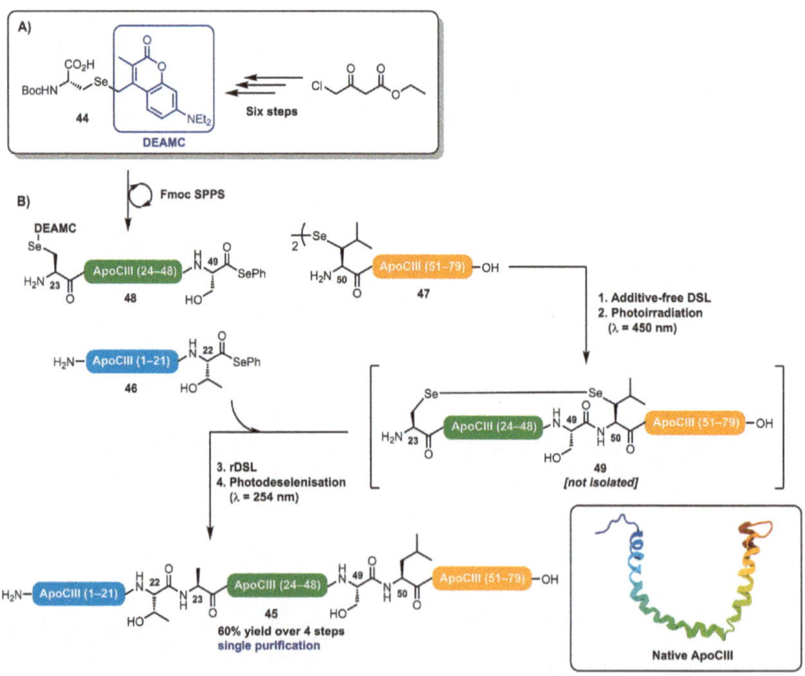

Fig. 10 (A) Synthesis of Sec(DEAMC) building block **44**. (B) Chemical synthesis of ApoCIII **45** through iterative DSL reactions [PDB 2JQ3 (Gangabadage et al., 2008)].

by inhibiting the activity of lipoprotein lipase, it also plays a crucial role in promoting the secretion and assembly of very low-density lipoproteins and triglyceride-rich lipoproteins in plasma. As such, modulators of ApoCIII activity hold great therapeutic potential toward mitigating the risks associated with dyslipidemia and cardiovascular disease (Ginsberg & Brown, 2011; Luo & Peng, 2016; Taskinen & Borén, 2016). The 79-residue protein **45** was disconnected at sterically demanding Thr22-Ala23 and Ser49-Leu50 junctions, necessitating synthesis of three suitably functionalized peptide fragments (as outlined in Fig. 10B). Fragments **46** and **47** were prepared using standard Fmoc SPPS, with the latter involving incorporation of a β-seleno Leu residue. A bifunctional fragment **48** bearing an N-terminal Sec(DEAMC) residue and a C-terminal aryl selenoester was accessed using a side-chain anchoring strategy (Hanna, Kulkarni, Watson, Premdjee, & Payne, 2017). With purified fragments in hand, the protein assembly began with the first DSL reaction between fragments **47** and **48**. While this ligation proceeded smoothly to completion in 30 min under additive-free conditions, the steric hindrance of Leu residue at the ligation junction meant that the ligation needed to be performed at elevated temperature (37 °C). Next, the reaction mixture was subjected to photoirradiation ($\lambda = 450$ nm, 36 W), which resulted in quantitative DEAMC removal in 10 min, generating the ligated product as an intramolecular diselenide **49**. Unfortunately, subsequent ligation could not be initiated, which can be attributed to the aggregation-prone and hydrophobic nature of N-terminal fragment **46** (Lins et al., 2002). Thankfully, this bottleneck could be resolved by switching to rDSL conditions, whereby the ligation was performed under high dilution (200 μM final concentration with respect to monomeric fragment **47**), in the presence of additives (20 mM TCEP and 30 mM DPDS). After removal of residual DPDS by Et_2O extraction, photodeselenization (200 mM TCEP, 40 mM GSH, pH 7.0, $\lambda = 254$ nm, 35 W, 5 min) and purification by HPLC, native ApoCIII **45** was isolated in an excellent 60% yield over four steps. This iterative ligation strategy using the DSL reaction manifold lays the foundation for the chemical synthesis of larger and more complex protein targets in the future.

7. Summary

By virtue of its speed, efficiency and unique features, the DSL technology has enabled access to numerous protein targets, including post-translationally modified proteins, selenoproteins, lipoproteins, and diversified protein libraries for potential therapeutic applications. Importantly, DSL has served as an important new tool for the field of chemical protein synthesis as it has provided (and will continue to provide) access to targets that were considered synthetically intractable a few years ago. With the availability of selenylated amino acids combined with recent advances facilitating iterative ligations, it is envisioned that this methodology will continue to be a central methodology for the synthesis of native and engineered proteins bearing native or designer modifications that cannot be accessed through recombinant technologies. As such this ligation methodology should contribute to our understanding of how PTMs and other modifications modulate the structure and function of proteins.

Acknowledgments

The authors acknowledge grant support from the Australian Research Council Centre of Excellence for Innovations in Peptide and Protein Science (CE200100012) and the NHMRC Investigator Grant (APP1174941 to R.J.P.). The authors also acknowledge support from the Research Training Program scheme (to M.J.B.).

References

Agouridas, V., El Mahdi, O., Diemer, V., Cargoët, M., Monbaliu, J. M., & Melnyk, O. (2019). Native chemical ligation and extended methods: Mechanisms, catalysis, scope, and limitations. *Chemical Reviews*, *119*(12), 7328–7443.

Agten, S. M., Watson, E. E., Ripoll-Rozada, J., Dowman, L. J., Wu, M. C. L., Alwis, I., et al. (2021). Potent trivalent inhibitors of thrombin through hybridization of salivary sulfopeptides from hematophagous arthropods. *Angewandte Chemie International Edition*, *60*(10), 5348–5356.

Antonczak, A. K., Simova, Z., & Tippmann, E. M. (2009). A critical examination of Escherichia coli esterase activity. *Journal of Biological Chemistry*, *284*(42), 28795–28800.

Bang, D., Chopra, N., & Kent, S. B. H. (2004). Total chemical synthesis of crambin. *Journal of the American Chemical Society*, *126*(5), 1377–1383.

Besse, D., Siedler, F., Diercks, T., Kessler, H., & Moroder, L. (1997). The redox potential of selenocystine in unconstrained cyclic peptides. *Angewandte Chemie International Edition*, *36*, 883–885.

Bieberich, E. (2014). Synthesis, processing, and function of N-glycans in N-glycoproteins. In *Glycobiology of the nervous system* (pp. 47–70). Springer.

Bock, A., Forchhammer, K., Heider, J., Leinfelder, W., Sawers, G., Veprek, B., et al. (1991). Selenocysteine: The 21st amino acid. *Molecular Microbiology*, *5*, 515–520.

Bondalapati, S., Jbara, M., & Brik, A. (2016). Expanding the chemical toolbox for the synthesis of large and uniquely modified proteins. *Nature Chemistry*, *8*(5), 407–418.

Brahma, R. K., Blanchet, G., Kaur, S., Kini, R. M., & Doley, R. (2017). Expression and characterization of haemathrins, madanin-like thrombin inhibitors, isolated from the salivary gland of tick Haemaphysalis bispinosa. *Thrombosis Research*, *152*, 20–29.

Bray, B. L. (2003). Large-scale manufacture of peptide therapeutics by chemical synthesis. *Nature Reviews Drug Discovery*, *2*(7), 587–593.

Calisto, B. M., Ripoll-Rozada, J., Dowman, L. J., Franck, C., Agten, S. M., Parker, B. L., et al. (2021). Sulfotyrosine-mediated recognition of human thrombin by a tsetse fly anticoagulant mimics physiological substrates. *Cell Chemical Biology*, *28*(1), 26–33.e28.

Cergol, K. M., Thompson, R. E., Malins, L. R., Turner, P., & Payne, R. J. (2014). One-pot peptide ligation–desulfurization at glutamate. *Organic Letters*, *16*(1), 290–293.
Chen, B., Sun, Y., Niu, J., Jarugumilli, G. K., & Wu, X. (2018). Protein lipidation in cell signaling and diseases: Function, regulation, and therapeutic opportunities. *Cell Chemical Biology*, *25*(7), 817–831.
Chen, J., Wan, Q., Yuan, Y., Zhu, J., & Danishefsky, S. J. (2008). Native chemical ligation at valine: A contribution to peptide and glycopeptide synthesis. *Angewandte Chemie International Edition*, *120*(44), 8649–8652.
Chin, J. W. (2017). Expanding and reprogramming the genetic code. *Nature*, *550*(7674), 53–60.
Chisholm, T. S., Kulkarni, S. S., Hossain, K. R., Cornelius, F., Clarke, R. J., & Payne, R. J. (2020). Peptide ligation at high dilution via reductive diselenide-selenoester ligation. *Journal of the American Chemical Society*, *142*(2), 1090–1100.
Cho, J. H., Park, C. B., Yoon, Y. G., & Kim, S. C. (1998). Lumbricin I, a novel proline-rich antimicrobial peptide from the earthworm: Purification, cDNA cloning and molecular characterization. *Biochimica et Biophysica Acta—Molecular Basis of Disease*, *1408*(1), 67–76.
Clayton, D., Kulkarni, S. S., Sayers, J., Dowman, L. J., Ripoll-Rozada, J., Pereira, P. J. B., et al. (2020). Chemical synthesis of a haemathrin sulfoprotein library reveals enhanced thrombin inhibition following tyrosine sulfation. *RSC Chemical Biology*, *1*, 379–384.
Crawley, J. T. B., Zanardelli, S., Chion, C. K. N. K., & Lane, D. A. (2007). The central role of thrombin in hemostasis. *Journal of Thrombosis and Haemostasis*, *5*(Suppl. 1), 95–101.
Crich, D., & Banerjee, A. (2007). Native chemical ligation at phenylalanine. *Journal of the American Chemical Society*, *129*, 10064–10065.
Crick, F. (1970). Central dogma of molecular biology. *Nature*, *227*(5258), 561–563.
Dardashti, R. N., Kumar, S., Sternisha, S. M., Reddy, P. S., Miller, B. G., & Metanis, N. (2020). Selenolysine: A new tool for traceless isopeptide bond formation. *Chemistry*, *26*(22), 4952–4957.
Dawson, P. E., Muir, T. W., Clark-Lewis, I., & Kent, S. B. H. (1994). Synthesis of proteins by native chemical ligation. *Science*, *266*(5186), 776–779.
Dery, S., Sai Reddy, P., Dery, L., Mousa, R., Dardashti, R. N., & Metanis, N. (2015). Insights into the deselenization of selenocysteine into alanine and serine. *Chemical Science*, *6*, 6207–6212.
Dery, L., Sai Reddy, P., Dery, S., Mousa, R., Ktorza, O., Talhami, A., et al. (2017). Accessing human selenoproteins through chemical protein synthesis. *Chemical Science*, *8*, 1922–1926.
Durek, T., & Alewood, P. F. (2011). Preformed selenoesters enable rapid native chemical ligation at intractable sites. *Angewandte Chemie International Edition*, *50*(50), 12042–12045.
Fan, C., Deng, Q., & Zhu, T. F. (2021). Bioorthogonal information storage in l-DNA with a high-fidelity mirror-image Pfu DNA polymerase. *Nature Biotechnology*. https://doi.org/10.1038/s41587-41021-00969-41586.
Figueiredo, A. C., de Sanctis, D., Gutiérrez-Gallego, R., Cereija, T. B., Macedo-Ribeiro, S., Fuentes-Prior, P., et al. (2012). Unique thrombin inhibition mechanism by anophelin, an anticoagulant from the malaria vector. *Proceedings of the National Academy of Sciences of the United States of America*, *109*(52), E3649–E3658.
Gangabadage, C. S., Zdunek, J., Tessari, M., Nilsson, S., Olivecrona, G., & Wijmenga, S. S. (2008). Structure and dynamics of human apolipoprotein CIII. *The Journal of Biological Chemistry*, *283*(25), 17416–17427.
Gieselman, M. D., Xie, L., & Van Der Donk, W. A. (2001). Synthesis of a selenocysteine-containing peptide by native chemical ligation. *Organic Letters*, *3*, 1331–1334.
Ginsberg, H. N., & Brown, W. V. (2011). Apolipoprotein CIII. *Arteriosclerosis, Thrombosis, and Vascular Biology*, *31*(3), 471–473.

Griffiths, R. C., & Mitchell, N. J. (2021). The chemical synthesis of site-specifically modified proteins via diselenide-selenoester ligation. In W. M. Hussein, R. J. Stephenson, & I. Toth (Eds.), *Peptide conjugation: Methods and protocols* (pp. 231–251). New York, NY: Springer US.

Guenther, W. H. H. (1967). Methods in selenium chemistry. III. Reduction of diselenides with dithiothreitol. *The Journal of Organic Chemistry*, *32*, 3931–3934.

Gui, Y., Qiu, L., Li, Y., Li, H., & Dong, S. (2016). Internal activation of peptidyl prolyl thioesters in native chemical ligation. *Journal of the American Chemical Society*, *138*(14), 4890–4899.

Haase, C., Rohde, H., & Seitz, O. (2008). Native chemical ligation at valine. *Angewandte Chemie International Edition*, *47*, 6807–6810.

Hanna, C. C., Kulkarni, S. S., Watson, E. E., Premdjee, B., & Payne, R. J. (2017). Solid-phase synthesis of peptide selenoesters via a side-chain anchoring strategy. *Chemical Communications*, *53*, 5424–5427.

Harpaz, Z., Siman, P., Kumar, K. S. A., & Brik, A. (2010). Protein synthesis assisted by native chemical ligation at leucine. *ChemBioChem*, *11*, 1232–1235.

Hondal, R. J., Nilsson, B. L., & Raines, R. T. (2001). Selenocysteine in native chemical ligation and expressed protein ligation. *Journal of the American Chemical Society*, *123*, 5140–5141.

Italia, J. S., Peeler, J. C., Hillenbrand, C. M., Latour, C., Weerapana, E., & Chatterjee, A. (2020). Genetically encoded protein sulfation in mammalian cells. *Nature Chemical Biology*, *16*(4), 379–382.

Jiang, W., Zhang, B., Fan, C., Wang, M., Wang, J., Deng, Q., et al. (2017). Mirror-image polymerase chain reaction. *Cell Discovery*, *3*, 17037.

Kambanis, L., Chisholm, T. S., Kulkarni, S. S., & Payne, R. J. (2021). Rapid one-pot iterative diselenide–selenoester ligation using a novel coumarin-based photolabile protecting group. *Chemical Science*, *12*, 10014–10021.

Kent, S. B. H. (2009). Total chemical synthesis of proteins. *Chemical Society Reviews*, *38*(2), 338–351.

Kent, S. B. H., Sohma, Y., Liu, S., Bang, D., Pentelute, B., & Mandal, K. (2012). Through the looking glass—A new world of proteins enabled by chemical synthesis. *Journal of Peptide Science*, *18*(7), 428–436.

Kryukov, G. V., Castellano, S., Novoselov, S. V., Lobanov, A. V., Zehtab, O., Guigó, R., et al. (2003). Characterization of mammalian selenoproteomes. *Science*, *300*, 1439–1443.

Kulkarni, S. S., Sayers, J., Premdjee, B., & Payne, R. J. (2018). Rapid and efficient protein synthesis through expansion of the native chemical ligation concept. *Nature Reviews Chemistry*, *2*, 0122.

Kulkarni, S. S., Watson, E. E., Premdjee, B., Conde-Frieboes, K. W., & Payne, R. J. (2019). Diselenide–selenoester ligation for chemical protein synthesis. *Nature Protocols*, *14*(7), 2229–2257.

Kumar, K. S. A., Bavikar, S. N., Spasser, L., Moyal, T., Ohayon, S., & Brik, A. (2011). Total chemical synthesis of a 304 amino acid K48-linked tetraubiquitin protein. *Angewandte Chemie International Edition*, *50*(27), 6137–6141.

Lins, L., Flore, C., Chapelle, L., Talmud, P. J., Thomas, A., & Brasseur, R. (2002). Lipid-interacting properties of the N-terminal domain of human apolipoprotein C-III. *Protein Engineering Design & Selection*, *15*(6), 513–520.

Liu, J., Chen, Q., & Rozovsky, S. (2017). Utilizing selenocysteine for expressed protein ligation and bioconjugations. *Journal of the American Chemical Society*, *139*(9), 3430–3437.

Liu, C. C., & Schultz, P. G. (2010). Adding new chemistries to the genetic code. *Annual Review of Biochemistry*, *79*(1), 413–444.

Liu, J., Zhang, Z., & Rozovsky, S. (2014). Selenoprotein K form an intermolecular diselenide bond with unusually high redox potential. *FEBS Letters*, *588*, 3311–3321.

Lu, J., & Holmgren, A. (2009). Selenoproteins. *Journal of Biological Chemistry*, *284*, 723–727.
Luo, M., & Peng, D. (2016). The emerging role of apolipoprotein C-III: Beyond effects on triglyceride metabolism. *Lipids in Health and Disease*, *15*(1), 184.
Maity, S. K., Jbara, M., Laps, S., & Brik, A. (2016). Efficient palladium-assisted one-pot deprotection of (acetamidomethyl)cysteine following native chemical ligation and/or desulfurization to expedite chemical protein synthesis. *Angewandte Chemie International Edition*, *55*(28), 8108–8112.
Malins, L. R., Giltrap, A. M., Dowman, L. J., & Payne, R. J. (2015). Synthesis of β-thiol phenylalanine for applications in one-pot ligation–desulfurization chemistry. *Organic Letters*, *17*(9), 2070–2073.
Malins, L. R., Mitchell, N. J., McGowan, S., & Payne, R. J. (2015). Oxidative deselenization of selenocysteine: Applications for programmed ligation at serine. *Angewandte Chemie International Edition*, *54*, 12716–12721.
Malins, L. R., & Payne, R. J. (2012). Synthesis and utility of β-selenol-phenylalanine for native chemical ligation–deselenization chemistry. *Organic Letters*, *14*(12), 3142–3145.
Malins, L. R., & Payne, R. J. (2014). Recent extensions to native chemical ligation for the chemical synthesis of peptides and proteins. *Current Opinion in Chemical Biology*, *22*, 70–78.
Malins, L. R., & Payne, R. J. (2015). Synthetic amino acids for applications in peptide ligation-desulfurization chemistry. *Australian Journal of Chemistry*, *68*, 521–537.
Maxwell, J. W. C., & Payne, R. J. (2020). Revealing the functional roles of tyrosine sulfation using synthetic sulfopeptides and sulfoproteins. *Current Opinion in Chemical Biology*, *58*, 72–85.
McGinty, R. K., Kim, J., Chatterjee, C., Roeder, R. G., & Muir, T. W. (2008). Chemically ubiquitylated histone H2B stimulates hDot1L-mediated intranucleosomal methylation. *Nature*, *453*(7196), 812–816.
Merrifield, R. B. (1963). Solid phase peptide synthesis. I. The synthesis of a tetrapeptide. *Journal of the American Chemical Society*, *85*(14), 2149–2154.
Metanis, N., Keinan, E., & Dawson, P. E. (2010). Traceless ligation of cysteine peptides using selective deselenization. *Angewandte Chemie International Edition*, *49*(39), 7049–7053.
Mitchell, N. J., Kulkarni, S. S., Wang, S., Malins, L. R., & Payne, R. J. (2017). One-pot ligation–oxidative deselenization at selenocysteine and selenocystine. *Chemistry*, *23*(4), 946–952.
Mitchell, N. J., Malins, L. R., Liu, X., Thompson, R. E., Chan, B., Radom, L., et al. (2015). Rapid additive-free selenocystine–selenoester peptide ligation. *Journal of the American Chemical Society*, *137*(44), 14011–14014.
Mitchell, N. J., Sayers, J., Kulkarni, S. S., Clayton, D., Goldys, A. M., Ripoll-Rozada, J., et al. (2017). Accelerated protein synthesis via one-pot ligation-deselenization chemistry. *Chem*, *2*(5), 703–715.
Monigatti, F., Gasteiger, E., Bairoch, A., & Jung, E. (2002). The sulfinator: Predicting tyrosine sulfation sites in protein sequences. *Bioinformatics*, *18*(5), 769–770.
Muir, T. W., & Kent, S. B. H. (1993). The chemical synthesis of proteins. *Current Opinion in Biotechnology*, *4*(4), 420–427.
Muir, T. W., Sondhi, D., & Cole, P. A. (1998). Expressed protein ligation: A general method for protein engineering. *Proceedings of the National Academy of Sciences of the United States of America*, *95*(12), 6705–6710.
Neumann, H., Peak-Chew, S. Y., & Chin, J. W. (2008). Genetically encoding Nε-acetyllysine in recombinant proteins. *Nature Chemical Biology*, *4*(4), 232–234.
Newberry, R. W., & Raines, R. T. (2017). The n→ π* interaction. *Accounts of Chemical Research*, *50*(8), 1838–1846.
Nurk, S., Koren, S., Rhie, A., Rautiainen, M., Bzikadze, A. V., Mikheenko, A., et al. (2021). The complete sequence of a human genome. *bioRxiv*. 2021.2005.2026.445798.

Peeler, J. C., Falco, J. A., Kelemen, R. E., Abo, M., Chartier, B. V., Edinger, L. C., et al. (2020). Generation of recombinant mammalian selenoproteins through genetic code expansion with photocaged selenocysteine. *ACS Chemical Biology*, *15*(6), 1535–1540.

Pollock, S. B., & Kent, S. B. H. (2011). An investigation into the origin of the dramatically reduced reactivity of peptide-prolyl-thioesters in native chemical ligation. *Chemical Communications*, *47*, 2342–2344.

Premdjee, B., Andersen, A. S., Larance, M., Conde-Frieboes, K. W., & Payne, R. J. (2021). Chemical synthesis of phosphorylated insulin-like growth factor binding protein 2. *Journal of the American Chemical Society*, *143*(14), 5336–5342.

Quaderer, R., & Hilvert, D. (2002). Selenocysteine-mediated backbone cyclization of unprotected peptides followed by alkylation, oxidative elimination or reduction of the selenol. *Chemical Communications*, 2620–2621.

Quaderer, R., Sewing, A., & Hilvert, D. (2001). Selenocysteine-mediated native chemical ligation. *Helvetica Chimica Acta*, *84*, 1197–1206.

Sakamoto, I., Tezuka, K., Fukae, K., Ishii, K., Taduru, K., Maeda, M., et al. (2012). Chemical synthesis of homogeneous human glycosyl-interferon-β that exhibits potent antitumor activity in vivo. *Journal of the American Chemical Society*, *134*(12), 5428–5431.

Sayers, J., Karpati, P. M. T., Mitchell, N. J., Goldys, A. M., Kwong, S. M., Firth, N., et al. (2018). Construction of challenging proline-proline junctions via diselenide-selenoester ligation chemistry. *Journal of the American Chemical Society*, *140*(41), 13327–13334.

Shchedrina, V. A., Everley, R. A., Zhang, Y., Gygi, S. P., Hatfield, D. L., & Gladyshev, V. N. (2011). Selenoprotein K binds multiprotein complexes and is involved in the regulation of endoplasmic reticulum homeostasis. *Journal of Biological Chemistry*, *286*(50), 42937–42948.

Shieh, H.-S., Mathis, K. J., Williams, J. M., Hills, R. L., Wiese, J. F., Benson, T. E., et al. (2008). High resolution crystal structure of the catalytic domain of ADAMTS-5 (aggrecanase-2). *Journal of Biological Chemistry*, *283*(3), 1501–1507.

Stone, M. J., & Payne, R. J. (2015). Homogeneous sulfopeptides and sulfoproteins: Synthetic approaches and applications to characterize the effects of tyrosine sulfation on biochemical function. *Accounts of Chemical Research*, *48*(8), 2251–2261.

Swatek, K. N., & Komander, D. (2016). Ubiquitin modifications. *Cell Research*, *26*(4), 399–422.

Tan, Y., Wu, H., Wei, T., & Li, X. (2020). Chemical protein synthesis: Advances, challenges, and outlooks. *Journal of the American Chemical Society*, *142*(48), 20288–20298.

Tang, S., Liang, L. J., Si, Y. Y., Gao, S., Wang, J. X., Liang, J., et al. (2017). Practical chemical synthesis of atypical ubiquitin chains by using an isopeptide-linked Ub Isomer. *Angewandte Chemie International Edition*, *56*(43), 13333–13337.

Tang, S., Si, Y. Y., Wang, Z. P., Mei, K. R., Chen, X., Cheng, J. Y., et al. (2015). An efficient one-pot four-segment condensation method for protein chemical synthesis. *Angewandte Chemie International Edition*, *54*(19), 5713–5717.

Taskinen, M.-R., & Borén, J. (2016). Why is apolipoprotein CIII emerging as a novel therapeutic target to reduce the burden of cardiovascular disease? *Current Atherosclerosis Reports*, *18*(10), 1–8.

Thompson, R. E., Chan, B., Radom, L., Jolliffe, K. A., & Payne, R. J. (2013). Chemoselective ligation-desulfurization at aspartate. *Angewandte Chemie International Edition*, *52*(37), 9723–9727.

Thompson, R. E., Liu, X., Alonso-Garcia, N., Barbosa Pereira, P. J., Jolliffe, K. A., & Payne, R. J. (2014). Trifluoroethanethiol: An efficient additive for one-pot ligation-desulfurization chemistry. *Journal of the American Chemical Society*, *136*(23), 8161–8164.

Townsend, S. D., Tan, Z., Dong, S., Shang, S., Brailsford, J. A., & Danishefsky, S. J. (2012). Advances in proline ligation. *Journal of the American Chemical Society*, *134*(8), 3912–3916.

Tulloch, L. B., Howie, J., Wypijewski, K. J., Wilson, C. R., Bernard, W. G., Shattock, M. J., et al. (2011). The inhibitory effect of phospholemman on the sodium pump requires its palmitoylation. *Journal of Biological Chemistry, 286*(41), 36020–36031.

Veber, D. F., Milkowski, J. D., Denkewalter, R. G., & Hirschmann, R. (1968). The synthesis of peptides in aqueous medium, IV. A novel protecting group for cysteine. *Tetrahedron Letters, 26*(9), 3057–3058.

Veber, D., Milkowski, J., Varga, S., Denkewalter, R., & Hirschmann, R. (1972). Acetamidomethyl. A novel thiol protecting group for cysteine. *Journal of the American Chemical Society, 94*(15), 5456–5461.

Verma, S., Hoffmann, F. W., Kumar, M., Huang, Z., Roe, K., Nguyen-Wu, E., et al. (2011). Selenoprotein K knockout mice exhibit deficient calcium flux in immune cells and impaired immune responses. *Journal of Immunology, 186*, 2127–2137.

Vijay-Kumar, S., Bugg, C. E., & Cook, W. J. (1987). Structure of ubiquitin refined at 1.8Å resolution. *Journal of Molecular Biology, 194*(3), 531–544.

Wan, Q., & Danishefsky, S. J. (2007). Free-radical-based, specific desulfurization of cysteine: A powerful advance in the synthesis of polypeptides and glycopolypeptides. *Angewandte Chemie International Edition, 46*, 9248–9252.

Wang, X., Ashhurst, A. S., Dowman, L. J., Watson, E. E., Li, H. Y., Fairbanks, A. J., et al. (2020). Total synthesis of glycosylated human interferon-γ. *Organic Letters, 22*(17), 6863–6867.

Wang, D., Bresnahan, W., & Shenk, T. (2004). Human cytomegalovirus encodes a highly specific RANTES decoy receptor. *Proceedings of the National Academy of Sciences of the United States of America, 101*(47), 16642–16647.

Wang, M., & Casey, P. J. (2016). Protein prenylation: Unique fats make their mark on biology. *Nature Reviews Molecular Cell Biology, 17*(2), 110–122.

Wang, X., Corcilius, L., Premdjee, B., & Payne, R. J. (2020). Synthesis and utility of β-selenophenylalanine and β-selenoleucine in diselenide – selenoester ligation. *The Journal of Organic Chemistry, 85*(3), 1567–1578.

Wang, X., Sanchez, J., Stone, M., & Payne, R. J. (2017). Sulfation of human cytomegalovirus protein UL22A Enhances binding to the chemokine RANTES. *Angewandte Chemie International Edition, 56*(29), 8490–8494.

Watson, E. E., Liu, X., Thompson, R. E., Ripoll-Rozada, J., Wu, M., Alwis, I., et al. (2018). Mosquito-derived anophelin sulfoproteins are potent antithrombotics. *ACS Central Science, 4*(4), 468–476.

Watson, E. E., Malins, L. R., & Payne, R. J. (2021). Expanding native chemical ligation methodology with synthetic amino acid derivatives. In *Total chemical synthesis of proteins* (pp. 119–159).

Watson, E. E., Ripoll-Rozada, J., Lee, A. C., Wu, M. C. L., Franck, C., Pasch, T., et al. (2019). Rapid assembly of a library of anticoagulant sulfoproteins. *Proceedings of the National Academy of Sciences of the United States of America, 116*(28), 13873–13878.

Wilkinson, B. L., Stone, R. S., Capicciotti, C. J., Thaysen-Andersen, M., Matthews, J. M., Packer, N. H., et al. (2012). Total synthesis of homogeneous antifreeze glycopeptides and glycoproteins. *Angewandte Chemie International Edition, 51*(15), 3606–3610.

Yan, L. Z., & Dawson, P. E. (2001). Synthesis of peptides and proteins without cysteine residues by native chemical ligation combined with desulfurization. *Journal of the American Chemical Society, 123*, 526–533.

Yang, R. L., Pasunooti, K. K., Li, F. P., Liu, X. W., & Liu, C. F. (2009). Dual native chemical ligation at lysine. *Journal of the American Chemical Society, 131*, 13592–13593.

Yin, H., Lu, D., Wang, S., & Wang, P. (2019). Development of powerful auxiliary-mediated ligation to facilitate rapid protein assembly. *Organic Letters, 21*(13), 5138–5142.

FSC
www.fsc.org
MIX
Papier | Fördert gute Waldnutzung
FSC® C083411

Kolibri 360 GmbH
Ferdinand-Jühlke-Straße 7
99095 Erfurt, Deutschland
produktsicherheit@kolibri360.de